ELECTRONIC BAROQUE

Inside Technology

Wiebe Bijker, Rebecca Slayton, and Edward Jones-Imhotep

A list of books in the series appears at the back of the book.

ELECTRONIC BAROQUE

BUILDING A HISTORICAL ORGAN FOR THE PRESENT

PETER F. PETERS

The MIT Press
Cambridge, Massachusetts
London, England

The MIT Press
Massachusetts Institute of Technology
77 Massachusetts Avenue, Cambridge, MA 02139
mitpress.mit.edu

The MIT Press would like to thank the anonymous peer reviewers who provided comments on drafts of this book. The generous work of academic experts is essential for establishing the authority and quality of our publications. We acknowledge with gratitude the contributions of these otherwise uncredited readers.

This book was set in Bembo Book MT Pro by Westchester Publishing Services. Printed and bound in the United States of America.

Library of Congress Cataloging-in-Publication Data

Names: Peters, Peter Frank, 1960– author.
Title: Electronic baroque : building a historical organ for the present / Peter F. Peters.
Description: Cambridge, Massachusetts : The MIT Press, 2025. | Series: Inside technology | Includes bibliographical references and index.
Identifiers: LCCN 2025000543 (print) | LCCN 2025000544 (ebook) | ISBN 9780262553636 (paperback) | ISBN 9780262385183 (epub) | ISBN 9780262385190 (pdf)
Subjects: LCSH: Organ (Musical instrument)—Netherlands—Amsterdam. | Organ (Musical instrument)—Construction—History—21st century. | Organ (Musical instrument)—Construction—History—18th century. | Hildebrandt, Zacharias, 1688–1757. | Elbertse, Hans. | Yokota, Munetaka. | Elbertse Orgelmakers. | Hermann Eule Orgelbau. | Sinua (Firm)
Classification: LCC ML582.8.A47 P47 2025 (print) | LCC ML582.8.A47 (ebook) | DDC 786.5/19492352—dc23/eng/20250128
LC record available at https://lccn.loc.gov/2025000543
LC ebook record available at https://lccn.loc.gov/2025000544

10 9 8 7 6 5 4 3 2 1

EU Authorised Representative: Easy Access System Europe, Mustamäe tee 50, 10621 Tallinn, Estonia | Email: gpsr.requests@easproject.com

CONTENTS

PREFACE

The journey that led to this book began when I first heard about the Orgelpark in Amsterdam in 2007. At the time, I was fascinated by organs as musical machines with a long history. The Orgelpark had recently opened its doors as a concert venue that aimed to integrate the pipe organ in musical life. It took another six years before I visited the former church in the spring of 2013. The Orgelpark's artistic leader, Johan Luijmes, whom I knew as artistic programmer of a center for new music in Maastricht, where I worked, had asked me if I would want to become a researcher in a project that aimed to build a new baroque organ for the Orgelpark. His idea was that the documentation of this ambitious project by a researcher from the field of science and technology studies (STS) would be interesting not just for the organ world but also for researchers, professionals, and artists in other fields.

Soon after, I was introduced by Johan to Loek Dijkman and Sylvia de Munck, president and vice president of the Utopa Foundation, which funds the Orgelpark. After the board of the Utopa Foundation had decided to build the new organ, I received a research grant from the foundation that allowed me to follow the building process as it unfolded as a member of the core group that oversaw project. It was the start of five years of participating in meetings, study trips, visits to organ workshops, and concerts and symposia in the Orgelpark. The former Parkkerk (Park Church), right at the edge of the Vondelpark in the center of Amsterdam, became a familiar place to me.

I learned to appreciate the organs that the Orgelpark houses as characters in a novel with their own histories and habits. Many times, I listened to the musicians who performed their music on or together with these instruments. The concerts and yearly symposia in June brought me into contact with organists, organ scholars, musicians, and researchers from all over the world. During study trips, I had the chance to listen to and learn about historical organs in the Netherlands and in Germany. Regular visits to organ workshops taught me how complicated the making of an organ and its pipes is. Following the construction of an organ unfold became a way for me to ask philosophical questions as well.

That I have been able to write this book, I owe to Johan Luijmes and his unwavering enthusiasm for what it could become. I am grateful to Loek Dijkman and Sylvia de Munck for their interest in my perspective on the project, their generosity, and their patience when finishing the book took longer than I hoped. What makes the Orgelpark so special is that it is not only a concert venue but also a research center. Hans Fidom has been director of the Orgelpark research program from the start, combining this position with a chair in Organ Studies at the VU University Amsterdam. Throughout the project and during the writing of this book, I benefited greatly from his knowledge of organs and their history. His precise and constructive comments on versions of the manuscript have been invaluable.

Where organs are usually built by one company, the Utopa Baroque Organ was built by the Orgelpark in collaboration with four companies. Elbertse Orgelmakers has been responsible for the case, the windchests, the wooden pipes, the bellows, and the wind channels as well as the smaller parts that connect the keys to the pipes. Hans Elbertse has welcomed me many times into his organ workshop in Soest, the Netherlands, where we discussed the work as it progressed. Our discussions on how he and his coworkers articulated and solved problems have helped me to understand more about how an organ is built. I also thank Bert van den Heuvel, Jos Elbertse, and their colleagues for sharing their design ideas and experiences as craftspeople with me. The pipes of the Utopa Baroque Organ were made by Hermann Eule Orgelbau in Bautzen, Germany. During study trips in 2015, I met Dirk and Anne-Christin Eule, Helmut Werner, Ingolf Huth, Andreas Baum, Konrad Dänhardt, and Thomas Habelock, whom I thank for what I learned about organ pipes, their anatomy, and their making. Once made, organ pipes must be given their own sound character; in organ terms, they must be voiced. In the baroque organ project, this was the responsibility of Munetaka Yokota. He is one of the leading experts in historically inspired organ building and pipe voicing, and it was a privilege to follow his work and that of his assistant, Mariya Kato. For generously sharing their knowledge and expertise with me, I am grateful. The electronic parts of the organ and the software needed to use its digital console were developed by Sinua in Dussseldorf, Germany. I thank Benedikt Aufterbeck and Thomas Stöckl for explaining their work to me. Thanks also to Gerard de Jongh, who introduced me to historically inspired organ painting.

Visiting the concerts and symposia at the Orgelpark was one of the joys of participating in the project. I thank the staff of the Orgelpark—Sonja Duimel, Trevor Grahl, Martje van Loenen, Jacqueline Oskamp, and Karlijne Swart—for their kind hospitality and assistance during symposia.

From the start, the Orgelpark's new baroque organ project was seen as a research project, and organists, organ scholars, organ builders, musicians, artists, and scholars from various fields were invited to share their ideas about what the new organ could, or should, be. Research into historical organs was necessary to develop the final plans for the instrument. During a study trip with the members of the project's core team, we met with Carl Johan Bergsten, Hans Davidsson, Johann Norrback, Ibo Ortgies, Paul Peeters, and Joel Speerstra at the Göteborg Organ Art Center (GOArt) of the University of Gothenburg, Sweden. I thank them for what I learned about historically informed organ building and their interest in my perspective on their work as an STS scholar. Building the new baroque organ with historical examples in mind was one of the goals of the project, and the core team visited organs built or inspired by Arp Schnitger and Zacharias Hildebrandt in Germany. We were welcomed in their churches to learn more about the historical organs they play, restored, or built by Sebastian Bartsch, Martin Böcker, Frits Elshout, David Franke, Thiemo Janssen, Annette Herr, Martin Hofmann, Rudolf Kelber, Jörg Kraemer, Espen Melbö, Marko Mitzscherling, Thomas Müller, Martina Pohl, Johannes Trümpler, and Jan Werner. I thank them all for what they shared about the unique histories and characters of the instruments they take care of.

Building a new baroque organ was the subject of debate and reflection throughout the project during symposia organized by Hans Fidom in the Orgelpark. Often, these events also involved concerts. The expertise and artistry of the participants of these symposia has shaped the arguments in this book in many ways. I am thankful to Anne la Berge, Jos de Bie, Wim Diepenhorst, David Franke, Hans-Ola Ericsson, Clare Gallagher, Alfredo Genovesi, Robert van Heumen, Erik de Jong (Spinvis), Piet Kee, Jacob Lekkerkerker, Wouter Snoei, Henk Verhoeff, Ansgar Wallenhorst, and Kristian Wegscheider for taking the time for interviews or for sharing their thoughts, knowledge, and artistry with me through their presentations and concerts. Randall Harlow introduced me to his concept of the "hyper organ," which became a source of inspiration during the project, and I also thank him for his detailed comments on my manuscript. I want to thank Stef Tuinstra for introducing me to the historical Hinsz organ in Roden, the Netherlands, in 2007, which sparked my interest in historical organs, and Anja Hendrikx and Erwin Wiersinga for teaching me how to play an organ.

My colleagues at the Faculty of Arts and Social Sciences of Maastricht University have followed my work on the new baroque organ with interest throughout the years, and without them I would not have been able to write this book. I want to thank Karin Bijsterveld, René Gabriëls, Jacqueline Graff, Anna Harris, Ruud Hendriks, Anique Hommels, Ike Kamphof, Jens Lachmund, Harro van Lente, Darian Meacham, Jessica Mesman, Maaike Meijer, Cyrus Mody, Wiebe Nauta, Christine Neuhold, Bernike Pasveer, Janosch Prinz, Jan de Roder, Vivian van Saaze, Geert Somsen, Renée van de Vall, Sophie Vanhoonacker, Marlies Vermeulen, Jo Wachelder, and Sally Wyatt for their support, advice, and reading tips. Darryl Cressman, Maarten Doorman, Louis Stiller, Tsjalling Swierstra, Tullio Viola and Rein de Wilde commented on earlier versions of chapters and the manuscript, corrected the English, and helped me to shape my arguments. I thank the members of the Maastricht University Science, Technology, and Society Studies research program and of the Department of Philosophy for their comments during presentations. I also want to thank the students of the research master's program CAST (Cultures of Arts, Science, and Technology) who reflected on recreating musical cultures of the past for the twenty-first century during their joint research project and presented their papers during a symposium at the Orgelpark in 2014 and 2015. I am also grateful to the Faculty of Arts and Social Sciences of Maastricht University for funding a sabatical that allowed me to analyze the empirical material that I collected during my fieldwork.

My collaborations with colleagues from other faculties and organizations have helped me to develop my thinking, in particular, Henk Borgdorff, Hannah Bosma, Sven Dupré, Julia Kursell, Harro Maas, Frank Mineur, Stephanie Pitts, Hans Radder, Linnea Semmerling, and Floris Schuiling. Margaret Meredith gave me valuable feedback in the early stages of my writing.

Shortly after the inauguration of the Utopa Baroque Organ, in March 2018, I became director of the Maastricht Centre for the Innnovation of Classical Music, a structural collaboration between what is now Philzuid, the orchestra of Southern Netherlands, Maastricht University, and the Conservatory Maastricht of Zuyd University and its research center, What Art Knows. Translating what I learned about innovating organs to the world of symphonic music turned out to be an inspiring and fruitful endeavor. Our discussions on the concept of the "musical work" and how it still regulates classical music practice, nineteenth-century idealist esthetics, and twentieth-century music mediation sociology were an important intellectual background while writing this book. But even more important was what we learned together in our projects and concert experiments. I am grateful to Ruth Benschop, Imogen Eve, Joachim Junghanss, Mette Laugs, Karoly Molina, Denise Petzold, Stefan Rosu, Neil Tòmas Smith, Veerle Spronck, and Ties van de Werff for being a wonderful and inspiring team. I sincerely thank them all for being so invested in the project of innovating music. I also thank Neil for his helpful comments and correcting the English of chapters in this book.

My colleague at the Faculty of Arts and Social Sciences Wiebe Bijker told me about the Orgelpark back in 2007. Sharing our interest in music and both being amateur musicians, we talked about organs in the years that followed. In his role as editor of the Inside Technology series at the MIT Press, Wiebe encouraged me to write this book, and without his continued interest and his gentle deadlines in the final stages, it would not have been finished. His coeditor of the series, Trevor Pinch, came to the Orgelpark in 2015, after I invited him to present his work on analog synthesizers and circuit bending. I will remember him as a warm and generous colleague and as one of the most inspiring and engaged voices in the dialogues among art, music, and STS. I thank the series editors, Rebecca Slayton and Edward Jones-Imhotep, for reflecting with me on the title of the book. I am grateful for the detailed comments of the three anonymous reviewers. For their support and advice in the publication process of this book, I am thankful to Justin Kehoe, Jitendra Kumar, and Sheena Meng.

Finally, I wish to thank my partner, Bernike Pasveer, and my daughter, Stijn Pasveer, for always finding ways to break new ground in our lives, and, most of all, for making music together.

INTRODUCTION

It was during a visit to the Organ Museum in Borgentreich, not far from Kassel in Germany, that I understood what this book is about. The museum is dedicated to the pipe organ in the nearby St. Johannes Baptist Church. Whereas a layperson might see an ordinary historical organ, an expert would know why this instrument deserves its own museum. What makes the organ in Borgentreich special is the way its windchests are constructed. Windchests are large wooden boxes on which the organ pipes sit. They lead the air—or "wind"—from the bellows into a pressurized chamber. After that chamber, two devices keep the wind from entering the pipes: stops and keys. If an organist opens a stop, a rank of pipes on the windchest will sound when keys on the manual are pressed.

The most common type of windchest is the slider chest. When a stop is pulled, a wooden slat in a windchest slides into position; the holes in it align under the pipe feet, and wind can flow into the pipes. Instead of slider chests, the organ in Borgentreich has six spring chests. This type of windchest uses a different mechanism to open or close the pipe ranks of a stop (see figure 0.1). It does not have a slat with holes that slides below the pipe feet but, instead, has small valves under each pipe (see figure 0.2). When a stop is drawn, all of the pipe valves in a rank are pushed down to allow the wind to flow into the pipes. Whereas slats in a slider chest need space to move, causing them to leak wind, spring chests are airtight and, therefore, more reliable. They are also notorious among organ builders because they are complicated to build. More importantly, however, the first spring chests, appearing in Italian organs from the fifteenth century, were hard to maintain: If something went wrong, one had to take out all the pipes to repair them. The solution to this maintenance problem can be seen in the organ in Borgentreich. The small valves are mounted on a wooden beam, one corresponding to each key of the keyboard, that can easily be pulled out of the windchest without touching the pipes. Even though it was technically superior to the slider chest, the

Figure 0.1
Model of the spring chest in the St. Johannes Baptist Church in Borgentreich, Germany (photo taken by the author).

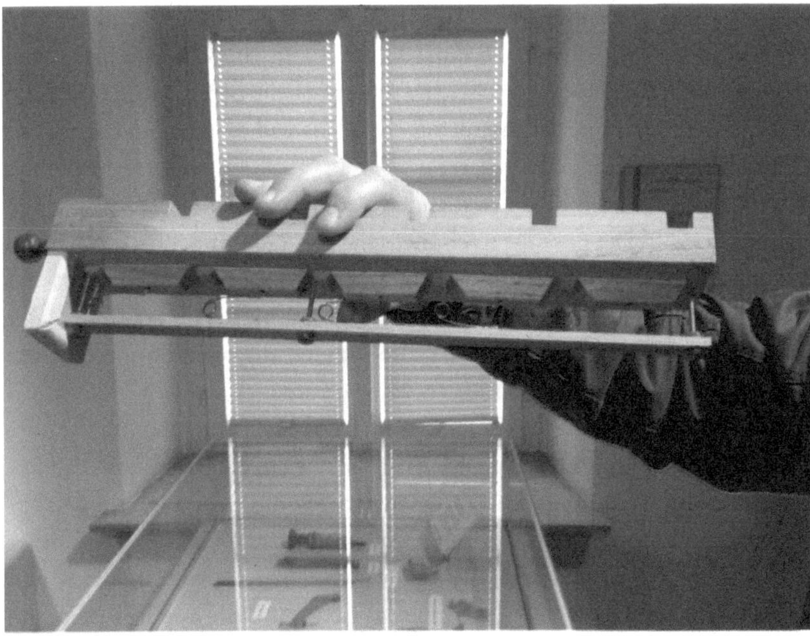

Figure 0.2
One of the wooden beams with its small valves taken out of the spring chest model (photo taken by the author).

spring chest ultimately went into oblivion. After 1700, slider chests worked just as well and were significantly more affordable.

It took until 2013 for the archaic organ technology of the spring chest to be revived. In that year, it became the focus of a new project carried out by the Orgelpark in Amsterdam. This concert venue's mission is to integrate the pipe organ into musical life, and it had decided to build a new organ. This instrument was to have two faces. On the one hand, it would be a baroque organ on which music by Johann Sebastian Bach and his contemporaries could be played according to historically informed performance practices, where "baroque organ" refers to organs built between 1650 and 1775 in Northern and Western Europe, notably in the Netherlands, Germany, and Scandinavia. On the other hand, it would allow access to the pipe sounds through a digital console. This would allow musicians to use the instrument to create new music, as well as open new perspectives on historically informed performance practices.

Key to this dual design was the spring chest of which I had seen an example in the St. Johannes Baptist Church and the Organ Museum in Borgentreich. To be more precise, the success of the project would depend on the small pipe valves in the spring chest. Instead of opening them with a wooden slat with metal stickers, as in the seventeenth-century original, each pipe valve would be opened and closed by its own electromagnet. In this way, it would be possible to connect the age-old windchest mechanism to a digital console, offering new forms of control of the instrument and its sounds to musicians. Using the digital console, musicians could basically create any combination of pipes. By also varying the opening speed of the valves, yet-unheard sound worlds could be created.

I visited the museum in Borgentreich in 2016, as a member of the core team that was responsible for the new baroque organ project. We were making trips to historical organs in Germany that could be used as models for the new instrument, organs built by famous builders such as Arp Schnitger (1648–1719) and Zacharias Hildebrandt (1688–1757). Although it would be impossible to copy these instruments, or even build replicas of them, they were an important inspiration. How might one build a new organ that would have the same sound quality as the historical organs and would also offer access to these sounds in new ways? I was invited to follow the project that aimed at building a new baroque organ from the start in 2013 as an STS scholar working on music and its technologies. What fascinated me about this project, and what became clear to me in Borgentreich, was that the unassuming pipe valves in the spring chest were not only relics of the past, interesting from an organological perspective. They also had the potential to bring something new into the world of music.

This book is about the project that created the Utopa Baroque Organ, which now resides in the hall of the Orgelpark in Amsterdam. Even though it is brand new, it embodies the long history of pipe organs, dating back to the Middle Ages and even earlier. These instruments have always incorporated new knowledge, materials, techniques, crafts, and ideas. Studying how the new baroque organ project mobilized this long history to create a pipe organ for the twenty-first century fits into the tradition of "object lessons" in science and technology studies. Bicycles, automated subway trains, bridges, air pumps, bush pumps, and synthesizers—studies of these artifacts and innovations are often post hoc reconstructions.[1] Stabilization of artifacts and innovations in networks and practices cannot be explained only from their intrinsic properties or qualities but should consider local circumstances and contingencies. Offering detailed descriptions and analyses, these reconstructions share the argument that things could have been otherwise.

In the case of the Utopa Baroque Organ, I was able to witness its design and building as it happened. In fact, when I started writing the first chapters of this book, the organ was still in the workshop in Soest. This means that I have been part of the story that I narrate in this book. My own learning has coevolved with that of the people who built the organ. Interest in the organ project originated in part from my own experiences as an amateur organist, fascinated by the sound, materiality, and historicity of these instruments. In their complexity, they embody musical and technical knowledge that makes them a strategic research site for anyone interested in the role of science and technologies in musical practices. And finally, there was the curiosity to see as an insider how the ambitious project would develop. My initial question was straightforward: Will the actors be able to design and build an organ that is considered important by the organ world? In other words, how will the new instrument be endowed with musically interesting qualities? Looking back, this question was too simple since it assumes that these esthetic qualities are unambiguously clear and can actually be planned, designed, and built. Not only did the process of creating the Utopa Baroque Organ turn out to be much more complicated, but my own ideas of what constitutes esthetic qualities and how they relate to knowledge and making practices evolved as a result of participating in the project.

Through my investigations into the making of the Utopa Baroque organ, I aim to contribute to three interrelated academic fields: new or relational musicology, science and technology studies of sound and music, and the emerging field of artistic research in music. What connects these fields in how they study musical cultures can be summarized as the shift from an internalist focus on the "musical work" and how it is analyzed, to a more inclusive perspective on "music-in-action" (DeNora 2016, xi; Acord and DeNora 2008), or the empirical study of the conditions under which music is produced, performed, and consumed. Central to this practical or performative turn in music studies is the reflection on how music exists, how to study music in action, and how to consider the esthetic, normative aspects of musical practices. Before outlining the chapters of this book, I will briefly contextualize these three issues.

In her book *The Imaginary Museum of Musical Works* (1992), Lydia Goehr approaches the question of musical ontologies in two steps. She first critically discusses the attempts of analytical philosophers to describe and define how we should understand the reality of musical works. She shows how any fixed definition has to establish a set of either essential properties or identity conditions. This approach, she argues, does not account for the historical, contingent, and possibly changing character of these properties and conditions. Goehr suggests that instead of describing what kind of *object* a musical work is, we should study how the *concept* of a work emerged in classical music and how it has functioned therein.[2] Her strategy is to trade thought experiments in music ontology for historical research. Drawing on archival sources, she shows how the idea of the stable musical work emerged only around 1800. In what Goehr calls the Beethoven paradigm, the musical work regulated how composers notated their music, how performers were expected to be true to the score to give authentic performances, and how audiences listened in silence to hear the beauty of the work itself.

Since the nineteenth century, music history, musicology, and music sociology have been closely aligned to the Beethoven paradigm. Scholars made a distinction between the study of music as writing and music as practice. Musicologists analyzed scores and their histories; sociologists, historians, and anthropologists focused on how music works and how its practices evolve. Recent research in music sociology and new or relational musicology, however, has questioned this gap between page and stage, between the musical work itself and the practical work that has to be done to make music happen (Aguilar et al. 2021; Clarke and Cook 2004; Cook 2014; DeNora 2003; Devine and Boudreault-Fournier 2021; Martin 2006). Music in the making is understood here as a social, material, and situated practice. Examining musical practices in these situations of "action" can draw on interdisciplinary practice research. To give an example from the field of science and technology studies: If we think of musical instruments as technological artifacts, we can study how their meaning and use developed over time in different practices, or how certain aspects were kept stable while others changed (Pinch and Bijsterveld 2003).[3]

Coming from a background in the sociology of innovation, Antoine Hennion offers another example in his argument that music from the past can appear among us here only through a collective process of assembling mediators such as musical instruments, scores, norms, languages, institutions, interpreters, and teachers. As ethnographers of musical practices, we can traverse these assemblages. Working in the context of actor-network theory in the 1980s and the 1990s, a theoretical and methodological approach to social theory that understands reality as assemblages of relations between entities (Blok, Farías, and Roberts 2020; Latour 2005), Hennion emphatically criticized the opposition between the esthetic analysis of the musical work as an isolated art object and the analysis of the social conditions through which it exists (Hennion 2015). In the past decades, post–actor network theory and mediation theoretical approaches that eroded the dualism between the work itself and its social conditions have informed the study of musical practices (Born 2005; Hennion and Levaux 2021; Mattozzi and Parolin 2021; Piekut 2014; Schuiling 2020).

The concept of the musical work as it originated in the nineteenth century had practical consequences; it led to a division of artistic and reflexive labor. Composers create, musicians perform, audiences listen. Musicologists took up the task of scholarly reflection on music. They analyzed scores as representations of musical structures and meanings. Music criticism became the domain of experts who had the knowledge and experience to make value judgments about works and performances. Even though the practical or performative turn in music studies has challenged this division of labor, the issue of the music's normative and esthetic value remains relevant in relational musicology (Born 2010). The question of how empirical studies can consider the normative aspects of musical practices is taken up not only in science and technology studies but also in research in the arts (Borgdorff, Peters, and Pinch 2020).[4]

Music, as a matter that concerns us, must be arranged, interpreted, rehearsed, worked upon, listened to, put to the test.[5] In our understanding of it, knowing and making go hand in hand. In musical practices, this insight is not new, of course. To create new music, musicians and composers have always reworked and adapted existing music, studied sources or the contexts in which they were written, and developed new skills and technologies. Recently, however, artist-researchers in the field of artistic research have made this work more explicit.[6] Some artists present their art not only as works or performances that are acknowledged and evaluated in art worlds.[7] They also document the research that their art making requires in ways that allow academic communities to assess its epistemic value. They thus expand the ways in which their artworks can exist and be made relevant. Artistic research in music shows how we can think of "music in action" as experiments. Setting up these experimental situations combines reflection and creation, and the new baroque organ project that the Orgelpark initiated is an example. My aim in this book is to describe and analyze the building of the new baroque organ to understand how working with the "pastness" of pipe organs can open up new futures for the instrument and its music.

In chapter 1, I attempt to situate the Orgelpark, the new baroque organ project, and my own research by exploring three interrelated issues. The first reveals the Orgelpark as a site of disruption. The building in which it is housed used to be a church, and its German Romantic organ was mainly used to accompany the singing of the congregation. After the Park Church had been refurbished as a concert hall, it became a place where music was performed and researched as a form of art. The use and qualities of the music and the organ sound itself were now mainly defined and assessed according to esthetic criteria. The second issue that I introduce regards the long history of evolution and adaptation of pipe organs. Throughout the centuries, they have reflected and embodied the state of the art in music, architecture, handicraft, and science. This raises the question of how the new baroque organ project related to this rich and complex historicity, and what can be learned from it about the transmission of musical culture through time. And, finally, there is the issue of how entering the Orgelpark opens a world in which questions of knowledge, artisanry and esthetics, historical significance and cultural transmission, tradition and innovation, and renewing crafts and musical performance are deeply entangled.

Chapters 2 and 3 concern the first two years of the project, in which decisions were taken about the historical reference organs and the design of the windchest of the new organ. The focus in this period was on mobilizing organological and other technical knowledge to finalize the plan for the instrument.

The new baroque organ project took historical organs in Germany as models, more specifically, organs that Johann Sebastian Bach had played. Chapter 2 analyzes the way the new baroque organ project can be related to the historically inspired performance and organ-building practices that developed in the twentieth century. Three "expeditions" to some of the organs of Bach were seen as a necessary step in the design process of the new organ because they helped members of the core team to situate themselves historically not only in terms of organological and mechanical aspects of the instruments but also in their sound characteristics. Joining the other members of the core team on these journeys enabled me to follow the discussions on practical and technical problems that emerged along the way, and to understand the background of the choices the team had to make during the first stages of the project. The choice between organs built by Arp Schnitger and Zacharias Hildebrandt as the historical reference is an example. Referring to the work of Hans-Georg Gadamer on historical hermeneutics, or the engagement with the question of how to read and give meaning to the past, I analyze how the team members negotiated a hermeneutical situation in which they balanced the goal of reconstructing a historical sound world with the realities of contemporary musical life in the Orgelpark.

In chapter 3, I focus on the design process of the hybrid windchest that started with the making of a first test organ. The core issue in the discussions and debates that the prototype triggered was how to combine a convincing baroque organ concept with the possibility of accessing its sound through electronic means, thus inspiring organists, improvisers, and composers to make new music. To design the hybrid windchest, the project team worked with organ-building traditions from various periods, resulting in a final design of the windchest that combined knowledge and understanding, as well as practical solutions and techniques ranging from the sixteenth to the twenty-first century. To understand the historical references embodied in the windchest, I draw on the work by Bruno Latour on the hybrid ontology of things, and I discuss how this work is helpful to redefine what it means to design an artifact, such as the hybrid windchest, that uses elements from different historical periods.

In chapters 4 and 5, I recount the actual making of the new baroque organ in the two-year period between the acceptance of the final plan and the transport of the organ from the workshop in Soest to the Orgelpark in Amsterdam in the summer of 2017.

Whereas the previous chapters showed how the final design drew together elements with various histories, in chapter 4, I focus on the practical work to "fit" these heterogenous elements into a pattern that corresponded with the project's mission. Analytically, I use concepts from Richard Sennett's work on craftsmanship and artisanry. Sennett studies crafts in a pragmatist register that draws on John Dewey's ideas about education, experience, and inquiry and highlights the intimate relations between problem-solving and problem-finding, the iterative process of making and understanding, of shaping objects and learning from how they behave in our hands. In following the making process of the organ, I am interested in how problems were not given but articulated in situations in which inquiry implied trying out and taking risks. Throughout the chapter, I focus on examples of such problematic situations regarding the technologies of transmission in the new baroque organ: transporting the wind from the bellows to the windchest, connecting the movement of the stop drawknob to the windchest, and transmitting the movement of the finger on the key to the pallet in the windchest. In the new baroque organ, these transmission mechanisms combine historical organ-building conventions with present-day organ technologies and the use of electronic components and software to control the movement of organ parts. The central question in the chapter is how the builders succeeded in making these many parts fit together.

Chapter 5 explores the work that was done to recreate or reenact the sound of Hildebrandt's pipes. The work started by determining the design of the pipes of the various registers of the new organ. Measurements of Hildebrandt pipes, along with methods of pipe making in his time, were taken as a starting point. Ideas about the importance of reenacting traditional pipe-making practices as closely as possible played an important role in how the pipes for the new organ were actually made. Whereas the previous chapter analyzed how practical problems and puzzles were found and solved to make things fit, in this chapter I focus on how the making of the pipes coevolved with the ability to value their sound. In doing so, I look at the role of materials such as alloys, leather, wood, and cloth, as well as the use of machines and tools in manipulating these materials. Theoretically, I draw on the work of anthropologist Tim Ingold about materiality and skill in making and sociologist of music Antoine Hennion, who uses the concept of instauration to express the idea that things in the making, such as organ pipes, are calling us. We feel a responsibility to bring them to "more" existence through our actions. Making is understood here as "helping becoming," an active and reciprocal process, defining at the same moment what we make and what makes us.

Chapters 6 and 7 concern the work that was done to give the new baroque organ the interfaces that would allow musicians to interact with it and describe the first moments it was presented and used in concerts.

Chapter 6 focuses on how combining a traditional and a digital console would make the sound that was crafted by the pipe makers accessible in novel ways. At both the mechanical and digital console, an organist can follow baroque conventions in touch, articulation, and registration. The digital console also enables performers to link the organ to computers and other electronic devices so it can serve new musical purposes. Whereas the traditional console is grounded in a conception of music as tones, intervals, and harmonies, the digital console connects the organ to the world of electro-acoustic sound. After outlining the tradition of electro-acoustic music, I show how ideas in this tradition provided an important benchmark in the design of the digital console. Concerts offered opportunities to experiment and demonstrate ways of interacting with the organ. For me as a researcher, these concerts were an important way to understand how the new baroque organ could be positioned in the landscape of twenty-first-century music.

In chapter 7, I continue the exploration of how the new baroque organ became an instrument endowed with musical qualities. This happened not only in the design and making but also in the placement of it in the hall of the Orgelpark. A recurring theme in the debate on the appearance of the organ was whether it should present itself as a historical instrument or as an instrument of the twenty-first century. Not only in its appearance but also in its sound and temperament—that is, the way it is tuned—the organ had to be integrated in the Orgelpark. The debate on the temperament of the new organ was settled with the choice for Hildebrandt as the main historical reference. After the organ had been shaped, painted, voiced, and tuned, it was ready to present to its audiences. This was done during a series of concerts that started on March 21, 2018, the 333rd birthday of Johann Sebastian Bach. Combining performances of Bach's music with improvisations on the digital console, these concerts anticipated the future programming that the new organ would make possible. To understand how we can think of the Utopa Baroque Organ as an "art thing," I revisit the esthetic and hermeneutic philosophy of Hans-Georg Gadamer that I introduce in chapters 1 and 2. Rather than being experienced in an alienated esthetic consciousness, art exists through its acquirement of new uses and adaptation to changed circumstances. Being presented under the new conditions of the contemporary, artworks are fundamentally unfinished and their meaning is never exhausted. I show how Gadamer's concept of art as play allows us to understand the work that was to be done to achieve the task of building the organ.

In the conclusion, I pull the preceding chapters together in an argument about how working with the "pastness" of pipe organs and their music in present situations can lead to the imagination of the new and unfamiliar. Making a new organ did not mean executing a plan or copying a historical model; rather, its qualities emerged in the very making process through which they could be recognized and put to the test. Building the Utopa Baroque Organ required working with history in three ways: by being aware of the strangeness and familiarity of the historical organ worlds that inspired the project, by reassembling technologies or *things* from various historical periods and, in doing so, adding to their varied ontologies, and, finally, by engaging in problematic situations that emerged in the shaping of these new assemblages through trying things out, taking risks, and dealing with the future through the imagination.

1

A NEW BAROQUE ORGAN

FAREWELL TO THE VAN LEEUWEN ORGAN

People are playing a ball game in the Vondelpark in the center of Amsterdam. Cycling through the park, I follow my usual route, passing joggers on the main path that curves around one of the ponds. To my right are the stately houses built for the bourgeoisie of the late nineteenth century, when the city's elite took the initiative to erect a concert hall that would rival the best in Europe. In front of me, partly covered behind the trees, I see the pinnacle on the roof of what used to be called the Parkkerk (Park Church). Taking a right turn, I leave the park through an iron gate and lock my bicycle in front of the former church. Above the entrance is the name of the building: *Orgelpark*. Crowning the white letters are nine vertical bars in red, green, purple, yellow, blue, violet, orange, dark green, and pink. The abstract composition may refer to the pipes of an organ. After a substantial renovation, the former church opened its doors back in January 2007 as a concert venue dedicated to organists who want to give their instrument a new role in musical life.

On this first day in June 2016, my aim is to have a last look at the Van Leeuwen organ inside and to play on it. In a few days, the organ builder will start to take it apart. After dismantling the organ case and taking out the hundreds of pipes one by one, the windchest on which the pipes stand will be disconnected from the wooden channels through which the air flows from the bellows. Packed in cases, the parts of the instrument will be transported to the Music School in Nowy Targ, near Krakow, Poland, where they will be reassembled.

Today the staff of the Orgelpark are busy in their offices preparing for the three-day research symposium that is organized every year. There are no rehearsals in the main concert hall, so there is some time for me to play. I walk up the stairs and open the balcony door. The sight of the sharp-angled, V-shaped case of the Van Leeuwen organ and its symmetrical rows of front pipes ascending from the middle to the sides has become familiar to me in the years that I have been visiting the hall (see figure 1.1). The two manuals and the pedal are located at the front of the organ. On either side of the manuals are vertical rows of stop drawknobs that allow the organ player to change the sound and timbre of the instrument. I turn on the blower, a machine that provides the wind for the pipes. Inside the case, I hear the almost imperceptible sound of an electrical motor gaining speed. I sit down on the organ bench and look at the stop drawknobs. They have names such as Prestant 8', Speelfluit 4' and Nachthoorn 2'. The Dutch spelling indicates that the organ was built in the Netherlands.[1] The term 8' means eight foot and indicates the size of the longest pipe of the respective stop, producing a frequency of about 256 Hz. A four-foot stop sounds an octave higher, two-foot stops two octaves higher. Drawing a stop means that the corresponding row of pipes will sound, each row with a different sound character. Combining the stops, I can create a variety of sounds. One of the stops sounds beautiful, I know—the Roerfluit 8'—so I decide to draw its stopknob.

I play a short chorale variation by Johann Sebastian Bach, *Wer nur den lieben Gott lässt walten* (BWV 691). Being an amateur organ player, I have often practiced and played this piece. It has a two-part structure characteristic of the Lutheran chorales that afforded the basic musical material for Bach and so many other composers of church music before and after him. When I play the first embellishment, a rather long trill, I realize how little touch I need to press the keys. Hardly any force is needed to transfer the movement of the key via a system of trackers, through the organ from the manual to the valves in the windchest just below the pipes. When I press a key, a valve opens and pressured air, or "wind," flows into the pipes for which stops have been pulled. While playing, I try to pay attention to my articulation. One of the first things my organ teacher taught me was that, on an organ, the trick is to minutely control the length of a note and the duration of the gap between the notes. She also made me listen to the sound of the key action, the mechanism that links the keyboards to the organ pipes: If you can hear the movement of the trackers inside the case, this means that your fingers, hands, and arms are probably not relaxed enough. In a way, the instrument teaches me how to play it. Listening to the sound of the single stop makes me aware of the acoustic in the hall. For the French organist Aristide Cavaillé-Coll the acoustic was the most important stop on any organ.

Figure 1.1
The Van Leeuwen organ (1954) on the north balcony in the hall of the Orgelpark in Amsterdam, June 2016 (photo taken by the author).

The day after my visit, the Van Leeuwen organ took center stage during a concert given by Viennese organist Peter Planyavsky, marking the opening of the symposium.[2] Before the concert, Loek Dijkman gave a short speech as chair and founder of the Utopa Foundation that initiated the Orgelpark and sponsors it. Dijkman described how the Van Leeuwen organ had entered the Orgelpark. When it arrived, the church's own instrument, a German romantic Sauer organ that was inaugurated in 1922, had already been restored and renovated. In addition to the Sauer organ, a new French romantic organ had been built by the Dutch organ builder Verschueren in 2009.[3] By commissioning this big new organ, the Foundation followed its mission to support aspects of organ culture such as organ building.

Although music from the baroque era could well be played on the two romantic organs, one of the organ specialists who advised the Orgelpark at the time suggested they buy the neo-baroque organ built by Willem van Leeuwen in 1954, for a church in The Hague. Its objective, unromantic sound was said to reflect the sober spirit of the postwar era of rebuilding the country (Fidom 2006b). Like many other pipe organs from this period, its sound ideal was derived from organs built in the seventeenth and eighteenth century, hence their stylistic label of neo-baroque. In his speech, Dijkman talked about the Van Leeuwen organ, but he also spoke of the "labor pains" of a new baroque organ in the making, one that would replace the Van Leeuwen organ. Even though he was glad that it would be used by music students in Poland, Dijkman went on, saying farewell to the Van Leeuwen organ was not easy: "So with pain in my heart, this organ is leaving us."[4]

To some in the audience the speech may have been puzzling. Why was an organ that had been acquired only a few years earlier going to be replaced by a brand-new instrument that basically has the same function: to enable the performance of organ music from the baroque and, more specifically, the compositions by Johann Sebastian Bach? This question is at the heart of this book. It is about the building of the instrument that replaced the Van Leeuwen organ and what this project can teach us about how, in musical practices, the new comes into being. Paraphrasing the American pragmatist philosopher John Dewey, I hope to find out what the organ builders retained from the past and what they expected from the future, and how this directed their actions in the present (Dewey 2005 [1934], 18).

My visits to the Van Leeuwen organ and its farewell concert were just two of many in the years that I was engaged in the research for this book. The Orgelpark became a familiar place to me, a place that fascinated me both as a researcher and an amateur of music. Its mission to give the organ a new role in musical life makes it a studio for the study of the transmission of musical cultures. I use the term transmission to refer to the work that has to be done to reiterate and innovate these cultures, both historical and contemporary. My aim in describing and analyzing the building of the new baroque organ is to understand how working with the "pastness" of pipe organs and their music can lead to the imagination that anticipates the unfamiliar.

In this chapter, I will situate the Orgelpark, the new baroque organ project, and my own research by exploring three interrelated issues. The first reveals the Orgelpark as a site of disruption. For many years, the Sauer organ accompanied the singing of the congregation. After the Park Church had been refurbished as a concert hall, it became a place where music was performed and researched as a form of art. Organs became concert instruments, their sounds to be listened to attentively by a silent audience. The use and qualities of the music and the organ sound itself were now mainly defined and assessed according to esthetic criteria that are no longer linked to a religious practice. The changing use and valuation of the organs in the Orgelpark highlights a second issue. Of all musical instruments, organs have the longest history of evolution and adaptation. Throughout the centuries, they have reflected and embodied the state of the art in music, architecture, handicraft, and science. As "mirrors of their time," organs have a historical and cultural significance (Snyder 2002). This raises the question how the project to build a new baroque organ relates to this rich and complex historicity. And finally, there is the issue of how a concert venue can be a site of interdisciplinary research on contemporary musical practices. Entering the Orgelpark opens a world in which questions of knowledge, artisanry and esthetics, historical significance and cultural transmission, tradition and innovation, and renewing crafts and musical performance are deeply entangled.

In what follows, I will first describe how the Orgelpark came into being as a concert venue. Focusing on one of its organs, a replica of the oldest remaining Dutch organ, the Peter Gerritsz organ from 1479, I will explore how organs can have long histories in which they are continuously adapted to changing needs and values. The inauguration of the Van Straten organ, as the replica was called, turned out to be an important step toward the plans for building a new baroque organ and, in particular, the radical idea that is at the heart of its design. I end the chapter with some reflections on the theoretical grounding and methodological approach of my research.

BRINGING THE ORGAN BACK TO MUSICAL LIFE IN THE ORGELPARK

First-time visitors to the Orgelpark tend to overlook the two brass sculptures of a cobra that wind along the two stairs to the cloakroom. They function as handrails but they also refer to a chapter in Protestant church history in the Netherlands. In 1923, the Park Church became the focus of a theological controversy after its vicar, J. G. Geelkerken, gave sermons in which he claimed that the story in Genesis 3 about the snake talking to Eve should not be taken literally but rather as a metaphor. The disagreements led to a secession in the Reformed Churches in the Netherlands. Although suspended, Geelkerken continued to draw thousands of people to the Park Church. Every week their psalms resounded under its elegantly curved ceiling.

Situated on the edge of the Vondelpark, the church in the Gerard Brandtstraat opened its doors for the first time on Christmas, 1918. It had been designed by the architect E. A. C. Roest in a Romanesque style, which shows in the rounded windows that were originally fitted with stained glass. The square ground plan gives the building a symmetrical aspect. Its interior is sober and sturdy. The geometric ceiling adds a sense of spatiality. Upon entering, the church is set as a theater with three balconies situated around the church floor facing the pulpit. Above the pulpit is the Sauer organ, built by what is now considered to be one of the most important organ factories in the late nineteenth and early twentieth century, established in the mid-nineteenth century. The building functioned as a church until 1994. One year later, the municipality of Amsterdam took over the building.[5] Although not given a definite new purpose, it was used intensively, first as a dance studio and later as a rehearsal space for the Dutch Wind Ensemble. Various initiatives were taken, but the costs of rent and renovation of the building were too high to permanently accommodate any nonprofit organizations. Like many other churches in the Netherlands, the deteriorated building waited for a new use.

In the early 2000s, Loek Dijkman thought of a way to support organists and their practice. As owner and general director of a group of companies

Figure 1.2
The Park Church, around 1940, with the Sauer organ and the pulpit below it, and the north balcony to the right on which the Van Leeuwen organ was later placed (source: Stadsarchief Amsterdam/N.V. W. ten Have).

operating in transport packaging, he had launched the Utopa Foundation in 1988. In a book that he published in 2013, Dijkman explains the ideals behind the initiative (Dijkman 2019 [2013]). In his view, a company's responsibility to the society in which it operates goes beyond providing employment and making profit. Instead of paying "surplus profit" to shareholders, it should be used for the greater good. The objective of the Utopa Foundation became to "stimulate and promote the creative talent of people, particularly those whose potential goes unrecognized, for whatever reason."[6]

During an interview in April 2016, Dijkman and his wife, Sylvia de Munck, who is a member of the board of the foundation, recounted the events that led to the establishment of the Orgelpark. Comparing the organ to a multifaceted diamond, Dijkman reflected on the instrument as "an impressive sound machine, a bit pompous even, that evokes thoughts of death, eternal life and sacred moments."[7] But his personal interest in the organ was not the main reason that the board of the foundation decided to invest in a new future for the pipe organ. The focus was on the organist, the musician, the artist, not primarily on the instrument. Coinciding with the closing down of churches, the profession of not only organists and organ builders was under threat but also composers of organ music. Support of artisanry being one of the focal points of the foundation, it was necessary to ensure that they could continue their craft by creating a new role for their instrument in musical life.

The foundation started to finance the restoration of organs in the Netherlands and initiated a project called "The Organ in the Classroom." This featured a mobile model of a pipe organ through which pupils could learn how these instruments work. The restoration projects to which his foundation contributed brought Dijkman into contact with the department of Sounding Monuments at the Cultural Heritage Agency, which oversees the restorations of historical organs and carillons in the Netherlands. He developed a network in the Dutch organ world and first met the people who were to play an important role in the Orgelpark: Johan Luijmes, who later became artistic director, and Hans Fidom, who became the Orgelpark sponsored chair in Organ Studies at the VU University Amsterdam and director of the Orgelpark Research Program.

The Orgelpark was established in 2003. After Dijkman learned that the former Park Church was available, the Utopa Foundation decided to renovate the building to use it as a concert venue.[8] The name referred to the nearby Vondelpark, but also to the plan to have a "park of organs." A restoration plan for the structure of the building was made, as well as that for the design of the entrances and the interior. At the same time, the restoration of the Sauer organ was prepared. Because the Park Church had been given the status of national monument, the renovation was done with great care to keep the original building intact as much as possible, while creating new space for the cloakroom, the box office, the foyer, and the restrooms. Whereas the original church had brown furniture and grayish-white walls and ceiling, the new concert hall was painted in a gamut of colors, ranging from various tints of gray and dark and lighter green in the lower segments to combinations of yellow and orange on the balcony. The ceiling was painted in a light pastel tone of yellow.[9] The Romanesque windows on three sides of the hall were covered by translucent images from the history and the renovation of the building. The row of name plates with organ composers along the balcony should be seen as a playful reference to the nearby Concertgebouw, which also has name plates (see figure 1.3).[10]

Figure 1.3
Interior of the hall of the Orgelpark with the Van Leeuwen organ in front, the Sauer organ to the left, and the Verschueren organ to the right. Below the Van Leeuwen organ is the Molzer organ (source: Orgelpark/Joyce Vanderfeesten).

To fulfill its mission, the Orgelpark needed other organ sound worlds than the Sauer organ it already possessed. Being one of the remaining three by this important builder in the Netherlands, it is an official sounding monument. Elbertse Orgelmakers, an organ builder based in Soest, the Netherlands, who had extensive experience with nineteenth- and early twentieth-century organ technology, restored the instrument.[11] The Utopa Foundation commissioned a French Romantic organ from the organ builder Verschueren and bought a salon organ built by Molzer in 1913 in Vienna, transported it to the Orgelpark, and restored it. The choice of these instruments reflected Dijkman's initial preference for organ music—and art in general—from the Romantic and early Modern period. Another reason to focus on the nineteenth and early twentieth century was that organ music from the seventeenth and eighteenth century is frequently performed in the Netherlands:

> I discussed this with Johan [Luijmes—PP] and we said: let's just pick 1850. If someone asks for Bach, we will say: Bach? Never heard of [him]. That was

> nonsense of course. But we wanted to bring the organ back to music and to contemporary concert practice, and I thought French organ culture was terrific: Franck, Widor, Langlais and Messiaen.[12]

The Utopa Foundation decided to not only finance the renovation of the former Park Church and the restoration of the Sauer organ but also to provide an annual budget for the concert hall. In the early years, around one hundred concerts and other events such as symposia were scheduled. The first concert in the new hall was given on January 20, 2007. In the first issue of the periodical *Timbres*, which was published twice a year, Dijkman and the team at the Orgelpark underlined their intention to create a venue where musicians, artists, and their audiences would feel welcome and at home.[13]

In the years following the opening, artistic director Johan Luijmes has developed a programming style that explores the organ in all its possibilities. The concerts not only focus on its own repertoire and use as an instrument on which to play and improvise but also present combinations of the organ with other musical instruments and genres that are not usually associated with organ music. Classical music, jazz, and improvised music are often linked with other art disciplines such as dance, mime, performance art, and film. From the start, the idea was to combine the organization of concerts with doing research. The Orgelpark's research program, led by Hans Fidom, organizes yearly symposia, publishes research reports on aspects of organ culture and history, and contributes to the programming of the venue.

The mission of the Orgelpark resonates with a broader cultural development in the Netherlands and beyond. Due to declining attendance, churches merge with others or close their doors. As a result, the instruments that accompanied the singing of the congregation or church choirs are no longer used for their original task. Some historical organs were given the status of sounding monuments and acquired a new role as concert instruments, as in the case of the Sauer organ in the Park Church. Others were sold and moved, like the Van Leeuwen organ. When the religious context of use disappears, the question of how to make the organ, its music, and its practice relevant in new ways becomes all the more urgent. The transformation of a church into a concert space underlines that this new relevance is often sought in highlighting the art character of organ music. In the case of the Orgelpark, this also shows in the use of composer name plates as a tongue-in-cheek reference to the nineteenth-century admiration for the genius-composer as creator of timeless musical works of art. Yet a large part of the repertoire performed at concerts was created in times when music was not seen as art.[14] To make matters more complex, the removal of the Van Leeuwen organ suggested that the organ sound itself was a crucial parameter. Its postwar rendering of baroque sound was to be replaced by an instrument that matched more recent sound ideals. As will become clear, these ideals are closely connected with the historically informed performance practice that emerged in the 1960s.

At the Orgelpark, new interrelations among the function of organ music, its performance practices, and its esthetic valuation become visible. From a historical perspective, however, these changes are not new: Organs have always evolved and adapted to new uses, tastes, cosmologies, scientific insights, and artisanry. They embody cultural histories that go back ages. Tailor-made for one place, they were designed for the architecture and acoustic of that place, itself part of a landscape and a culture that gave the organs their significance and meaning. As technological artifacts, they have been sites of exchange between the knowledge from the organ builder as an artisan and knowledge from other domains. One of the organs in the Orgelpark, the Van Straten orgel, a replica of the oldest organ in the Netherlands, is an exemplar of this long history of changes and adaptations.

A MIRROR OF ITS TIME

The Van Straten Organ was unveiled at the Orgelpark in 2012.[15] This organ was designed and built as a replica of an instrument originally built in 1479 by Peter Gerritsz for the Nicolaï Church in Utrecht. The historical Gerritsz organ produced its last sounds in 1885, a description of which was recorded by a member of the congregation:

> During a Sunday morning service in the month of May 1885 it could still accompany the singing of the congregation, but it was impossible for the organist to continue; the instrument was not even able to produce its usual shrieking and moaning. (Cited in Dijk 2009, 73)[16]

The Dutch State bought the Gerritsz organ in 1886 and transferred it to the recently opened Rijksmuseum in Amsterdam. The organ would be silent from that moment on, but the case and its interior parts were of interest to Dutch cultural historians, musicians, musicologists, and organ experts. After it stopped being a musical instrument, that is, after it stopped being an object that produced musical sounds, it became an artifact through which researchers could learn about fifteenth-century building techniques and musical culture. The period around 1500 was a time of astonishing innovations in the art of organ building, all of which assumed the creation and application of various kinds of knowledge, ranging from mathematics and natural history to knowledge of materials such as wood, leather, and various alloys (Bormann 1966). Organ builders learned and transferred knowledge and insights from previous generations while also experimenting with new designs, techniques, and materials.

Organ scholars were able to reconstruct the long history of the Gerritsz organ from various sources. These range from the descriptions of payments in old church accounts to a close reading of the characteristics of the remaining material, such as inscriptions and markings of pipes, or the traces of repairs and changes of the instrument.[17] The instrument that Peter Gerritsz built in 1479 had two divisions ("Hoofdwerk" and "Bovenwerk") and a pedal. A division is the collection of ranks of pipes controlled by a particular manual. The grandson of Peter Gerritsz, Cornelis Gerritsz, gave the organ a new Hoofdwerk as well as a Rugpositief (back positive) in 1547 (Diepenhorst 2009, 206). Additional pipes were added to the Hoofdwerk, and changes to the keyboard were made. In the early seventeenth century, the organ was given a major renovation. In later decades, the shutters of the main case and the positive were repainted, and stops were added or removed. This process of repairing, renovating, and changing the organ continued in the eighteenth century. In 1733, the organ builder Christian Müller gave the organ two new keyboards. After Müller, Gideon Thomas Bältz conducted numerous repairs, though without removing or changing the oldest parts of the organ (Diepenhorst 2009, 210). In the nineteenth century, the organ increasingly suffered from leakages in the wind channels and windchests, and there were defects in the pipework. From 1872 onward, the organ was no longer maintained (Dijk 2009, 61).

After having resided for decades in the Rijksmuseum, the Gerritsz organ's case was transported in 1952 to the Koorkerk in Middelburg, the Netherlands, where it was prominently displayed. Because of a lease-lend agreement, the Rijksmuseum remained the owner of the organ.[18] The inner parts such as the windchests and the remaining pipes led a nomadic existence, being stored at various places, such as "sounding monuments" within the Cultural Heritage Agency (CHA) of the Dutch Ministry of Education, Culture, and Science in Amersfoort.[19]

In 2009, the CHA decided to build a study copy of the Gerritsz organ to gain new insights about the old organ. This project was funded by the Orgelpark. Although a few parts of the original organ had deteriorated beyond repair, such as the bellows and a few of the pipes, the shape, size, and material constitution of most of the organ's parts could be replicated so that an organ in the style of the fifteenth century could be built. This plan to replicate the Gerritsz organ culminated in the building of the Van Straten organ, named after Rudi van Straten from the department of Sounding Monuments of the CHA, who took the initiative for the replication project. The goal of building this new instrument was to go back to the earliest stage in the history of the Gerritsz organ, the situation in 1479. In doing so, it would become possible to gain knowledge about the Gerritsz organ that could be helpful in case the original had to be restored.

The replica project also gains meaning in the context of the early music movement, and, in particular, this movement's interest in performing music on historically accurate instruments (Lawson and Stowell 1999). From the 1960s onward, more and more musicians undertook research that sought to recreate the original sound of medieval, Renaissance, and baroque music. Pioneers like Nikolaus Harnoncourt, Gustav Leonhardt, and Frans Brüggen played from critical text editions, performed on restored original instruments or replicas, and adjusted the size of choirs, orchestras, and ensembles to historical conventions. They based their interpretations on primary source materials ranging from instrumental and theoretical treatises to surviving instruments, iconography, historical archives, references in literature, journals, newspaper reports, and sometimes letters, diaries, catalogues, advertisements—and for post-1900 music, even recordings (Lawson and Stowell 1999, 17–41). Against this craving for historical accuracy and the establishment of what could be considered "authentic" early music, some critics of the early music movement argued that this research strategy resulted in performances that focused on mere sound and completely lacked the expressive qualities of the musician.[20]

Reviving interest in early music would have been impossible without the flourishing development in instrument building and restoration that makes it possible for musicians to play string, wind, and keyboard instruments from different time periods. Surviving instruments are used to study and experiment with matters of technique, style, and interpretation, and instrument builders take these old instruments as starting points to relearn old building practices.

The restoration and replication of old church organs is closely related to the reinterpretation of early music by musician-researchers and the corresponding development of instrument building (Fidom 2000). Organs, however, present a special case. They are very large and expensive instruments that, historically, were not simply replaced. As the biography of the Peter Gerritsz organ showed, they were continually extended and restored in different periods. As a result, these ancient instruments contain within them layers of material, scientific, artisanal, and artistic knowledge. In the making of the Van Straten organ these sediments of knowledge played a crucial role.

The replication process consisted of three phases: studying the remaining parts of the Gerritsz organ, actually making the instrument, and reconstructing its use in the fifteenth century. The RCA advisor to the project, Wim Diepenhorst, compared the old windchest to the Rosetta stone (see figure 1.4). Read by an expert like Diepenhorst, the windchest contains

Figure 1.4
Detail of the windchest of the Peter Gerritsz organ (photo taken by the author at the office of the Cultural Heritage Agency [CHA] of the Dutch Ministry of Education, Culture, and Science in Amersfoort, the Netherlands, 2016).

information about the number of keys of the original organ, the number of pipes that sound when a key is pressed, the circumference of these pipes, the width of the keys on the keyboard, and the dimensions of the organ. "Reading" the windchest, though, is a process of conjectures and refutations. The details of the windchest have to be interpreted using historical knowledge from descriptions and drawings found in archives that date back to the fifteenth century. There is a limit to what can be known from interpreting remaining parts. Relearning historical organ building skills through making a sounding organ was essential in deepening the understanding of the Gerritsz organ. In a process of trial and error that could not rely on experience or on just copying the old parts of the windchest and the pipes, mistakes were made and new knowledge was gained.[21] As the original fifteenth-century keyboard from the organ has been lost, the organ builders had to reconstruct it. Conjecturing from remaining materials and old drawings, Diepenhorst and the organ builders found that the keys were wider than those found in later organs. The greater width of the keys proved to be important for the understanding of musical performance practice at the time the Peter Gerritsz organ was built. Having the Van Straten organ enables scholars and musicians to test existing knowledge about historical performance practice by actually playing on a musical instrument of which no playable originals remain.

Because the last moans and shrieks that the Peter Gerritsz organ brought forward have been recorded only in words, it is impossible to know, with accuracy, how it sounded in 1885 or before. After having been silent for more than a century, the Gerritsz organ became an object through which knowledge about fifteenth-century musical culture is produced for twenty-first century musicians, composers, researchers, and listeners. Making a new organ old tells us much more about historical musical cultures than the old organ did. Rather than reading the past through studying organs from that past, the team engaged in creating an imagination of that past with its instruments and practices, where accuracy is not so much a matter of having the historical facts in place but of making that past work in the present. The most obvious aspect of this is the sound. Reading original artifacts such as the windchest, making a new organ and its pipes, and performing it as a musical instrument was what it took to make the new Van Straten organ sound "old" and to have an idea of what medieval organs and their music may have sounded like. As it turned out, however, soon its historical sounds were also used to make new music.

A RADICAL DESIGN IDEA

The inaugural concert program on April 21, 2012, presented two faces of the Van Straten organ: one looking back to the organ sound of the Middle Ages and the Renaissance, and one looking forward to the use of this sound material in new music. Before the interval, Gregorian chant was sung in alternation with Wim Diepenhorst playing a selection from the *Buxheimer Orgelbuch*, a collection of organ tablatures published in Munich in the 1460s. In his review of the concert in the daily newspaper *Trouw*, music journalist Christo Lelie observed that although the organ has no more than five separate sounds, it remained interesting during the whole concert. One reason for this, Lelie argued, were the four blacksmith bellows that blow directly into the windchest. When their levers are operated manually, the wind pressure varies slightly, which results in a lively tone (Lelie 2012). Next on the program were two new compositions that had been commissioned for the reconstructed gothic organ. One of these, *Organism* by the Dutch composer Wouter Snoei, combined the amplified sound of the windflow from the bellows with electronic manipulations of the organ sound.[22] In hindsight, this dual use of the replica was pivotal for ideas that shaped the design of the new baroque organ.

After the arrival of a medieval organ replica, the Orgelpark was able to cover the main style periods in the organ repertoire, ranging from medieval and Renaissance, to baroque, Romantic, and contemporary music. The situation was not ideal for Johan Luijmes, however, as the one responsible for the programming of the venue. He dreamed of having a baroque organ that excelled through the pure beauty of its sound. During his training as an organist, Luijmes played mostly on historical organs.

> For years, I had my organ lessons on an organ from 1650. This has shaped my sound ideal. It is maybe through habit that I got used to the idea that historical organs and my ideal organ sound go together. . . . I started playing the piano and got into the early music movement. In our experience a neo-baroque organ was nothing at all. For me it is very personal: I just don't like to play Bach on the Van Leeuwen organ.[23]

Even though it had been decided back in 2007 that Bach's music would not take center stage in the Orgelpark, Luijmes wanted to include it in the program more often. In a short email to the board of the Utopa Foundation sent in the fall of 2012, Luijmes asked for a new baroque organ. In the board's initial rejection of the proposal, they argued that the Netherlands already has many organs well suited to perform the music by Bach, both historical organs and new instruments, such as the organ built by Verschueren in the Grote Kerk in Dordrecht. Doubling the possibilities that existed elsewhere did not fit the mission of the Utopa Foundation.

A new argument was put forward by Hans Fidom, who is responsible for the Orgelpark's research program. Mentioning the spring chest as the linking pin, he came up with the idea to build a baroque organ that could also be an instrument for making new music. This would help to realize the foundation's mission of giving the organ a new role in musical life. It turned out to be a convincing idea. In the spring of 2013, the board of the Utopa Foundation decided that a new organ had to be built at the Orgelpark to fill the gap between medieval and Renaissance music sound concepts and Romantic and contemporary ones. It should be designed in such a way that music from the baroque period could be played on it, especially music composed by Johann Sebastian Bach. The organ should also make its sound material accessible through electronic means, thus giving composers the possibility to create new music for the instrument. This was the initial problem that the new organ was set to solve, yet how was such an organ to be built?

To understand the design principle behind the dual character of the new organ, it is necessary to have a closer look at its windchest. In general, any windchest supports the pipes, provides a space just below the pipes that can contain pressurized air, and has mechanisms that enable the organist to control which pipes will sound (Shannon 2009, 62). Like many other parts of the organ, the construction of windchests evolved over time. One of the most interesting aspects of the Peter Gerritsz organ is that, at the end of its active life in the Nicolaï Church in Utrecht, it incorporated three subsequent stages of the technical solutions to the problem of how to distribute the air from the bellows to the pipes after a key has been pressed. In his 1479 organ, Peter Gerritsz used a "*blokwerk*" chest—a piece of massive wood in which air channels, or grooves, have been cut out for every key on the keyboard. Each channel leads air to multiple pipes that sounded together on each key.[24]

During the time of Peter Gerritsz, organ builders were experimenting with ways to control different ranks of pipes individually. Instead of hearing several pipes when pressing a key, as with the *blokwerk*, it would be possible to decide which one, two, or more pipes—each having its own character—would sound. This allowed more variation in the organ sound. To enable shutting off ranks of pipes that belong to the same sound family or register, sixteenth-century organ builders developed two mechanisms. One mechanism uses wooden sliders with holes in them: When the holes are positioned directly under the pipes, they sound. When this is not the case, they are silent. The organist can manipulate the sliders by drawing stops.[25] A second system, called the spring chest, has small valves directly under each pipe. By drawing a stop, the organist opens the valves for a rank of pipes, allowing the air to enter. This relatively complex system can still be found, especially in historical organs in Italy and also in a small number of organs in Northern Europe, such as the Nieuwe Kerk (New Church) in Amsterdam and organs in some German churches (Bicknell 1998b; see also Peeters and Vente 1984, 16–17).[26]

Because of their complexity and the higher manufacturing costs compared to slider chests, spring chests have rarely been used since 1700. In the context of the Orgelpark and its mission to create new roles for the instrument in musical life, however, the archaic technology offered possibilities. The radically new element in the design of the new baroque organ, suggested by Hans Fidom, is the use of spring chests instead of slider chests. His idea was that if the small valves below the pipe feet could be manipulated individually, it would be possible to play any combination of pipes and, thus, to create as yet unheard sounds without having to invent new pipes or stops. At the same time, this design opened the possibility of building an organ according to any historical sound concept. The way to do this was to control the opening and closing, and even the speed at which this happens, of the pipe valves by electromagnets. When seated at the mechanical console of the organ, the organist would be able to use the traditional stop action of the spring chest. In addition, controlling the electromagnets through computer software would afford access to the sounds of the organ pipes by playing the organ from a remote digital console.

By the spring of 2013, the outlines of the project were clear. The Utopa Foundation had decided to build a new organ that aimed at recreating the sound qualities of those familiar to Johann Sebastian Bach. This ambition should be seen against the background of the early music movement and its plea for the historically informed performance of music from the baroque. Knowledge of seventeenth- and eighteenth-century organ-building practices, as well as knowledge about performance practices at that time, would be necessary to reach that goal. In addition, the organ would innovate various aspects of organ design such as the temperament or tuning system of the organ, tone color, and flexible wind pressure. Its digital interface would offer possibilities to performers, improvisers, and composers that no other organ in the world could match.

The dual character of the organ, facing both past and future, raised practical challenges for the people at the Orgelpark. What historical organs and organ builders should be used as a reference? How could they ensure the new instrument would match the sound quality of these historical instruments? How could they combine the principles and practices of historicizing organ building with the need to include electronic parts in the new organ? How would they design the mechanical and digital keyboards—the interface between organist and instrument? And they had more mundane questions as well: How big should it be? How many stops should it have? How should it look? Answering these questions required all kinds of work: research into historical instruments, their sound, and how they were made; designing and making the new organ; developing new electronic and digital technologies; and reflecting on how the new organ could be a way to perform music, both old and new, in artistically interesting and relevant ways. My aim is to tell the story of how the new baroque organ was brought into being and, in doing so, to examine and understand the practical, collaborative, and imaginative work of handing down the past into the present to make new things.

THE SUSPENSE OF NOT KNOWING THE OUTCOME

In June 2013, I became a member of the core team that would supervise the design and building of the new baroque organ. The board of the Utopa Foundation had decided that it would be beneficial for the project to have a researcher on the team who would follow the making process itself, in addition to the organological research needed to design and build the organ. Having a background in science and technology studies and an interest in musical practices, I fit the profile. My fascination for organs and contemporary music had brought me into contact with Johan Luijmes in the years before, and he had introduced me to Loek Dijkman. It was agreed that half of my research time in the years to come would be funded by my university and the other half by the Utopa Foundation. This offered me a unique position to follow the building of the new baroque organ from the very start—in real time, so to say—and from within. As a member of the core team, I could be present at meetings, make study trips with the other members to relevant places abroad, visit the various building sites such as organ builders workshop and the Orgelpark itself, and have regular conversations with the main characters in the project's story that I was to write.

For my research, it was an exciting prospect that I would be able to follow the making of the organ from the start. I imagined that the paradox at the heart of the project from the start—namely, that what we consider to be "baroque sound" was to be handled with electronics—would be fascinating to study and analyze as an STS scholar. I was especially interested in how the historical heterogeneity of the instrument would inform debates on the musical and esthetic value of the instrument. This angle would connect my own research to recent dialogues between science and technology studies and the arts and humanities. In addition, STS scholars have extensively studied innovation, and my question was how this work could inform my own analysis of the project.

Many case studies in the field of science and technology studies are post hoc reconstructions of technological projects or innovations, often with the aim of learning more about the reasons for their success or failure. A good example is Bruno Latour's book *Aramis, or the Love of Technology* about a high-tech automated subway in Paris in the 1980s. Framed as a detective novel, the book tries to find out who "killed" Aramis. The vehicle never made it from the drawing board to the rails, and the book attempts to explain this failure with the fact that Aramis succumbed under the weight of having to meet the often conflicting demands of too many actors and institutions, such as the unions of metro drivers, safety regulations, and train factories. In the end it turned out to be impossible to reconcile these tensions in a coherent design (Latour 1992). *Aramis* is actor-network theory in action. It explains the failure of the innovative technology by arguing that to bring it into being, a network of actors had to be assembled that was flexible enough to adapt to changing situations, yet stable enough to continue to exist.[27] The heterogeneity of the core design idea of the baroque organ, combining historical reference with futuristic possibilities, reminded me of the fate of Aramis and its incompatible requirements, but in my case study, there was the suspense of not knowing the outcome.

For a researcher interested in innovation, this ignorance might actually be a good thing. A central heuristic in science and technology studies is to not explain the success and failure of innovations from what we know in hindsight. As Trevor Pinch and Frank Trocco put it in their book *Analog Days* on the invention and development of the Moog synthesizer in the 1960s, "in hindsight, all arrows seem to point one way, and we forget that the picture was more confusing at the time, with myriad road signs pointing in many different directions" (Pinch and Trocco 2002, 11). When I became a member of the core group, there were indeed many directions in which we could go, as well as the "enthusiasm and uncertainties of *what it was like back then* [italics in the original]" (2002, 11) that Pinch and Trocco sought to recreate in their account of how the 1960s culture shaped the synthesizer. Central to their story are not only Bob Moog and Don Buchla, who made their first prototype synthesizers in 1964, but also the musicians who adopted them to make new music. One of them was Wendy Carlos, who on her album *Switched-On Bach* (1968) used the Moog synthesizer to perform Johann Sebastian Bach's music in creative new ways. What connected Bach and Carlos, Pinch and Trocco argue, was the chromatic keyboard that both used to play their instruments. Embedding his new sound machine into a culture of use that went back centuries, Moog's synthesizer reached users who were hesitant to operate the knobs and sliders on Buchla's modular synthesizer.

The plans for the new baroque organ were reminiscent of the Moog synthesizer in their aim to combine innovation with tradition. If the people at the Orgelpark wanted the new organ to be a success, that could well be a wise strategy. In their research on the innovation of classical music instruments, Bijsterveld and Schulp (2004) asked why they are so remarkably stable in their design and outlook compared to other artifacts. Through interviewing instrument makers and studying promotional literature, they found that successful new instruments often innovated and sustained tradition at the same time. Reflecting on the reception of new technologies in music, Pinch and Bijsterveld (2003) found that this reception is often accompanied by a debate on the boundary between "instruments" and "machines." Whereas musical instruments are associated with embodied skills, an experienced sense of hearing, a developed taste, and even art, a machine mechanizes the creative act of making music into something that is only repetitive: "new technologies sharpen the perennial issue of what makes for good music and 'art'" (2003, 538). Pinch and Bijsterveld argue that this perennial issue cannot be put to rest by reading an essence into musical instruments. Rather, they are interested in how the breaching of conventions and boundaries by the introduction of machines reveals how, in musical cultures, groups and actors establish the criteria for "good music."[28] In line with their arguments, one could expect that the plans for the new baroque organ would trigger debates about its qualities in the communities of organ builders and organists, and following these debates would offer a way to learn about contemporary musical practices that revolve around the organ as an instrument with a long history.

Instrumental innovations not only challenge esthetic conventions, they also change the ways in which music from the past can be made present. Or, as Antoine Hennion puts it in his book *The Passion for Music*, how it can "appear among us" (Hennion 2015, 1). Hennion emphatically rejects the dualism between the esthetic analysis of the musical work as an isolated art object and the analysis of the social conditions that are revealed in its production and reception. To overcome this dualism, he argues, the focus should be on the myriad of mediators that are assembled in the collective process of making music happen: "instruments, scores, media, languages, institutions, interpreters and teachers" (2015, 3).[29] Studying music means that we have to trace the mediations that are relevant to what we want to understand. An example is the debate between what Hennion calls the "Moderns" and the "Baroques," musicians who favored the modern performance style that emerged immediately after the Second World War and the proponents of a historically informed performance practice. The controversy—or quarrel, as Hennion puts it—could be explained as a difference in taste: The Moderns preferring modern, static, powerful instruments to the dynamic, feeble sound of period instruments; the Baroques being interested in restoring the "authentic" sound of baroque music. Hennion, however, renders the debate as an opportunity to distinguish between competing forms of mediation that bridge a gap of more than two centuries (Hennion 2015, 174). Contrary to the Moderns, the Baroques studied period instruments, paintings, treatises, texts, music programs, and manuscripts and relearned forgotten performance practices and temperaments. If performing music from past sources in the present requires mediation, the question then is how the organ builders at the Orgelpark would mobilize mediators to construct their instrument and how the new baroque organ itself would become a mediator to bring music into existence (Born 2005; Kronengold 2018).

For any entity to come into existence and continue to exist, it needs mediators, or "attachments," a concept that Hennion developed in later work.[30] His sociology of mediation allows us to rethink the ontology of music beyond given categories such as "the social" or "the esthetic." This also has consequences for how we can think of music as art. In her book *The Imaginary Museum of Musical Works* (1992), already mentioned in the introduction, Lydia

Goehr traced the genesis of the notion of "musical work." Rather than analyzing or defining the mode of existence of musical works as objects, Goehr studied how the work concept emerged historically around 1800 and regulated musical practice in what she calls the Beethoven paradigm (1992, 4). In trading philosophical thought experiments in music ontology for historical research, an approach that resonates with Hennion's account of the quarrel between the Moderns and Baroques, Goehr underlines that presenting music as works of art depends on practical achievements rather than conceptual analysis.[31]

In my use of the term "presenting" in the previous sentence, I draw on the work of Hans-Georg Gadamer who uses it in the sense of making present (in German: *Darstellen*). In his book *Truth and Method* (1989 [1960]), Gadamer develops a hermeneutic account of how art exists.[32] He criticizes nineteenth-century Romantic writers and philosophers such as Johann Gottlieb Fichte and Friedrich Schiller. They argued that art and esthetic experience are abstract because they derive from a realm of the beautiful that is timeless. The esthetic experience must focus on the artwork itself and abstain from non-esthetical aspects such as goal and function. By disregarding the original context of life, and the religious or secular function that gave it significance, it becomes visible as a pure work of art. Gadamer calls this process of abstraction "aesthetic differentiation."[33] The esthetic differentiation also created an external existence of art in places solely dedicated to it such as the universal library, the museum, the theater, and the concert hall. In Gadamer's hermeneutic approach, however, the work of art cannot be isolated from the contingencies of the conditions under which it presents itself. On the contrary, it can exist only in its relations to the lifeworld. Although every work of art has its own world, this does not mean that when this original world has vanished, it can exist only in an alienated esthetic consciousness. Rather, it exists through its acquiring of new uses and adapting to changed circumstances.

Since works of art can exist only when they are presented under the new conditions of the contemporary, they are fundamentally unfinished and their meaning is never exhausted. This open future for Gadamer applies not only to art but also to cultural transmission in a broader sense. We are not separated from the past but connected to it through tradition. The past is actively handed down in our situated understandings and applications of it. Bringing the new baroque organ and its music into existence thus requires bringing the traditions that it aims to hand down—of organ music, building and playing—into a "*thoughtful mediation with contemporary life*" as Gadamer phrased it (1989 [1960], 161, emphasis in the original).

LEARNING ALONG

Earlier in this chapter, I showed how the decision to replace the Van Leeuwen organ with a new baroque organ exemplifies two interrelated issues. After the Park Church had been refurbished as a concert hall, its organs became concert instruments, their sounds primarily evaluated in the frame of art music. The plans for the new organ can also be read as a new chapter in the long history of these "mirrors of their time" and the many ways in which they helped to transmit cultures and traditions. When studying these two issues in the Orgelpark, I draw on ideas and insights from science and technology studies, music studies, and hermeneutic and pragmatist philosophy. What these literatures have in common is that they help one understand what it takes to bring a musical instrument and its music into existence. This ontological focus, then, can be translated into a practical question: What work was to be done to actually make the new baroque organ? To be able to tell the story of this work, I had to become a participant observer.

In his book *Making* (2013), the anthropologist Tim Ingold reflects on the role of the participant observer when he sharply contrasts ethnography and anthropology. Ethnography's aim, Ingold claims, is documentary, in that it generates accounts based on learning *about* a practice. Anthropology, on the other hand, is transformational; it sets out to learn *from* a practice by corresponding to it. Whereas ethnographic research results in "field material"—qualitative data to be analyzed using an exogenous body of theory—anthropology seeks to work with its materials the way an artisan would work with wood, following what is going on and responding to it. Participant observation is the term for this second method, and it does not aim for descriptive fidelity as ethnography does, Ingold claims (2013, 4–5). The idea that it would be possible to collect data to construct knowledge after the fact goes against Ingold's conviction that we can never be "outside" the world that we study. Rather, Ingold's epistemological exemplar is the artisan who thinks through making and acquires knowledge through practical and observing engagement "with the beings and things around us" (2013, 6). This is what Ingold calls the "art of inquiry," in which we learn from interacting with the materials with which we work rather than establishing an opposition between mental ideas and practical facts: "Thus the art of inquiry moves forward in real time, along with the lives of those who are touched by it, and with the world to which both it and they belong. Far from answering to their plans and predictions, it joins with them in their hopes and dreams" (2013, 7). Joining the hopes and dreams of the organ builders is something I had to learn by doing.

In the daily practice of my own research, the clear lines that Ingold draws between the observing and documenting ethnographer and the participating and learning anthropologist were often more blurry. With hindsight, I recognize myself in both.[34] Following the project in real time meant that I would not know what events, or "road signs," as Pinch and Trocco (2002, 11) put it, would prove to be relevant. In general, I tried to be present as much as possible at moments that I expected to be important. Especially in the first years of the project, these included the meetings of the core group, symposiums and conferences in the Orgelpark, study trips, and listening experiments. During these moments, the members of the core group would normally be present. I made sound recordings and took photographs. Sometimes I was able to make field notes on the spot, but more often I wrote these shortly after I returned to my office. Over the years, I made hundreds of photographs and videos, collected tens of hours of recordings and interviews, and filled many notebooks.[35] I have analyzed interview texts, other documents such as memos and minutes of meetings, and travel plans, as well as some pictures and my own comments using qualitative data analysis software. This enabled me to code the material, which was helpful in at least two ways. First, it made it possible to quickly navigate through the texts, and second, it enabled me to make connections and experiment with relating concepts to the material.

Documenting my research, however, in no way meant that I kept myself "outside" the world that I studied. In fact, throughout the years, I had to negotiate a fine balance between involvement and distance. As a member of the core project team, I was a participating observer most of the time. I was considered to be "one of us" by the others, and I myself often felt the urge to not just listen and observe but to join the discussions in the heat of the moment. When doing research over a longer period of time, this balancing act becomes even more challenging. Making trips together, co-organizing conferences and symposiums, visiting working sites, and becoming a regular visitor at the Orgelpark meant that I did not just follow my "actors"; they became my travel companions, both literally and figuratively.

I came to the baroque organ project sensitized by my attachment to the instrument. At ten years old I began piano lessons, and making music became very important for me. Like many people, I associated the sound of the organ with church visits in my youth, and I did not particularly like their sometimes loud and harsh sound. During my years of study in Groningen, in the north of the Netherlands, I worked as a music journalist and became aware of the rich organ landscape that surrounds the city. Still, I was more interested in the cultural and historical value of these instruments and how they are embedded in the places than in how they sounded or who built them. Only many years later did I actually start playing the organ. Being used to playing music from the baroque, and especially Bach, on the piano, I was interested to know how it sounded on an organ.[36] Some aspects of organ playing proved to be difficult to learn, especially the pedals, but I soon became fascinated by the instrument and its endless musical possibilities. Being able to regularly play on a historical organ from 1788 in Roden, the Netherlands, I learned the feeling of interacting with an age-old instrument that is so unique. I cherished the ritual of getting the key, entering the dark church, climbing the stairs to the balcony, and switching on the lights and the blower. Over the years, some organs became as familiar as old friends, yet always somewhat strange in their static sound and their manuals that do not respond to touch the same way piano keys do. When studying organs, I not only draw on the knowledge that I acquired over the years of how they work but also on my fondness of them as musical machines with a life of their own.

Ingold's art of inquiry does not aim at the kind of understanding that requires generalized descriptions of facts that can be explained by applying theoretical concepts. Rather, he invites the researcher to set up a relation with the world that he calls *correspondence* (Ingold 2013, 7, italic in the original).[37] The new baroque organ project is connected to multiple contemporary and historical practices and, thus, creates a hands-on way of learning more about these practices. But my aim is not that of the organologist, the historian of technology, or the musicologist. I instead see myself as an organ anthropologist and an empirical philosopher in the domain of music. Following the actors is for me not only a way of understanding the practices that they embody but also of learning along with them, about how we give meaning and relevance to music and its historical objects in our society and, more importantly, about how we need history to imagine where we could go next.

JOINING THE DREAMS OF A NEW ORGAN

The core group—Loek Dijkman, Johan Luijmes, Hans Fidom, and I—of the new baroque organ project met the first time on June 22, 2013, in Leiden. In his starting memorandum, Loek Dijkman listed the main questions that, according to him, needed to be answered at this stage: What are the research goals? Who are the people who have to be involved? What role is there for the Cultural Heritage Agency? What are the roles of Hans, Johan, Peter, and Loek? What is the time frame for the project? How do we organize the communication about the project? In response to this memorandum, we all wrote our own short memorandums, in which we answered the questions by Loek and also, on his request, formulated our "dreams" in relation to the project.

Hans Fidom hoped that the new baroque organ project would revitalize the belief in progress of the early twentieth century. More specifically, he thought of the research necessary to realize the project as a case study in the influence of technology on culture, and one of the research goals was to position the new organ and its project in the history of technology and music.

In his text, Johan Luijmes pointed out the importance of having an organ that distinguishes itself "through the pure beauty of its sound," rather than through the repertoire that is played on it—much like the Van Straten organ, without disqualifying the two other main organs in the Orgelpark, the Verschueren and the Sauer.[38] He hoped for a new instrument that would make it possible to create even more varied programs; new music, he wrote, thrives best by placing it in a context of older music. For him, the research would have not only a technical instrument as its object but also an artistic one. Reflecting on how the new organ was built would thus contribute to the field of artistic research. This parallel research would show that "the specific ways of thinking and working of artists can be greatly beneficial to society," as he formulated it.[39]

For me, the opportunity to follow the innovative project from its first days was a dream in itself. From my background in science and technology studies, and having a strong interest in music and its technologies, I hoped that my questions and reflections during the project could contribute to an organ that would not only be built but also used in innovative ways. And I dreamed of writing a book like this.

Why did the Van Leeuwen organ have to be replaced by a new baroque organ? What exactly is "new," and how do you make it? The radical design idea of a dual windchest as the heart of an organ that faced both the past and the future set the project in motion. For the Utopa Foundation, it reflected the mission of the Orgelpark to give the organ, its builders, and its musicians a new role in musical life. During that meeting in June 2013, we started on a journey without knowing where it would lead. There was no blueprint—anything could still happen. The first step was to actually make trips to other places and, as with all travel, this assumed the ability to orientate oneself.

2

TRAVELING TO THE ORGAN WORLDS OF BACH

A STRANGE TIME

The hall of the Orgelpark is filled with the murmuring of an expectant audience. On the podium below the Sauer organ is a hybrid collection of instruments, some from Japan and Indonesia, others from European early music. An electric guitar is standing next to a harmonium. Musicians from the ensemble Serafin enter the space, followed by the Dutch singer and songwriter Spinvis. The program leaflet mentions the title of this concert in February 2016: *Mein junges Leben hat ein End* [My Young Life Has an End]. It refers to the set of six variations on a short choral melody written after 1605 by the Dutch organist and composer Jan Pieterszoon Sweelinck (1562–1621). The technique of variation and adaptation, demonstrated in Sweelinck's composition, is the central topic of the concert. Spinvis's own songs were adapted for organ and harmonium, and the Dutch composers Roderik de Man and Guus Janssen were asked to compose variations on Sweelinck's piece. Opening the concert, the organist Una Cintina played the original composition on the Van Leeuwen organ. Its theme consists of the eight notes of a descending octave with a simple rhythmic twist at the end, resulting in an "immortal motive," as one music critic observed the next day in her review of the concert (Luttmer 2016). As simple as they are, Sweelinck's notes proved to be a rich source for new music.

In his introductory remarks to the audience, Spinvis pointed out the relation between the Orgelpark and Sweelinck, who had been organist at the Oude Kerk (Old Church) in the nearby center of Amsterdam:

> Welcome to the Orgelpark. Organs, organs, organs, as far as the eye can see, and even further. Because also below us, in the basement, there are hundreds of organs. A thousand maybe. And below that, even more basements with even more organs. Nobody knows exactly how many organs there are. [In the background, the musicians are tuning their instruments and speak to each other in a low voice.] Once,

an expedition was sent down to find it out. [Laughter in the audience.] One man came back. [More laughter.] The rest still roam around between the thousands and thousands of organs and . . . between the thousands and thousands of organists who are still there, dead and forgotten, covered by cobwebs, fallen over their keys, over the last chord on their keyboards. One of them is called Sweelinck, Jan Pieterszoon Sweelinck, born in Deventer [a town with a long history of trading and book printing in the East of the Netherlands] in 1562. 1562. What a strange time this must have been. One had a God who could do anything. One ate thrushes. One just talked during concerts. [Laughter.] Because silence only came later. Jan Pieterszoon Sweelinck was a man in between. In between worlds. In between Renaissance and Baroque. In between the sixteenth and seventeenth centuries. In between Roman Catholic and Protestant. We do not know much about him, we know his music but apart from that almost nothing. He left his music and a chest. In that chest is everything we do not know about him. What we do know, and that is a fact, is that he once sat at a table somewhere in Amsterdam to write his beautiful music, four hundred years ago.[1]

The story of the Orgelpark founded on basements filled with organs and organists struck me as an apt metaphor. It exemplifies the past as both foreign and familiar, as supporting the present, as a place that can be visited by an expedition even though it is not certain if its members would return, and that has much that we do not know. Even though it was still in the making at the time of the concert, the new baroque organ was historical in the many ways that Spinvis hinted at. In this chapter, I trace how organ history was mobilized in the early days of the project to create references and criteria for the sound, design, and construction of the new organ. Analytically, I am interested in how the ambition to make a historically informed instrument had to be balanced with the need to solve practical problems that emerged as the project evolved. Creating convincing historical reference required complex negotiations and making hard choices.

Not only in a metaphorical sense did the first phase of the new baroque organ project start with Jan Pieterszoon Sweelinck. In the organ world, he is seen as the founder of what is called the North German organ school. From 1600 onward, his fame as an organist attracted pupils, many of whom came from the northern parts of Germany, hence his nickname "Organistenmacher," or maker of organists. The organ tradition that Sweelinck started would extend into the eighteenth century and reached its pinnacle in the music of Johann Sebastian Bach. Jacob Praetorius, Paul Siefert, Samuel

Scheidt, and Heinrich Scheidemann are among Sweelinck's most famous pupils. North German organ music was formative in Bach's musical development. As a young man of twenty, in 1705, he walked from his hometown of Arnstadt in Thuringia to Lübeck near Hamburg to listen to Dietrich Buxtehude, another important organist in the North German tradition (Wolff 2001). He stayed for three months and was impressed by the sound of the large organs he heard in Lübeck and Hamburg. The latter's Jacobi Church boasted an organ built by the influential north German organ builder Arp Schnitger (1648–1719) from 1689 to 1693. In 1720, Bach returned to Hamburg because he was interested in the vacant organist position at this same church, but most of his life, he played on instruments in the central German regions of Thuringia and Saxony. Important organ builders here were Gottfried Silbermann (1683–1753), Tobias Heinrich Gottfried Trost (1680–1759), and Zacharias Hildebrandt (1688–1757).

Any attempt to build a new organ on which Bach's music sounds convincing from a historical perspective starts with a choice between two organ styles. Should we think of the big instruments of the Hanseatic north German cities and towns such as Hamburg or Lübeck, or are the organs in Thuringia and Saxony that Bach played and tested during his life a better point of reference? The new baroque organ project was no exception, and in this chapter, I will reconstruct the debates and considerations in the core group about this choice. Three study trips were organized to visit organs that were thought to be relevant to the project. This relevance could be found not only in organological and mechanical aspects of the instruments but also in their sound characteristics. The "expeditions" to some of the organs of Bach (Wolff and Zepf 2012) were seen as a necessary step in the design process of the new organ. Joining the other members of the core team on these journeys enabled me to follow the discussions on practical and technical problems that emerged along the way, and to understand the background to the choices the team had to make during the first stages of the project. Following the "young" Bach to north Germany or staying in his home region of Thüringen would have consequences for what the new organ could become.

HISTORICIZING ORGAN BUILDING

Early September 2013, the core group met at the office of the Utopa Foundation in Leiden. Organ builder Hans Elbertse from Soest, the Netherlands, had now joined the team. His work on the restoration of the Sauer organ and his successful repairs of the new Verschueren organ were reasons for Loek Dijkman to invite him as the project's main contractor. During the meeting, the content of a draft press release, written by Hans Fidom, as well as the role of research in the project were discussed. To communicate the goals of the project to the outside world, its basic assumptions were validated again. According to Dijkman, the use of electronics was paramount to the project. Without this, he argued, there would be no real innovation: "Then we are making a replica of a historical instrument. I do not think that is the fun part."[2] The project was to be as open as possible, inviting a broad range of experts to comment on the plan. To build a network of expertise, the press release would be sent to a selection of international organ builders and other experts and institutions.[3] International organ experts would be invited to join a reference group. Colloquia and symposia offered additional possibilities to discuss the plan with experts and audiences who were not directly contributing to the project.

The press release summarized the basic design ideas. The new organ would connect the worlds of the replica of the medieval organ and the two Romantic organs in the Orgelpark. It would be suitable for performance of music by north German baroque composers like Dietrich Buxtehude and Georg Böhm, "and certainly also Johann Sebastian Bach."[4] The aim was to make an instrument that "will inspire the making of new music with 'spring chests', a seventeenth century invention where every pipe has its own pipe valve. By manipulating these valves through twenty-first century technology, the musician will get optimal access to the sound material: each one of the 3,500 pipes can be sounded in combination with any other pipe."[5] Concerning the size of the organ, the team believed that the balcony in the Orgelpark could be charged with a rather big organ. It was thought that there were no structural or architectural limits, which meant that the new organ could be a full-size baroque organ.

At the end of the meeting, a study trip to the organ research center GOArt at Gothenburg University, Sweden, was planned. As a first destination, GOArt made sense. The Swedish organ research center had built a new baroque organ modeled after organs by Arp Schnitger in the 1990s. As one of the outcomes of The North German Organ Research Project, the instrument in the Örgryte New Church in Gothenburg embodies the practice of historically informed organ building, which goes back to the period around 1900, when some organ builders became increasingly interested in the history of their practice. To better understand the significance of the Swedish project, and the study trip to the organ that came out of it, I will first describe how the interest in historical organ building emerged in the early twentieth century.

NEW UNDERSTANDINGS OF HISTORICAL ORGANS

Albert Schweitzer (1875–1965) was well known as a philosopher and a missionary, but he also had a lifelong fascination for the music of Bach as well as the organs on which the composer played. His father being a Lutheran-Evangelical pastor in Gunsbach, Alsace, Schweitzer developed a deep understanding of Bach's religious music and the meaning of the figures and motifs in his use of organ chorales. He became a devoted advocate for the study and restoration of historical pipe organs. As Stephen Bicknell notes in his overview of twentieth-century organ building, Schweitzer's critique of the organs built around 1900 as not suited for the performance of Bach's music echoed "a keen sense of disappointment at the fact that organ building was ruled by commercial considerations" (Bicknell 1998b, 82).

Schweitzer's ideas resonated with the German *Orgelbewegung*, or Organ Revival Movement, that arose in the 1920s. This movement, with Hans Henny Jahnn (1894–1959) and Christhard Mahrenholz (1900–1980) as important representatives, criticized the use of the modern organ as it had developed since the end of the nineteenth century as an instrument to perform organ music from the baroque. This "modern organ," which was generally labeled exactly that at the time in Germany, should have one and only one clear pitch, the same as that of the piano and other instruments. Nineteenth-century organs, therefore, have a variety of 8' stops that sound as the corresponding written note instead of a multitude of octaves that results from combining 16', 8', 4', and 2' stops, as in premodern organs. It also used new technologies for transferring the movement of the key to the windchest: Instead of a set of mechanical trackers, this was done with pneumatic and later electropneumatic systems. And finally it used "*Normalmensur*," or normalized measurements for the scaling of the pipes—the relation between the length and the internal diameter—developed by Johann Gotlob Töpfer (1791–1870) in the early nineteenth century.[6] In contrast to the modern organs, the builders and voicers of the *Orgelbewegung* favored a mechanical action, which is the mechanism that links the keyboards and stops to the windchest, and an organ design that mirrored the principles of North European baroque organs, for example, in having clearly separated divisions, or "choirs," of pipes and low wind pressures. Schweitzer's interest in the merits of historical organs and the ideas of the various movements for "organ reform" that developed in Europe and the United States before World War II were taken up in historicizing building styles until the 1970s, such as the postwar neobaroque style. What they shared was a focus on learning to understand "old" organs from the seventeenth and eighteenth centuries.

The interest in historical instruments can also be traced in organ restoration practices in the twentieth century. Until the late nineteenth century, organs were repaired and extended according to the technical and esthetic insights of their time. In most cases, it was seen as self-evident that the instrument should serve church music practice and thus be adapted to the kind of use this required. That the Peter Gerritsz organ became a museum piece in 1886 marked a shift in the valuation of historical organs from the baroque era and earlier. In the 1920s, organ enthusiasts such as Hans Henny Jahnn and his friend Gottlieb Harms (1893–1931) started to take care of antique organs such as the Schnitger organ in the Jacobi Church in Hamburg. The instrument had suffered during the First World War when its front pipes had been used to make weapons. Jahnn and Harms attracted attention to the organ and its builder by giving concerts and organizing conferences about the importance of restoring old instruments in German cities such as Hamburg and Lübeck (Lipski 1997). As valuable historical organs, they acquired the status of sounding heritage.

The interest in historical organs in the 1920s reflects what John Watson (2010) calls the "paradox of restoration and conservation of organs"—the need to restore historic artifacts to give them value through continued use versus the obligation to preserve their historical substance for future reinterpretation. Watson compares organs to buildings in the sense that they rarely can be retired from use. This means that they pass on from generation to generation through a series of changes. In making these changes, different values play a role, Watson argues. Quoting Alois Riegl (1858–1905), one of the founders of art history as a discipline, he distinguishes between use value in the present, which may result in repair and rebuilding of the organ, and commemorative value, which centers around the historical character of the organ and favors stabilization and conservation. Between these two values is the heritage value that "combines respect for an artifact's historical aspects with a claim to use it for present purposes" (Watson 2010, 29). During the twentieth century, the practice of organ restoration shifted in both directions on Watson's value continuum. Although there were some exceptions following the pleas of Jahnn for careful restoration, before the Second World War many important historical organs were modernized to enhance their use value.[7] After the war, a renewed interest in the historical values of antique organs sometimes led to the opposite approach, where the instrument was restored back to a supposed original state. As a result, later additions to the instrument were sometimes removed, deleting information about its genesis.

An international conference dedicated to Arp Schnitger in 1969 in Groningen, the Netherlands, marked a turning point in the history of organ building that cannot be separated from the revival of early music and performances on period instruments since the 1960s. According to Hans Davidsson in his overview of organ building in Northern Europe since 1969, the Schnitger conference was the start of a different approach to restoration as well as the actual copying of historical organs. The conference set new standards for the restoration of existing instruments and inspired builders in the Netherlands, Germany, Japan, and the United States to build replicas and new organs based on historical styles, such as the North German Schnitger style.

> The revival of the historical organ as a musical instrument and as an essential part of our cultural heritage, new principles for organ restorations, new ideals for organ building, and the application of historically informed principles of performance would be leading features in the development of the decades to come. (Davidsson 2003, 334)

This new paradigm meant that historical and organological research into baroque organs became even more important to the practice of organ building than it had been since the days of the *Orgelbewegung*. Documentation of relevant aspects of historical organs, as well as knowledge about their place and role in the organ landscape in which they functioned, became an important research resource. This research focused on systematically collecting information about organ cases and dispositions, windchests and wind systems, but especially the characteristics of the pipes, such as their dimensions and measurements, their metallurgical composition, and the ways in which the pipe mouths were shaped.

According to the German organist and organologist Harald Vogel, himself one of the pioneers of historically informed organ-building practice, making new organs using historical organs as a reference has led to new understandings that also changed the practice of organ restoration by offering new solutions to the paradox of use and conservation. Vogel underlines the importance of building replicas such as the Van Straten organ:

> We have actually come to the point now where replicas come closer to the original sound than the surviving originals themselves. Building new instruments in historical styles is the path that we must take in the future. There are two reasons why we cannot take our antique instruments closer to their original state then we are now. The first is that there are additions to the original material that we cannot take away. The second is that, in restorations, we have not gone far enough in many respects. I think, therefore, that the Göteborg project is a turning point in the history of restoration and replica in Europe: ideally, restoration should always be related to a replica. (Vogel 2003, 345)

THE NORTH GERMAN ORGAN RESEARCH PROJECT

The project that Vogel referred to originated in the 1990s at the University of Gothenburg, Sweden. Many of the ideas and ideals that shaped the paradigm of historically informed organ building culminated in this milestone undertaking. Hans Davidsson, then affiliated to the Gothenburg School of Music and Musicology as an organ teacher and musicologist, came up with the idea to build a seventeenth-century north German baroque organ.[8] From an artistic point of view, Davidsson argued, replicating antique organs became desirable as a reaction against modern organs produced by industrial methods. When organs became subjected to industrial production techniques, he wrote in 1993, "the main aim was no longer to attain the highest quality possible; instead factors such as capacity and profit became predominant . . . piece by piece, the accumulated experience of the skilled craftsmen disappeared. Thus the end result came to be determined more by the production process itself than by esthetical or stylistic aims" (Davidsson 1993, 9).

A central assumption in the North German Organ Research Project was that recovering the sound of historical organs required returning to artisanal building techniques that were displaced over a century before. These building techniques were thought to be intricately connected to the musical qualities of the organ, qualities that have been lost because of modern building techniques that efficiently rationalize production and disembedded the instruments from their original musical cultures. The ideas of Davidsson materialized in the building of a baroque organ in the Örgryte New Church in Gothenburg, Sweden. Built by the organ workshop of the GOArt research center, it was financed by grants from the Swedish government and the European Union, among others. The case of the instrument is a replica of the 1699 Schnitger organ in the Lübeck Dom, of which only photographs remain after it was destroyed during a bombing raid in 1942. The pipes of the replica were built after the pipework of Hamburg's St. Jacobi Church, which was removed from the organ case shortly before the church was also destroyed.[9]

The aim of the North German Organ Research Project was to gain the knowledge and experience necessary to construct, in a Swedish church, an organ in the way it might have been built by Arp Schnitger in late seventeenth-century northern Germany. Instead of asking one organ builder to build the organ, the project management assembled a team of organ-building experts and musicologists to conduct research on musicohistorical evidence, materials, acoustics, wind behavior, pipe making, and many other topics related to the building of a replica of an Arp Schnitger organ. The team collaborated with Chalmers University of Technology in Gothenburg to study wind behavior in the wind channels, the metallurgical properties of pipes, and the acoustics of the Örgryte Church. In addition, seventeenth-century artisanal methods for casting the metal sheets to build the pipes were reconstructed through a series of experiments (see also chapter 5).[10]

The North German Baroque Organ was inaugurated in 2000. In the years that followed, members of the team were involved in two other projects to build replicas of historical organs. One was carried out at the Eastman School of Music in Rochester, New York, designed after specifications of an instrument built by Adam Gottlob Casparini in 1776, in Vilnius, Lithuania.[11] The other project, a new organ for the Anabel Taylor Chapel of Cornell University in Ithaca, New York, aimed at reconstructing the tonal design of an instrument that Arp Schnitger had built for the Charlottenburg-Schlosskapelle in Berlin in the first decade of the eighteenth century.[12] All three replica projects reflect aspects of a more general trend in organ building since the 1969 conference in Groningen. First, the insight that cultural, artisanal, and esthetic contexts are particular to a certain period and style of organ building should be considered when building new instruments. Second, this contextualizing approach includes the use of artisanal techniques and aims at replicating the construction practices regarding woodwork and pipe making that were evidenced in historical reference organs. Finally, these instruments are multifunctional. As research organs, they enable the study of a variety of topics and questions; as teaching organs, they enable new generations of students to learn about historical performance practices; and as concert instruments, they contribute to innovating performance practices.[13]

During the visit to GOArt in September 2013, the core team met with the researchers at the organ research center: its director, Johan Norrback; Ibo Ortgies, a specialist on historical organ temperaments or tuning systems; Paul Peeters, an expert on historical organ building; Joel Speerstra, instrument builder and musician; and Carl Johan Bergsten, who created world-renowned organ databases. Hans Fidom introduced the Orgelpark's new baroque organ project. The GOArt experts responded by identifying a number of issues, such as choosing the temperament of the new organ, which, in their view, would require a solid research approach. A possible collaboration between the Orgelpark and GOArt was discussed, and it was decided that GOArt would not contribute to the project at an institutional level in a structural way but on the basis of input by individual members of the center.

After the meeting, we went to the Örgryte New Church to listen to the North German Baroque Organ. Joel Speerstra performed the partita *Ach wie nichtig, ach wie flüchtig* by Georg Böhm (1661–1733). The chorale and variations offered him the opportunity to demonstrate several stop combinations. To me, listening to the organ was somewhat confusing. I realized that I listened to a very young instrument, yet its sounds were said to closely approach those of age-old Schnitger organs. But, I wondered, what does it actually mean to make this comparison?

IN THE FOOTSTEPS OF ARP SCHNITGER

In the early stages of the Orgelpark's baroque organ project, the north German organ type and the sound concept of Arp Schnitger were regarded as the main historical reference. To learn more about Schnitger and the organs he built, and to study the spring chest in the organ in the north German town of Stade on which Schnitger worked as a journeyman of his uncle Berendt Hus (1630–1676), the core group scheduled a second excursion in October 2013. Sylvia de Munck, vice president of the Utopa Foundation, had at that time joined the core group.

In the small East Frisian town of Norden, our first stop, we planned to listen to a Schnitger organ in the Ludgeri church, built between 1688 and 1692. We drove through the empty landscape behind the dikes along the coast of the Wadden Sea to the small town. The church has a peculiar shape;

its Gothic choir is raised above the transept and the Romanesque nave, and its Romanesque bell tower is separated from the church by a street. This oddity is mirrored inside. The Schnitger organ is not positioned on a balcony just above the main entrance, as usual, but at the south side, halfway down the nave. All pedal pipes are placed in one pedal tower.

The organ in Norden is generally considered as one of Schnitger's most beautiful instruments. Thiemo Janssen, who shares his position as the church organist with his wife, Agnes Luchterhandt, explains that Schnitger was a visionary who was way ahead of his time. He worked on a number of existing instruments and built many new ones, all according to Janssen "on an incomparably high level, both in terms of their artisanal and artistic quality."[14] More than other organs, the Norden instrument shows how Schnitger adjusted the instrument to the building. Although he could have placed the organ at its traditional location at the north face, he chose to distance himself from his medieval and Renaissance predecessors. As a result, it was necessary to place all the large pedal pipes in one tower, where these were normally placed in two towers, one on each side of the organ.

In Stade, a town close to the river Elbe to the northwest of Hamburg, we visit an organ built between 1675 and 1688 in the Church of Saints Cosmas and Damian by Berendt Hus and Arp Schnitger. At age eighteen, Schnitger started to work for his uncle, Hus, first as an apprentice, and then, two years later, as a journeyman. As the organist on the instrument, Martin Böcker, explains to the group, Hus was still oriented to the past, to the sound world of the Renaissance, which explains the large number of high-pitched stops in the organ. He also chose to build a spring chest, presumably because it was more reliable in the wet sea climate, rather than a slider chest, which has long, thin sliders that can warp because of the humidity. The spring chest of the Stade organ is designed in such a way that the tone sliders that contain the small valves for each pipe can be taken out without first having to remove the pipes, as is the case in the Nieuwe Kerk (New Church) in Amsterdam.[15] This enables easy access to the valves in case of malfunction. It is this design that Hans Elbertse and Hans Fidom suggested to use for the spring chest of the new baroque organ in the Orgelpark. They measure the height and the width of the pallets, the tone channels, and the tone sliders, a term coined by Elbertse and Fidom (see figure 2.1).

After our visit to Stade, we pass through a another small town appropriately named Buxtehude and arrive at our hotel in Hamburg where we will

Figure 2.1
Hans Elbertse (right) and Hans Fidom measuring the dimensions of the tone sliders of the Hus/Schnitger organ in Stade, Germany. The tone sliders have round knobs and contain the small valves for each pipe. They can be taken out in case the valves malfunction (photo taken by the author).

visit two other organs the next day. During dinner, Loek Dijkman questions the idea of a dual temperament in one organ: "We do not want to make a 'monster,'" he says.[16] The hybridity of the new baroque organ, with its tuning system that allows two different temperaments and its mechanical and digital keyboards, becomes a point of discussion.[17] The organ in Norden can also be seen as a "monster"; it is constructed in the situation of that specific church, within certain degrees of freedom from the tradition, but also within certain limits. Within these limits, Schnitger built an organ that is an organic whole. Every old instrument is a product of the circumstances, we agree. What are the circumstances in the Orgelpark? How will they influence the design and sound of the new organ? What are the degrees of freedom?

The next morning, we walk to the St. Katharinenkirche to listen to the new organ built by the Dutch organ builder, Flentrop. It was completed in 2013, and has 520 pipes from the original organ from 1606 and 1647, which was destroyed in the Second World War. It was this organ that Johann Sebastian Bach played during his trip to Hamburg as a young man, and later, in 1720, when he applied for the job of organist in the nearby Jacobi church. The new Flentrop organ resulted from a historically informed building process and is, therefore, considered relevant for the Orgelpark's project. As Flentrop director Frits Elshout explains to us, the metal for the newly built pipes was cast according to seventeenth-century casting methods. In the group, we discuss whether the instrument's overall sound might be called "old," and what that actually means, and whether we think that the sound concept is artistically convincing.

In the nearby St. Jacobi Church, we listen to the organ that was restored between 1989 and 1993 by Jürgen Ahrend, who is well known for his early historically informed restorations of Schnitger organs. Although the case of the original organ built by Schnitger from 1689 to 1693 was destroyed during the war, the windchests and the pipes had been removed and, thus, survived. This pipework has been reinstalled during the restoration by Ahrend and was later used as reference material in the GOArt project in the Örgryte New Church in Gothenburg. After listening to organist Rudolph Kelber, who plays music by Mozart, we discuss the sound of Schnitger's organs. It is a warm autumn. In late October the weather is still good enough to sit outside and have a cup of tea. We discuss the pros and cons of taking his sound as our main reference for the new baroque organ.

From the descriptions, plans, and discussions at this stage, the new baroque organ emerges as a musical instrument that is hybrid in all its aspects. In terms of organ building, its main historical reference will be the craft of the Rembrandt among the seventeenth-century organ builders, Arp Schnitger. An argument supporting this is that Bach admired Schnitger's organs when he traveled to north Germany where he listened to Dietrich Buxtehude and studied with Georg Böhm. The sound of Schnitger organs is closely associated with the music of North German organ composers such as Buxtehude, Böhm, Bruhns, Krebs, Reincken, Scheidemann, and the brothers Scheidt and Tunder. The plan to make an organ with a dual-key action—both mechanical and electromagnetic—is at the heart of the project. Building spring chests and using magnets to manipulate the pipe valves realizes this design as it affords the creation of a baroque organ in an uncompromising way. Next to that, the organ will support live electronics, and even the use of lighting inside the organ is considered. These design pressures come from very different times and places.

Building an organ in the style of Schnitger means that it will have a quarter-comma meantone temperament. This system was commonly used in the seventeenth century and before. If one wants to give historically informed performances of the North German seventeenth-century repertoire and music from the Renaissance period, the quarter-comma meantone is ideal, but it is problematic for quite a few of Bach's organ compositions because he used the new temperaments that emerged in his time. To tackle this problem, a dual temperament could be a solution, but some pipes would need to be doubled, which would complicate the already intricate design of the spring chest. To understand the dilemma of this design problem, it is necessary to briefly explore historical tuning systems in more detail.

HISTORICAL ORGAN TUNING SYSTEMS

When listening to keyboard instruments, we are accustomed to what is called equal temperament. This tuning system became the standard in musical practices in the course of the nineteenth century. It more or less settled a longstanding theoretical and practical debate on how to solve a peculiar problem in music known as the "Pythagorean comma." Through experiments with sets of strings to study the properties of intervals in the harmonic series such as the octave, the fifth, and the major third, Pythagoras discovered that the interval of seven ascending octaves was not equal to twelve ascending pure fifths. In other words, twelve fifths do not fit in the seven octaves.[18] The solution to this problem we have today is to equally distribute the difference, called the Pythagorean comma, over all fifths by making by making them slightly impure. This tuning system is called equal temperament.[19] As a consequence, all other intervals, except the octave, are not pure either.

Before equal temperament became the standard for tuning keyboard instruments such as the organ, different temperaments were used, aiming to have at least a few pure intervals. In the sixteenth and seventeenth centuries, a system called quarter-comma meantone became a de facto standard in organs. In this system, eleven of the twelve fifths are slightly lowered, so that no less than eight thirds sound pure. This allows the triads, in which thirds have a prime role, to be purer as well. This temperament leaves one fifth that sounds very dissonant and was hence called the "wolf" interval. Overall, quarter-comma meantone allows pure thirds in some keys, and those were the keys that were mostly used.[20]

Around 1700, new temperaments were developed to give musicians the possibility of working in all twelve keys and not have triads that were too impure to use. Examples are the "well-tempered" systems that Andreas Werckmeister (1645–1706) developed. As Kent observes in his overview of keyboard tuning, at the time of Johann Sebastian Bach, there was a plurality in tuning systems (Kent 1998, 47). Bach's interest in innovative temperaments and the new possibilities these offered to composers to write harmonically complex music is reflected in the two sets of preludes and fugues that Bach published in 1722 and 1742, under the title *Das wohltemperierte Klavier* [*The Well-Tempered Clavier*]. Both volumes span the complete range of twenty-four major and minor keys.

The relation between temperaments and everyday musical practice becomes clear in four systems that were developed by Johann Georg Neidhardt (c.1685–1739), a contemporary of Bach. Neidhardt distinguished between systems "for the village and the small town" that were closer to meantone, and more irregular, well-tempered systems for "the large town and the court." These distinctions made sense, because large towns and courts were places of musical innovation that needed systems that enabled the use of many keys. Little is known about the preferences of Bach for the temperaments of his time (Ortgies 2014). A hint is given in the biography of organ builder Zacharias Hildebrandt by Ulrich Dähnert (1962). Bach was well acquainted with the work of Hildebrandt. In 1747, he tested an organ built by Hildebrandt in the St. Wenzel Church in Naumburg, Germany. Bach's son-in-law, Johann Christoph Altnikol (1719–1759), wrote about this occasion: "In tempering he follows Neidhardt and one can modulate very well in all keys without presenting the ear with anything repugnant—which for today's taste in music is the most beautiful" (Dähnert 1962, 115).

In the historically informed performance practices and historicizing instrument building that emerged since the 1960s, historical temperaments became an important issue. Recreating the conditions under which music was performed in pre-Romantic times included tuning them according to the principles at the time. It was argued that standardized equal temperament neglects the subtle differences in harmony that result from meantone and well-tempered systems and, hence, obscures the musical meaning of music from the baroque era.[21] The tuning issue became prominent in historicizing restoration practices of keyboard instruments, and organs in particular. Following the preference for equal temperament since the nineteenth century, many historical organs were retuned, which often meant that their original temperament was difficult to reconstruct. The question if and how historical organs should be brought back to their original temperaments became a source of extensive debates among organ experts, especially in the case of restorations of organs on which Bach's music was to be played. Given the fact that most of Bach's music is harmonically too complex to perform in a quarter-comma meantone temperament, a choice for a well-tempered system has to be made, yet without knowing which of the many well-tempered systems Bach would have preferred.

HARD CHOICES

The tuning issue figured prominently in the first months of the new baroque organ project. Most experts agree on the fact that the organs built by Arp Schnitger would have been tuned in quarter-comma meantone (Ortgies 2014). They are well suited for the music of the North German organ school, but much less so for Bach's harmonically complex compositions. If one chose Schnitger as the main historical model—which also made sense because his organ in Stade has a spring chest—and one wanted to be historically accurate in terms of tuning, it would be difficult to play Bach. In the spring of 2013, Hans Fidom had suggested solving this problem by giving the new baroque organ two temperaments, one meantone and one well-tempered.

During an Orgelpark symposium in May 2014 that focused on the temperament of the new organ, four experts presented their views. Jos de

Bie, an expert in historical organ temperaments, reflected on the question whether it might be possible to equip the new baroque organ with the ability to "switch" between temperaments that would favor both "old" and "new" music (Bie 2014, §140). De Bie identified four options: to alternate between equal and unequal temperaments, to combine a well-tempered and a meantone temperament, to build multiple manuals each with its own temperament, or to find a single and necessarily compromised temperament for the whole organ (2014). Generally speaking, combining two different temperaments in one organ requires sufficient common notes; otherwise, it would be better to build two organs. The first two options start from the principle of having both shared and extra pipes per octave to allow switching between the two temperaments. It would be necessary to build at least six extra pipes for every octave, which meant that the organ would become more expensive, would have more complex windchests, and also would become heavier (Bie 2014, §141–§172). In the third option, each of the manuals—and, therefore, the corresponding windchests—would have its own temperament. De Bie moves this option aside as being expensive and impractical, because coupling the manuals would produce bizarre results (2014, §173). The simplest and cheapest option, De Bie argued, would be the fourth option, to have a single compromise temperament for the whole organ (2014, §174–§179). The question of which temperament to choose became even more complicated, given the fact that in the "digital mode," pipes could be combined with any other pipe. This sound concept would work best with equal temperament.

Following in the footsteps of Arp Schnitger inevitably led to complex choices when it came to the temperament of the organ. This was one of the reasons that Hans Fidom suggested at the end of the trip to north Germany that the group broaden the scope of historical models and listen to organs built by Zacharias Hildebrandt. These organs represent a middle German sound ideal, one with which Bach was very familiar. A new study trip to the Thuringia region was scheduled for December 2013.

BRINGING HILDEBRANDT TO THE ORGELPARK

In the early morning of December 4, I met Hans Fidom and Johan Luijmes at the train station in Groningen, from where we travel by car to Borgentreich, Allstedt, and Naumburg in Germany. We have time to discuss the project as well as our impressions of the organs we had visited so far. In Borgentreich, we met Loek Dijkman, Sylvia de Munck, and Hans Elbertse.

The baroque organ in the Pfarrkirche St. Johannes Baptist has forty-five stops in total, three keyboards, and a pedal. Originally built in the sixteenth century for a nearby monastery church, the organ was moved to the Pfarrkirche in 1803. It was restored between 2003 and 2011 by Hermann Eule Orgelbau from Bautzen, Germany.[22] For us it is interesting because it has a spring chest (see the introduction). This characteristic of the organ is reflected in the organ museum associated with the church, which offers models of spring chests that are helpful to understand the mechanical aspects of this chest type.

We continue our journey to the Schloss Kapelle in Allstedt, built in the eleventh century and restored in 1974. The weather is cold and damp. The small chapel of the Schloss houses a small organ built by Kristian Wegscheider in 1990. The instrument offers the possibility to switch between two temperaments: a meantone and a well-tempered one. Listening to the meantone temperament, we are struck by the "openness" and "rest," as Hans Fidom puts it, in the sound when playing in keys such as C, G, and F. We also notice that the pipes that have a dual function—namely the six per octave that need to sound in both temperaments—have a clearly audible, to the brink of distractive, different sound quality. This leads to a discussion on whether having two temperaments in the new baroque organ is actually a good idea.

Provincial roads take us to Naumburg. In the small towns we pass, we see many houses that remind us of the atmosphere of the former German Democratic Republic. Later that day, in the hotel in Naumburg, I reflect on what I saw in my notebook: "How to write about these places and organs? The trips are gaugings into a strange world, beyond the two world wars, beyond the nineteenth century and Wagner. It's a world that has Weimar, Goethe, and, of course, Bach a stone's throw away. Searching for a sound that offers past and future at the same time. It is the sound of an outdated technology, a sound that we no longer understand how it was once heard" (Field notes, December 4, 2013—translation PP). I realized that this attempt to recreate a sound world of the past in the present fascinates me, without really understanding why.

THE HILDEBRANDT ORGAN IN NAUMBURG

The next morning, we walk to the St. Wenzel Church to listen to the organ that Zacharias Hildebrandt built in 1746, and that was restored between 1993 and 2000 by Hermann Eule Orgelbau from the nearby town of Bautzen. When entering the church, I am impressed by the first view of the organ, positioned relatively high in the large church. Organist David Franke welcomes us. He recounts how the organ was built, the hard work by an underpaid Hildebrandt, and the story of the assessment of its quality in 1746 by the organ builder Gottfried Silbermann, once the employer of Hildebrandt, and Johann Sebastian Bach. The commissioning city council chose Silbermann and Bach to inspect the new organ because they were "two well-known capable masters of their crafts and experts in the field" (Wolff and Zepf 2012, 148). The short report they wrote stated that the organ contains all the parts that were listed in the contract, and that all parts had been made with appropriate care, using the materials that were promised. "Nevertheless, it will be necessary to encourage Herr Hildebrandt to go through the entire instrument once more, stop by stop, in order to achieve more evenness in the voicing as well as in the key and stop actions" (Wolff and Zepf 2012, 148). The two men had a good time, David Franke tells us: "We know that from the fact that the bill of their visit to the café was substantially longer than the assessment report."[23]

The sound of the organ in the large church is overwhelming. When Franke plays Bach's *Fantasia in G major* (BWV 572) and *Nun komm, der Heiden Heiland* (BWV 659), I am deeply moved, and I later hear that the others had had similar experiences. Both Johan Luijmes and Hans Fidom play the organ. We listen to it, both up on the organ balcony and down in the church. After saying goodbye to David Franke and thanking him for his performances, we have lunch in a nearby *Konditorei*. Loek Dijkman jokingly summarizes his great admiration and enthusiasm for this organ: "If possible, I would say: just package it and deliver it to the Orgelpark."[24] We discuss the possibility of taking the Hildebrandt organ in Naumburg as the reference organ. A dual temperament would no longer be necessary because the instrument has a Neidhardt tuning from 1724, which makes it perfectly suited to play all the organ music by Bach. Hans Fidom points out another argument for Hildebrandt. The sound of his organs is characterized by a large number of 8' stops, which fits the acoustic in the Orgelpark better than the more "pointed" and "brilliant" sound of the Schnitger organs, and it also makes it easier to use in twenty-first century compositions and performances. What would be lost, however, is the possibility to play the North German organ repertoire in meantone temperament, one of the main reasons to focus on Schnitger. We think that the organ of Naumburg with its fifty-three stops would fit the Orgelpark, even though the number of stops is high. An alternative would be to study the specifications of smaller organs built by Hildebrandt. We leave Naumburg with the feeling that we might want to listen to the Hildebrandt organ again soon and head for Goslar-Grauhof to visit the Treutmann organ from 1737 in the Klosterkirche St. Georg. This organ is special because it combines influences from northern and middle Germany. When Johan Luijmes plays the instrument, he is very positive about the sound.

A week after the study trip, the core team met at the Utopa office in Leiden. Before the meeting starts, Loek Dijkman plays a recording of Bach's *Clavierübung III* made by David Franke on the Naumburg organ. He asks everyone to look back on the organs we visited and evaluate them. Johan Luijmes starts the discussion by saying that he was touched most by the organ in Naumburg, yet he wondered if this experience was not also an effect of its surroundings, the space of the church, and the fact that he is used to playing historical instruments since his conservatoire studies. "I think that, maybe, I not only listen to an organ when I listen to an organ."[25] Hans Elbertse thinks that the Hildebrandt sound concept would fit well in the Orgelpark. Hans Fidom summarizes the choice between Schnitger and Hildebrandt: A Schnitger organ is good for the North German school that preceded and influenced Bach and has a characteristic temperament, whereas a Hildebrandt organ is well suited to play Bach's music but is less characteristic in terms of its tuning. He agrees with Elbertse that "Naumburg is a bit friendlier to the space in the Orgelpark than Schnitger."[26]

Loek Dijkman challenges the idea that the project should take a historical model because, in his view, it is impossible to transfer the experience of the original instrument:

> I have the same feelings as Johan. What do I see? What do I hear? For me, the conclusion of our trips is that all these organs have special qualities. That may have to do with the organist playing the instrument, the light, the air, the building, the history, the religion. In short, an organ is more than just a sound. It is a product of sound, time and surroundings. If that is true, that could be a guideline for our project. It would mean that we should not replicate something, because

> that always leads to disappointment. Automatically. It is not the place, it is not the year, Bach did not sit behind the instrument, the organist was not there, the room temperature is different. Maybe we should skip the idea to make a copy altogether. By moving an organ to the Orgelpark it has become a different instrument.[27]

Questioning the possibility of an unequivocal historical reference, Dijkman argued, had consequences for the type of research that was necessary. In his view, research should focus on what one needs to know to build the organ, and it should not be a goal in itself. Hans Elbertse added that from his experience as an organ builder, he does not need scientific research to know why certain things work or not. Hans Fidom, on the other hand, argued for the idea of "process reconstruction" that had informed the GOArt project in Gothenburg. Here, reconstructing historical ways of working as well as using scientific methods and instruments to study aspects of the organ such as wind behavior was the way to get as close as possible to Schnitger's sound characteristic.

The study trip to Thuringia led to two important and interlinked changes in the design of the new baroque organ. First, Zacharias Hildebrandt replaced Arp Schnitger as the reference organ builder and his organ in Naumburg had become the main reference. A reason for this was the team was impressed by its sound. But there are additional arguments. The organ was assessed by Bach himself, who closely collaborated with Hildebrandt. Its sound character is characterized by the "*Gravität*," "*Brillianz*," and "*Poesie*" that Bach thought to be the most important elements in the sound of an organ.[28] With its large number of 8' stops, the sound fits the acoustic of the Orgelpark's hall. Second, the choice for Naumburg as a reference organ meant that the new baroque organ would have a single temperament, possibly a well-tempered Neidhardt temperament, because Bach's son-in-law J. C. Altnikol, who became organist in Naumburg in 1748, wrote that Hildebrandt "followed Neidhardt." As a result, the windchest would not have to be adapted to a larger number of pipes per octave, which would reduce the hybridity or "monster" character of the new organ. A new problem was identified, however. Hildebrandt did not build organs with spring chests, and this technology was crucial to having an electronic interface next to its mechanical keyboard. The next step in the design process was to explore in more detail whether the Hildebrandt's sound could be brought to the Orgelpark in combination with the spring chest technology.

STOPS AND WEIGHTS

Packaging the Hildebrandt organ in Naumburg and delivering it to the Orgelpark was not possible, of course. In what way could its sound still be a historical reference for the new organ? Even if it could fit in the hall, it was clear that the number of stops—a total of fifty-three, including some very big 16' and even 32' pedal stops—would be too large for the Orgelpark. Given that the assemblage of stops basically determines almost all aspects of an organ's design, the first task was to determine the specification, that is, the list of stops. Already during the meeting on December 11, 2013, Hans Fidom had proposed following the specification of an organ that Hildebrandt built in the Dreikönigskirche in Dresden between 1754 and 1757. This instrument was destroyed during the bombing of the city in 1945.[29] It had thirty-eight stops on two manuals and a pedal, and according to Fidom, it would better fit the hall in the Orgelpark. Having two instead of three manuals need not be a problem, because the digital console would enable the organist to make stop combinations freely.

The question of what types of research were necessary to design and build the new organ were raised again in a meeting on February 5, 2014. Hans Fidom had written an overview of the organological research that he thought should be carried out. The sound quality of an organ depends on choices made about the making of the pipes, the windchests, the bellows, and wind channels, as well as the key action. Expert knowledge about Hildebrandt and his practice would have to be mapped through contacts with organ builders, literature research, and measurements of Hildebrandt organs and pipework. In addition, it would be necessary to map the state-of-the-art knowledge on the magnets that were to be used in the spring chest. A second type of research, he argued, would focus on the question of whether the electronic parts in the windchest would influence the sound of the pipes. For this empirical research, he suggested using prototypes of the windchest and organizing listening sessions with experts. Additional research should focus on the wishes of composers and musicians concerning the use of live sampling of the organ; microtonality, or the use of intervals smaller than the semitones on a piano keyboard; flexible wind pressure; and the layout of the digital console. Finally, the design of the case was a topic for research. Given the preference in the team for something contemporary, he suggested the creation of an inventory of contemporary case designs from the past decades as a source of inspiration.

A month later, on March 12, the specification and the design of the organ were discussed again to see if the spatial, financial, and artistic conditions in the Orgelpark are met. Based on a specification proposal of forty-four stops, organ builder Hans Elbertse calculated that an organ this size would fit the balcony, even though it is unclear at this point if it could support its weight of approximately forty-four tons. The bellows would be positioned in a bellow case on the left of the balcony; the pedal pipes would be combined in one pedal case. The costs exceed the budget that was determined in the spring of 2013. The magnets and motors in the new organ will use a lot of electricity, Elbertse says, especially when they are switched on for a long period of time. This could raise the temperature in the tone channels up to five degrees Celsius. To prevent damage to the wood, he suggests building in a safeguard that would limit the amount of time that the magnets are switched on. Sylvia de Munck notices that this would affect the artistic possibilities available to musicians. The acoustic properties of the hall are discussed. An electronic system to control the reverberation times is seen as an option to fit the sound of the large organ in the space. At the end of the meeting, Johan Luijmes expresses his doubts about the organ that emerges from the first sketches and descriptions. He wonders whether having all pedal pipes in one tower will affect the sound negatively. Another worry he has is if it would still be possible to have a small orchestra and choir on the balcony to perform from there. "We should not have a big organ if that means that musicians cannot stand next to it."[30]

Only a few weeks later, the situation changed completely. The architectural engineer finally studied the original drawings of the Park Church and concluded that the north balcony can only support a weight of twenty tons, including chairs and members of the audience. Suddenly, an organ the size of Naumburg's or even Dresden's is out of the question. Quick calculations estimate the maximum weight of the organ around fifteen tons. During a short meeting on April 8, 2014, the team reconvenes to discuss the consequences. A solution is proposed by Hans Fidom, who recommends looking at other organs by Hildebrandt to see if they could be a reference model. The Hildebrandt organ in Lengefeld, Germany, recently restored by Kristian Wegscheider, has twenty-two stops and, thus, would be an option. An organ this size weighs 14.5 tons. According to Johan Luijmes, a middle-sized organ could still perform its task in the programming of the Orgelpark. The original idea to build an organ on which Bach would sound well, he says, would not be compromised by the smaller size. Solving the problem the other way, by strengthening the foundations of the building, is not an option for Loek Dijkman; he reminds the team of the heritage status of the building, which impedes large interventions in its architecture. Sylvia de Munck sees possible benefits for a smaller organ. "It might even sound more intimate and beautiful in that space."[31]

Hans Fidom and Hans Elbertse made detailed calculations of the weight of an organ with twenty-two stops and a larger variant with twenty-nine stops. Both options are discussed on April 22. They stay within the limit of fourteen tons, a weight mentioned by Hans Fidom. Fidom also made a new stop list based on Hildebrandt organs in Naumburg, Lengefeld, and Sangerhausen. It is clear that a historical reference to a single existing Hildebrandt organ such as in Naumburg will no longer be possible. How problematic is this? Hildebrandt's own practice might give an answer, Hans Fidom argues. When he had to build an organ in a small space, he used the specification of a larger one and adjusted the scaling of the pipes, thus staying within the same sound concept. "Naumburg," therefore, could still be the sound reference, even though the organ itself would be smaller. The team agrees on the benefits of an organ with twenty-nine stops and decides to calculate the extra costs of this specification.

Another important practical question at this stage is who will build the pipes for the new organ and who would voice them. Because we take Hildebrandt pipework as a reference, the pipe maker would ideally have knowledge of this pipework and practical experience in its restoration. The Hermann Eule Orgelbau company in Bautzen clearly is knowledgeable about Hildebrandt's pipes, given that they restored a number of his organs, including the organ in Naumburg. We also discuss the option of buying the pipe measurements of the Naumburg organ from Hermann Eule Orgelbau and choosing another pipe maker. This could be Munetaka Yokota, who voiced the GOArt organs in Gothenburg, Rochester, and Cornell. More than any other pipe maker and voicer, Hans Fidom says, Yokota has been able to create and voice pipes in such a way that they sound "old."[32] He also extensively studied Hildebrandt organs. During the meeting, Yokota is invited to come to Leiden and discuss a possible collaboration. Johan Luijmes suggests traveling with Munetaka Yokota to Naumburg to learn from his way of listening to the organ and to develop a shared understanding of the sound ideal within the core group.

On May 16, the core group meets with Munetaka Yokota in Leiden. After Dijkman introduced the Utopa Foundation, its mission, and the plans for the new baroque organ, Yokota talks about his work for the Göteborg Organ Art Center (GOArt). He explains that in his organ-building projects, he prefers to work with talents from the local communities, something that goes back to the way medieval organ builders worked. He underlines the importance of "process reconstruction": Instead of copying a historical organ, in his view, it is more important to understand how it was built in a specific situation and what the problems were that organ builders had to solve at a certain location. Only thus can we hope to achieve the sound quality of their organs, he argued. Engaging Yokota in the project gave a new impetus and direction to the debate within the core group on how to create a convincing reference to the organs that were familiar to Bach.

NEGOTIATING A HERMENEUTICAL SITUATION

A central question in the first year of the project was how the new baroque organ would be related to the history of the instrument and its music. The story that singer-songwriter Spinvis told about basements below the hall of the Orgelpark, each filled with organs and organists, evokes this history as a place that can be visited, yet is unfamiliar to us: "What a strange time this must have been." This chapter recounted how the members of the core team worked with history to better understand their project and to develop an idea of what the new organ could be. This appears paradoxical: How does one understand something that does not yet exist? And how does history play a role in this understanding?

At first sight, the story about the excursions by the core team can be read as an attempt to find a historical organ that could be used as a reference for the new instrument in terms of its construction and sound. But as the discussions during the study trips show, the questions how and to what extent historical reference should actually guide the project also figured prominently. The conversations during the team meetings focused on the practical problems and dilemmas that resulted from a historicizing approach, such as the choice of temperament, the design of the windchest, and the weight and specification of the new instrument. Not only did members of the core group have different opinions on these issues, their arguments echo positions in the debate on historically informed performance practice in the postwar period.

In 1951, the German composer Paul Hindemith wrote about the performance of early music:

> All the traits that made the music of the past lovable to its contemporary performers and listeners were inextricably associated with the kind of sound then known and appreciated. If we replace this sound by the sounds typical of our modern instruments and their treatment, we are counterfeiting the musical message the original sound was supposed to transmit. Consequently, all music ought to be performed with the means of production that were in use when the composer gave it to his contemporaries. (Hindemith, cited in Lawson and Stowell 1999, 10–11)

When Hindemith formulated his performance ideal, music from the baroque and classical era was mainly played on modern instruments. Contrary to the prevailing Romantic esthetic that grounded musical meaning in the expressive qualities of the performer, the historically informed performance practices that burgeoned since the 1960s saw the performance's fidelity to the original sound of historical instruments as an important precondition for its artistic value (Haynes 2007; Wilson 2014).

With the new focus on historical origins and sound fidelity came a debate on the concept of authenticity.[33] Although many musicians in the early music movement were hesitant to claim that their performances were more "authentic" than those on modern instruments, the term took center stage in an academic reflection on historicism in musical practice (Burstyn 1995; Davies 1991; Fabian 2001; Keller 1984; Kenyon 1988; Leppard and Pauly 1988). According to the analytic philosopher Peter Kivy (1998), we can distinguish at least four ways to link the notion of authenticity in a meaningful way to musical performance practice, in its faithfulness to the composer's performance intentions, to the performance practice of the composer's lifetime, to the sound of a performance during the composer's lifetime, and finally to the performer's own self, not in mimicking others. Kivy distinguishes between the practical question—if these forms of authenticity are possible—and the normative question—whether they are desirable. Disqualifying the first three types of authenticity as impossible, and recognizing the fourth as desirable, he welcomes "any mix of authenticities, that withstands the only relevant test there is: the test of listening" (Kivy 1998, 285).[34]

Following a similar line of argument, the American musicologist Richard Taruskin writes about the problem of historicity in musical performance. In his essay "The Pastness of the Presence and the Presence of the Past"

(1995), he claims that the authenticity of a musical performance has little to do with being faithful to anything that happened in the past, be it the composer's intentions, the actual sounds, or the practices of music making.

> I am convinced that "historical" performance today is not really historical; that a thin veneer of historicism clothes a performance style that is completely of our own time, and it is in fact the most modern style around; and that the historical hardware has won its wide acceptance and above all its commercial viability precisely by virtue of its novelty, not its antiquity. (Taruskin 1995, 102)

Opposing the two positions sketched here—Hindemith's plea for fidelity to historical sound by performing early music on period instruments and Kivy's and Taruskin's arguments against the desirability or even possibility of "authentic" or "historical" performance—has been typical for the historical performance debate. As Butt (2004) has shown, this debate raises fundamental questions about how performing music from the past is always an encounter with history that teaches us not only how things were different, but also how they could have been different: "It helps us create imaginary worlds, just like those of fiction, that chime with our own while revealing crucial differences" (Butt 2004, 46).

The new baroque organ project was intended as an encounter with history from the start. Yet, the goal of this encounter was not to replicate a historical instrument or sound, if possible at all, but, rather, to create possibilities for musicians to perform music from the past and the present in new ways. A central problem in this approach was how, in the building of the new organ, to convincingly "make" history while at the same time giving it meaning in the present. To analyze this problem in more detail, I draw on the historical hermeneutics of Hans-Georg Gadamer.[35]

In *Truth and Method* (1989 [1960]), Gadamer develops the outlines of a philosophy of hermeneutics. His goal is to regain the concept of truth for the humanities. Our experience of society and history, Gadamer argues, cannot be raised to a science by following the methodological procedures of the natural or social sciences. A historian does not see a concrete phenomenon as an instance of a universal rule.[36] The historical tradition becomes the object of research, but the understanding of this tradition is far from a construction of principles. Rather, it is the continued development of a history that comes from afar. We do not place ourselves outside this history but hand it down in our situated understandings and applications of it. Gadamer presents this handing down as hermeneutic work that he compares to an ongoing conversation.[37] What drives the conversation is the polarity of familiarity and strangeness of a historical text or object as "being a historically intended, distanced object and belonging to a tradition. *The true locus of hermeneutics is this in-between*" (Gadamer 1989 [1960], 295, italic in the original).

Historicism, Gadamer argues, conceives of time as a gulf that needs to be bridged because it separates us from the past. To overcome temporal distance, and to move toward historical objectivity, "we must transpose ourselves into the spirit of the age, think with its ideas and its thoughts, not with our own" (1989 [1960], 297). The idea that this would be possible, however, is naïve, according to Gadamer. Historical thinking can perform the task of understanding only if it takes its own historicity into account: "*Understanding is, essentially, a historically effected event*" (299, italic in the original). Being conscious of the fact that our understanding is affected by history means that we are aware of the hermeneutical situation. We are conscious that, as interpreters, we are not separated from the past but connected to it through tradition.

Gadamer further defines the concept of hermeneutical situation through the metaphor of "horizon" to indicate that every situation represents a standpoint that limits what can be seen. Applied to the thinking mind, horizons can be narrow, widened, or opened (301). Our horizon of understanding in the present is continuously being formed through our encounters with historical horizons. Gadamer calls this the "fusion of horizons": "In a tradition this process of fusion is continually going on, for there old and new are always combining into something of living value, without either being explicitly foregrounded from the other" (305).

One could now argue that a fusion of horizons leads to the formation of one horizon delineated by what Gadamer calls "the depths of tradition" (305). It is crucial, however, that within the historical consciousness of the hermeneutical situation, a tension between the historical object and the present is experienced. Hermeneutic understanding assumes that this tension is not covered up by attempting to assimilate it in the present but instead is seen by projecting a historical horizon that is different from the present (305). Only then can we mediate between historical ideas and our own thinking, which is what determines the historical life of a tradition. Understanding, Gadamer writes, "is to be thought of less as a subjective act than as participating in an event of tradition, a process of transmission in which past and present are constantly mediated" (291).

Gadamer's historical hermeneutics provide a framework to analyze how the work that was done in the first year of the new baroque organ project shaped a hermeneutical situation in which the members of the core team had been projecting different historical horizons to learn more about what the organ could, or even should, be. When shortly after the study trip to Thüringen Loek Dijkman said that making a copy would lead to disappointment because after "moving" a historical organ to the Orgelpark it would have become a different instrument, he was keenly aware of the impossibility of reconstructing the past without having to absorb it in the present. This led to dilemmas. Choosing Arp Schnitger as the reference builder had the advantage that he used spring chests in his organs, which made it possible to mediate a historical windchest design with electronic means to operate the organ. A disadvantage, however, was that his organs were tuned in quarter-comma meantone, which meant that not all of Bach's music could be played. Solving this by giving the instrument a dual temperament might give it a "monster" character. Choosing Zacharias Hildebrandt as the reference builder had as an advantage that the sound concept of his organs better fit the Orgelpark's acoustics and that the instrument could have a more versatile well-tempered temperament. A drawback would be that Hildebrandt did not build spring chests, whereas these were crucial to the dual design of the new organ, and that the preferred historical reference organ in Naumburg would be too big for the Orgelpark. Understanding what the new organ could be thus implied mediating between past and present hypothetically and evaluating the pros and cons of the different possibilities that resulted from this.

The study trips and the discussions during the meetings of the core team show how the hermeneutical situation of the project was negotiated during the first months. Although the basic design premise—an organ with a convincing baroque sound that can be made accessible through electronic means—was never challenged, the members had their own ideas, opinions, and preferences about how to realize this design idea. What had become clear, however, is that through the project they aimed to *participate* in traditions of organ building that spanned centuries. Following Gadamer's dialogical conception of tradition, rather than a form of antiquarianism, it is a continuing debate on questions, problems, and issues to which we ourselves contribute. Being at the heart of the new instrument, the design of the windchest of the new baroque organ, more than anything, shows how the members of the core team joined this debate.

3

DESIGNING AN ARCHAIC TECHNOLOGY

A COLLECTIVE THOUGHT EXPERIMENT

The organ workshop of Elbertse Orgelmakers is located in an industrial area on the outskirts of Soest, a small town in the center of the Netherlands. From the unassuming exterior, one would not expect the building to house a company that builds and restores musical instruments, some of them centuries old. Hans Elbertse, who is the third generation in a family of organ builders, moved to this building in 1996, from the workshop in the backyard of the house where his father, Jan, and grandfather, Johannes Josephus, lived and worked.[1] Elbertse's grandfather started as a young man in the workshop of the Dutch organ builder Michaël Maarschalkerweerd in 1895. At that time, Maarschalkerweerd had just finished the new organ (1891) for the Concertgebouw in Amsterdam.

The members of the core team visit the workshop in October 2013, to see a test organ that Elbertse built to experiment with the design of the new baroque organ's windchest. He shows us around. We start in a spacious stockroom filled with stacks of wood—mahogany, oak, pitch pine, Oregon pine, ebony, palisander. Some of the trees were felled eighty years ago. Elbertse explains that this old wood is used to build or restore windchests; because it dried naturally for a long time, it will not warp. Ebony, the most expensive wood he has in the workshop, is used for the black keys and for small mechanical parts in the organ. The workshop itself has two large rooms with workbenches and machines. Elbertse introduces us to his employees. They are working on the restoration of a cabinet organ from the eighteenth century that belonged to the owner of a mansion in the nearby village of Amerongen. We meet Bert van den Heuvel, the master organ builder of the firm, who takes us upstairs to another storage room for wooden parts and organ pipes to show the prototype. The sight of it evokes enthusiasm; the new baroque organ has its first physical appearance.

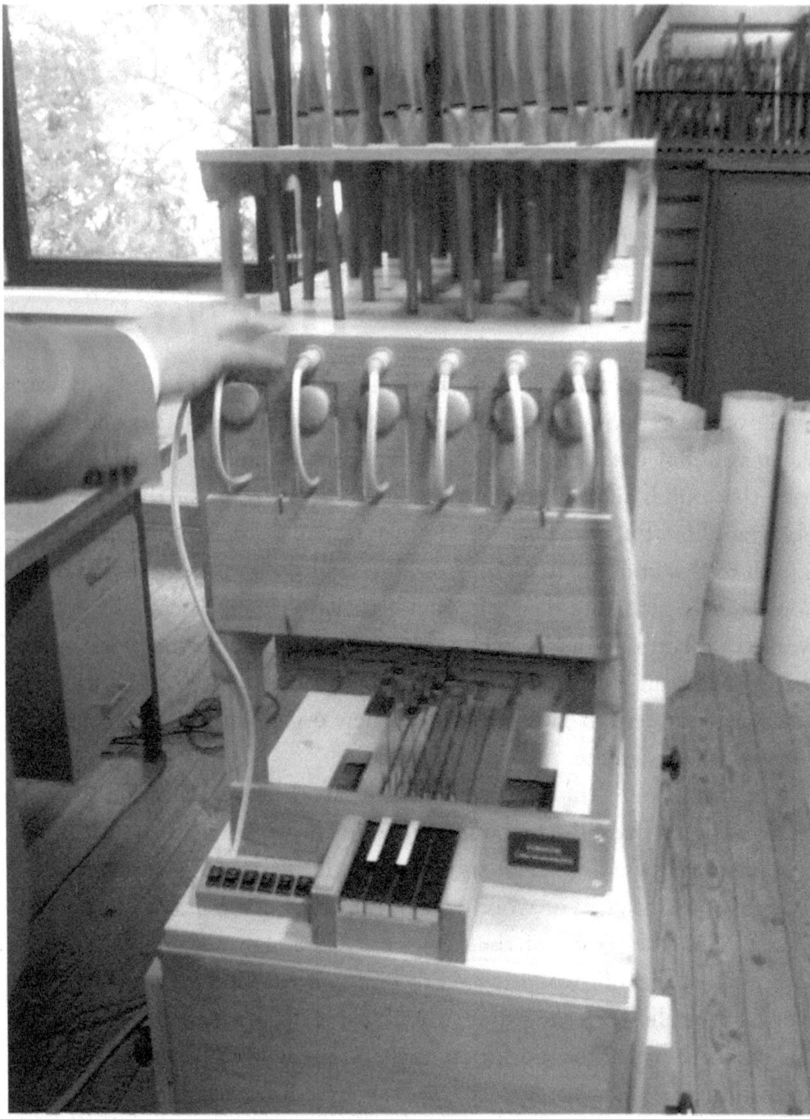

Figure 3.1
First test organ in the organ workshop of Hans Elbertse, October 2013 (photo taken by the author).

The strange structure is approximately two meters high (see figure 3.1). What we see is a wooden box on four legs that has a small organ manual with six organ keys and six on/off switches lying on top of it. Above the manual is a second wooden box with gray electronic cables coming out of it, leading into other parts of the structure. A set of organ pipes of various lengths and shapes, placed in a wooden rack, sits on top of the second wooden box. Elbertse explains the goal of the machine. Its second box, the windchest, materializes the central design idea of the new organ: the combination of a historical spring chest mechanism with electronically manipulated magnets that open small valves directly below the pipe feet. Right below the pipes are six tone sliders, each with a round knob to remove them. They house the pipe valves and the electromagnets (see figure 3.2) and can slide in and out of the windchest, analogous to the spring chest of the Hus/Schnitger organ in Stade, Germany (see chapter 2).

Wind flows from the small bellows inside the first box to the second box, from where it is led into the pipes in various ways. Using the switches, it is possible to shift between a mechanical mode (see figure 3.4a-b) and an electronic mode (see figure 3.5a-b). After the test organ is switched on and keys are pressed, it is possible to hear how the two modes affect the sound of the pipes. Not only does the apparatus allow Elbertse and Van den Heuvel to experiment with the mechanical and electronic construction of the hybrid windchest, it also makes it possible to evaluate the sound qualities that result from these differences.

The test organ and its sounds trigger an intense discussion that takes the shape of a collective thought experiment. Loek Dijkman hears differences between pipes played in the various modes. Elbertse explains how these result from slight changes in wind pressure, the type of pipe, and how it is constructed. "How many pipes do you want to use in the organ? 3,600?," Dijkman asks. "That means that there are 3,600 chances that it can go wrong."[2] Elbertse replies that the test organ allows the team to listen for the best way to make the pipe valve mechanism and thus to reduce the risks. Johan Luijmes points out that the spring chest mechanism only affects the organ sound when it is played in its twenty-first-century electronic mode. What counts, he says, is how the pipes sound when the organ is played mechanically. Hans Fidom explains that the anatomy of the windchest has impact on both "worlds," but that the historical sound has priority: "We will voice the pipes for the seventeenth-century sound. What comes out in the twenty-first-century

Figure 3.2
Tone slider with electromagnets that manipulate the pipe valves (photo taken by the author).

version is a present. We will see."[3] Dijkman agrees: "The electronic variant will have its own character."[4] He asks if Elbertse has old pipes in his workshop that he could place in the test organ. Elbertse returns with a pipe from a restoration project of a sixteenth-century organ. After he places the pipe in the test organ and listens to its sound, he takes it out again because he is not happy with it: "Do you know the problem with these old pipes? This is a Quintadeen that is made into a Holpijp. You can see the seams here. They wanted a different sound and adapted it."[5] Everyone agrees that in order to have valid listening tests, the pipes have to be newly constructed.

In this chapter, I focus on the design process of the hybrid windchest that started with the test organ and the discussions it evoked. This process involved meetings of the core group, experiments with prototypes of the windchest, listening tests, and a new study trip to Germany, as well as discussions with organ experts during symposiums in the Orgelpark. The core issue in this period was how to combine a convincing baroque sound concept with the possibility of accessing this sound through electronic means, and, thus, inspiring organists, improvisers, and composers to make new music. From an actor's perspective, this meant finding clever design solutions and making tough decisions, knowing that these would be intensely debated in the organ world. Theoretically, I focus on how to conceptualize the role of history in the design of the hybrid windchest. Following Gadamer, whose historical hermeneutics I introduced in the previous chapter, we will see how the project team participated in organ-building traditions from various periods. Working with these traditions resulted in a final design of the windchest that combined knowledge and understanding, as well as practical solutions and techniques ranging from the sixteenth to the twenty-first century.

Gadamer's hermeneutic understanding results from acknowledging the strangeness of the historical text, situating it in its own time, and giving it meaning by mediating it in our own time. To analyze the design of the windchest, however, the historical hermeneutics of Gadamer have limits that become clear when we think back to the old pipe that Hans Elbertse placed in the test organ, only to conclude that it did not work as he had hoped. How can Gadamer's horizon metaphor help understand the pipe as a historical object? Hans Elbertse made it clear that the pipe had been changed many times since it was made in the sixteenth century. Which horizons merge in our interpretation and understanding of it? And is interpretation actually what Elbertse is after, or is he using the pipe to demonstrate something that cannot be said in words? What would be the difference?

Answering these questions requires an approach of historicity and temporality that recognizes the old pipe as an object with a history. For this approach, I draw on the work by Bruno Latour on things, such as organ pipes, and how we can conceive of their history as everything that affected their *existence* (Latour 1993; Vries, de 2016, 132). When Elbertse placed the old pipe in his test organ, he changed its history. As will become clear, Latour's approach of the hybrid ontology of things is helpful to redefine what it actually means to design an artifact, such as the hybrid windchest, that uses elements from different historical periods.

In the previous chapter, I showed how the new baroque organ project draws on the practice of historicizing organ building as it emerged since the late 1960s. The design of the new baroque organ, however, also relates to nineteenth-century organ-building practices as well as recent innovative organ-building projects. These projects aimed at enhancing various aspects of the contemporary organ such as temperament, wind behavior, pipe sounds, and interfaces. After I analyze these traditions and innovations by mobilizing Latour's view on things and their histories, I will show how the restoration of the Sauer organ in the Orgelpark and its new digital console provided the germs of the design of the windchest for the new baroque organ. I then recount how the team tackled the problem of integrating electronic devices in the windchest. To experiment with possible solutions, two test organs were built by Hans Elbertse and Bert van den Heuvel, the second one designed by Hans Fidom. These were used not only to learn more about the technical aspects of the design but also to determine the effects of various design solutions on the sound of the pipes through listening tests. The final plan for the organ was ready in the early months of 2015. It drew together the various elements that would make the new baroque organ. In the next section, I will focus on the modern organ of the nineteenth and twentieth centuries, and the ways in which they incorporated new technologies to accommodate changing musical esthetics preferences.

ORGANS HAVE NEVER BEEN MODERN

Starting with the Organ Revival Movement in the 1920s, historicizing practices in organ building and organ music performance have often contrasted baroque organs to modern instruments. The term "modern" is then used in relation to the design of late-nineteenth-century organs, their production, and the technologies that were used inside them, as well as the esthetic criteria that allowed the assessment of their sound quality. As a pioneer of historically informed performance practice, Paul Hindemith contrasted the sound of modern instruments to that of historical instruments. One of the initiators of the North German Organ Research Project, Hans Davidsson, advocated the replication of historical organs as a reaction against modern building techniques. How, then, did the organ become modern, and more importantly, how does the term "modern" frame our understanding of how the new baroque organ project mobilized history in its design phase?

THE MODERN ORGAN OF THE NINETEENTH CENTURY

In the early 1800s, the basic principles that guided the building of organs had remained virtually unchanged since the days when Peter Gerritsz and his son worked on the organ in the Nicolaï Church in Utrecht (see chapter 1). Throughout the nineteenth century, however, the question of how to adapt the organ to contemporary scientific insights and technological developments as well as new esthetic ideals was taken up by organ builders and organ theorists. One of them was Johann Gottlob Töpfer (1791–1870) whose book *Die Theorie und Praxis des Orgelbaus [Theory and Practice of Organ Building]*, published in 1833, had a strong influence in the decades that followed. From the theoretical debates and practical experiments it sparked, a new organ esthetic emerged that can be summarized in three rules, as Hans Fidom puts it in his doctoral dissertation on German organ building between 1880 and 1918 (Fidom 2002a). The first is a unity of pitch. Like other instruments and the human voice, the organ should have a clear pitch and, therefore, had an abundance of 8' stops instead of a multitude of octaves that results from combining 16', 8', 4', and 2' stops, as in premodern organs (see chapter 2). A second rule is unity of timbre: Within a certain register, all individual pipes should sound the same. This was realized by standardizing the scaling of pipes using a mathematical formula suggested by Töpfer.[6] The work of Hermann Helmholtz (1821–1894) on the physics of sense perception indicated that fundamental tones have series of overtones that influence the way the timbre of a tone is perceived. Adding high overtones results in a sharper, brighter sound, and vice versa. This insight gave an additional impetus to the strategy to have many 8' stops in relation to higher stops. Finally, the design of the modern organ of the nineteenth century strives for a unity of dynamics, contrary to the terraced dynamics of the baroque organ. Instead of the contrasts of the baroque, both in terms of timbre and dynamics, Romantic esthetic ideals favored fluent transitions between the various stops as well as between the manuals.

The new esthetic ideals regarding the organ sound coevolved with the use of new technologies. These changed the basic design of the key action, the stop action, and the wind system (Owen 2002). The introduction of pneumatic technology in the organ disconnected the force that was needed to open a pallet in the windchest from the force of the finger. Instead, pressurized air from the bellows was used to power the transmission from the keyboard to the windchest. This enabled organ builders to include more 8' stops in the organ. Whereas in baroque music frequent changes of stops were discouraged by the fact that one needed assistants to draw and close all the stops by hand, pneumatic technology also allowed builders to equip the instrument with switches to quickly change registration. Organists could now plan in advance a whole range of stop combinations. This allowed smooth transitions in dynamics and tone color. In the last decades of the nineteenth century, organ builders also started to use electric wiring and electromagnets in parts of the action (Fidom 2006b).

The use of pneumatics and electricity in organ-building practices to transcend the limits of earlier instruments mirrored the artistic promises of technology in other musical practices at the start of the twentieth century. In 1907, the Italian composer, pianist, and writer Federico Busoni wrote in his *Sketch of a New Esthetic of Music* that the development of music was "impeded by our instruments. . . . In their scope, their sound, and their performative possibilities, our instruments are constrained, and their hundred chains shackle the would-be creator as well" (Busoni cited in Patteson 2015, 1). Busoni's idea that truly modern music could be achieved only by changing its technologies heralded a wave of initiatives to create new musical instruments. One example is the theremin, developed by Léon Theremin in the 1920s. It produces its tone when the electromagnetic fields around two antennas are manipulated by the moving hands of the player. The theremin exemplifies the search for new tone colors and sounds as well as a move away from the fixed twelve tones of Western tonality toward microtonality. Others experimented with compositions for automated musical machines (Bijsterveld 2008). In addition, other types of machines such as cars, trains, and factory machines entered the domain of music, or to be more precise, their noise did. Futurists such as Luigi Russolo took their fascination for the "noise and dynamics of mechanized, urban life" as a starting point for creating an "Art of Noises" and thus challenged the idea that music consists of tones (Bijsterveld 2008, 157; Patteson 2015). Where Busoni and Russolo were oriented toward the future, however, others saw the new musical machinery as a consequence of the very modernism that threatened a culture of humanistic values rooted in the past.

New musical instruments in the first half of the twentieth century offered seemingly limitless potential for new artistic creation, yet as Patteson observes, they were also unpredictable and uncontrollable in their consequences:

> Neither imperiously dictating musical reality nor obediently channeling their masters' wishes, they were "tricksters" whose mercurial nature belied their apparent fixity as material objects. Rather than expanding compositional possibilities in a linear and predictable way, the instrumental innovations of the early twentieth century scrambled conventional esthetic categories, destabilized the boundaries between the arts, and reshaped the relationship between past, present, and future in artistic consciousness. (Patteson 2015, 17)

How far was the pipe organ affected by this destabilizing of boundaries and esthetic categories? Even though large pipe organs offered an astonishing range of sounds, the organ never became the instrument of the musical avant-garde,[7] and organ builders produced their instruments for conventional contexts of use such as churches, theaters, and radio studios. As James Wallman observes in his overview of the organ in the twentieth century, the mainstream in organ building focused either on developing the modern, universal organ (and in the United States and Great Britain, the symphonic organ) or on building new instruments along more or less strict historical lines, expanding the tradition that began with the Alsatian organ reform movement in the early decades of the twentieth century (Fidom 2014a).

TRANSLATION AND PURIFICATION

Now that we know how and why organs became modern, we can turn to the question how this term frames an understanding of the role of history in the new baroque organ project, specifically in the design of its hybrid windchest. To answer it, I draw on Bruno Latour's book *We Have Never Been Modern*

(1993) and the concept of history that he develops there. Modernity comes in many versions, Latour writes, yet all of these are linked to the passage of time:

> The adjective "modern" designates a new regime, an acceleration, a rupture, a revolution in time. When the word "modern", "modernization", or "modernity" appears, we are defining, by contrast, an archaic and stable past. Furthermore, the word is always being thrown into the middle of a fight, in a quarrel where there are winners and losers, Ancients and Moderns. "Modern" is thus doubly asymmetrical: it designates a break in the regular passage of time, and it designates a combat in which there are victors and vanquished. (1993, 10)

Latour critically analyses this modern temporality and its understanding of time as a progression of historical periods and revolutions, as well as the idea of a general framework that organizes events in a before-and-after succession.

Instead of time as a singular line, Latour conceives of it as polytemporality where any situation has elements that come from different periods. A carpenter might use an electric drill that is thirty-five years old and a hammer that has a much longer history. Human activities, habits, and gestures are also far from homogeneous in a temporal sense, ranging from a few days to several thousand years (1993, 75). Rather than moving backward to a past that is the origin of our stable traditions, or forward into a future that promises radical innovations, Latour argues, "we have always actively sorted out elements belonging to different times" (1993, 76), and it is the sorting that makes the times, not the other way around. In his analysis of modernity as a concept, Latour does not follow in the footsteps of critical theorists or postmodernists. Instead, he denies that we have ever been modern at all. To make sense of this bold claim, I will summarize some of the arguments behind it.[8]

To explain what he means by being modern, Latour recounts his reading of a daily newspaper. One of the stories is about the hole in the ozone layer. It describes how upper-atmosphere chemists share their findings with political leaders who must come up with regulations for companies who have to phase out the use chlorofluorocarbons in their products. The article mixes chemical reactions and political reactions: "The horizons, the stakes, the time frames, the actors—none of these is commensurable, yet there they are, caught up in the same story" (1993, 1). The study of these strange situations, Latour writes, has been taken up by scientists, historians, and sociologists of science and technology who used notions such as translation, networks, or systems to show how nature and culture, knowledge and power had been tied together in "sociotechnical imbroglios" (1993, 7). The first meaning of the word *modern* thus designates these practices of "translation" that create entirely new types of beings that are hybrids of nature and culture (1993, 10).

The practices of translation are complemented by a second set of practices, those of "purification." These create the two entirely different "ontological zones . . . of human beings on the one hand; [and] that of nonhumans on the other" (1993, 10–11). Whereas the practice of translation brings about hybrids and mixtures, the practice of purification "establishes a partition between a natural world that has always been there, a society with predictable interests and stakes, and a discourse that is independent of both reference and society" (1993, 10–11).

Latour then shows how these two practices became shielded from criticism and change by what he calls the "modern Constitution." In a breathtaking overview of centuries of Western philosophical ideas, he shows how the work of philosophers such as René Descartes, who contrasted "thinking" to "things," and Immanuel Kant, who distinguished between the phenomenal world of the categories of the mind and the noumenal world of things-in-themselves, led to a whole series of dualisms: mind and matter; facts and values; knowing subjects and known objects; human society, culture, and politics versus nature and science; tradition and modernity; Us (moderns) versus Them (premoderns; Vries, de 2016, 124–129). Being modern thus depends on considering the two practices of translation and purification as separate, but as soon as we ask how the two are related and focus on the *work* that needs to be done to stabilize the dualisms, we stop being modern and realize that both our future and our past, as well as our relation to other "nature-cultures" (1993, 11), begin to change.

A PIPE WITH A HISTORY

My reading of Latour suggests alterations to the story of the nineteenth-century modern organ, as well as the rise of historically informed performance and building practices in the twentieth century. The insights and innovations that shaped the modern organ can no longer be presented as a rupture in time, separating it from its archaic predecessors. And we also have to rethink the arguments that the Baroques and the Moderns exchanged in their quarrel about the need for musical performance to be historically informed.[9] Instead, we have to conceive of organs as hybrids, or as Latour puts it, as "quasi-objects" (1993, 51), that exist through the multiple relations that bring them into being.

Contrary to a calendar conception of time where objects can be situated in the empty framework of time, Latour argues that the relations and connections between entities make time. In other words, beings such as organs exist in their networked relationships to other beings, and these networks construct irreversibility and reversibility (1993, 77).

His focus on the ontology of entities allows Latour to distinguish between "time" and "history." To make this more concrete, we can go back to the old pipe that Hans Elbertse used in the test organ. The pipe itself is a hybrid, a quasi-object, because it exists in its relations: to the knowledge of pipe-making practices in the sixteenth century, to the artisanal skills of those who changed its structure during a restoration, to the esthetic preferences that required it to sound differently, and more recently, to the brand-new test organ on which it was placed by Elbertse and quickly discarded as unhelpful. Whereas the modern Constitution and its practice of purification would present the pipe as an object, or even a brute fact of nature that has a singular *essence*, the practice of translation allows us to see how it exists in a trajectory of events that, taken together, constitute its *existence*. What the pipe is follows from what happened to it, the events that changed its trajectory. Tracing its history as everything that affected its existence, we can see its variable ontology, as Latour puts it: "History is no longer simply the history of people, it becomes the history of natural things as well" (1993, 82).

This inclusive conception of history allows us to make sense of the hybridity of Elbertse's test organ and to reframe the question of what it means to design with artifacts that have very different historical trajectories. This is the topic of the next section, in which I will show how working with the pneumatic console of the Sauer organ had an impact on the design of the windchest of the new baroque organ. In the years before he made the test organ, Elbertse and his colleagues had restored the Sauer organ in the Orgelpark in 2006, and added a second console to the instrument in 2011. Knowing about and working with the technological history of the Sauer resulted in insights and expertise that were key to the design of the windchest of the new baroque organ: the possibility of manipulating small valves directly below the pipe feet using the sixteenth-century apparatus of the spring chest.

A DIGITAL CONSOLE FOR THE SAUER ORGAN

The original organ in the Park Church in Amsterdam was built by the Sauer company under the responsibility of Karl Ruther. The man whose name graces the firm, Wilhelm Sauer, was born in 1832, and died six years before the organ was placed in situ. He founded what was recognized at the time as one of the most important organ manufacturers in Germany. The facade of the new Park Church organ with its ranks of green metal pipes is integrated in the west wall of the church, right above the round wooden pulpit. For many years, the console of the organ stood outside the hall of the church in a small room behind a round window to the left of the organ. It was moved to the hall in 1951, following an overhaul of the instrument by the organ builder Ernst Leeflang, who made the original pneumatic key action electropneumatic (Fidom 2002b). After the organ was no longer used during services in the 1990s, it deteriorated.

Preparing for the opening of the Orgelpark, the Utopa Foundation commissioned a restoration of the Sauer organ by Elbertse Orgelmakers, which was carried out in 2005 and 2006. The original pneumatic console was now placed in the pulpit and was made to fully function again. In addition, a completely new pneumatic action was installed, its design following the examples of Karl Ruther's other surviving organs. In the summer of 2011, a second console was inaugurated, designed by the Orgelpark and built by Hans Elbertse. In contrast to the original console, the second console can be positioned anywhere in the hall since it is connected to the organ through an Ethernet cable. For the Orgelpark, the moveable console has the advantage that the audience can see the organist play. In addition, organists can hear the organ better because the sound is not interrupted by the pulpit's large wooden sounding board between the original console and the organ. How was it possible to connect the electronic console to the pneumatic Sauer organ? Answering this seemingly simple question is crucial to understanding the hybrid design of the windchest of the new baroque organ. It requires a brief overview of organ innovations such as pneumatic and electropneumatic organs as well as experiments with new organ designs after the Second World War.

ORGAN INNOVATIONS IN THE NINETEENTH AND TWENTIETH CENTURIES

During the second quarter of the nineteenth century, organ builders in England, France, and Germany worked on the problem of how to build key actions that could control large numbers of 8' pipes, yet prevent the key action from becoming too heavy to play. A new pneumatic device, patented by the British engineer and organ builder Charles Spackman Barker

(1806–1879), aimed to solve this problem. Instead of transferring the pressure of the finger via a tracker to the windchest, as in a mechanical organ, Barker's "machine" used air pressure from the main bellows to power the key action (Fidom 2002b). Barker's pneumatic lever system did not become a success in his own country. In 1839, he offered it to the French organ builder Aristide Cavaillé-Coll, who was quick to recognize its advantages and started using pneumatic transmissions not only for the key action but also for the stop action, the "*moteurs à double effet*" (Fidom 2006a; Owen 2002).

Parallel to this innovation of the key action, organ builders in Germany experimented with new types of windchests. The rationale behind these experiments was the esthetic desire to create larger organs featuring more stops and manuals, to make them resemble orchestras and be as dynamic as pianos. This pushed the age-old technology of the slider windchest to the limit, as Owen (2002) writes, and cleared the way for a radical innovation ascribed to the German organ builder Eberhard Friedrich Walcker (1794–1872).

In slider chests, every key on the keyboard has its own channel. Through this channel, air can flow to several pipes belonging to different registers. These pipes will sound depending on what stops are drawn. Walcker came up with the idea to replace the note channels with stop channels.[10] In this concept, if a stop is drawn, the corresponding stop channel is put under wind pressure. All pipes now have individual valves to admit wind to the pipe.[11] If a key is pressed, valves will open in all of the channels. These valves were cone shaped, and since the German word for cone is *Kegel*, the new stop channel windchest type was called *Kegellade* (cone-valve chest). Cone-valve chests were initially played through mechanical key action, but soon the cones were operated by pneumatics.

In the second half of the nineteenth century, organ builders started to use electricity. It took several decades before the electromagnet, invented by William Sturgeon in 1825, was successfully used in the organ, where it could operate various types of pneumatic valves. Organ builders now were able to simplify the pneumatic cone-valve windchest organ. Each key on the keyboard became a switch that opened or closed pneumatic valves. Even though it was technically possible to construct windchests that replaced the pneumatic pipe valves with valves that were operated electromagnetically, these electric windchest designs were thought to have serious disadvantages by many organ builders. Because an electric magnet opens the pipe valve very quickly, the wind enters the pipe without a delay, "like a pistol shot," as became the expression around 1900 (Fidom 2006a; Owen 2002).[12]

New experiments with organ designs were undertaken after the Second World War, particularly since the last two decades of the twentieth century. In his overview of organ design innovations in recent years, the American organist and musicologist Randall Harlow discusses examples of these experiments. His goal is to show how they offer elements for what Harlow calls a new hyperorgan, "an organ with extended capabilities that seamlessly blend the electronic and acoustic worlds" (Harlow 2011, 3). The innovations described by Harlow, as well as the concept of the hyperorgan, were an important source of inspiration for the new baroque organ project and its design, and Harlow was invited to several of the international symposiums at the Orgelpark to give presentations and discuss the new instrument.

The first example Harlow presents is the thirty-one-tone organ that the Dutch physicist Adriaan Daniël Fokker (1887–1972) designed.[13] During the Second World War, Fokker became interested in the writings on musical tuning from the seventeenth-century physicist Christiaan Huygens, who had developed a system in which he divided the octave into thirty-one tones instead of the usual twelve. According to Fokker, realizing Huygens's tuning system in a thirty-one-tone organ would enable musicians to play music not only in historical temperaments but also from non-Western cultures and new music that uses microtonality. The instrument was built by the Dutch organ builder Pels and inaugurated in the Teyler's Museum in 1951. Its most salient characteristic, as Harlow remarks, is its keyboard. Instead of the twelve black and white keys on a conventional keyboard, the Huygens-Fokker organ also has blue keys, and all the keys are ordered in a roof tile pattern.[14]

Whereas the design of the Huygens–Fokker organ tackled the problem of just intonation, the organ built according to the plans of the German organist and composer Peter Bares in the Kunst-Station (art station) in Cologne, Germany, aimed at offering maximum tone color in the context of liturgical music.[15] For this church, which became a center of contemporary art, he designed an organ that has over a hundred stops, many of which are not to be found in ordinary organs. Built by the German organ builder Willi Peter in 2004, it is meant to be an instrument for new music, which is underlined by the contemporary design of the case.

A third example of innovative organ design that Harlow describes focuses on the wind system. Together with the Swiss organist Daniel Glauss, the Swiss organ builder Peter Kraul developed "an organ on which the player could manipulate not only the sounds and the stops, but the very wind itself. The wind as a third dimension after the keys and the stops should turn the inflexible organ into a dynamic, vibrant instrument" (Kraul 2006, 70). To realize flexible wind, the builders had to redesign not only the wind system but also the pipes and the mechanical keyboard. The latter was extended with mechanical levers to control the wind and with keys that are sensitive to touch. Remarkably, these prototypes do not have electric components.[16]

The French organ builders Daniel Birouste and Michaël Fourcade approached the problem of the static organ sound in their "Modulorgue." Normally, an organist can either open or close a valve by pressing and releasing a key. It is not possible to regulate the amount by which the valve opens. Birouste and Fourcade solved this problem by using digital valve technology, which enabled organists to produce sound effects that are not possible on conventional organs. In addition, the valve technology allows the combination of any pipe with other pipes. Thanks to this individual pipe control (IPC) system, organists have sounds at their disposal that are closer to the esthetics of contemporary composers (Harlow 2011).[17]

Although all examples link innovative organ technology to new musical possibilities, for Harlow it is the blending of acoustic and electronic worlds that is most promising in offering extended possibilities to create new sounds:

> In contrast to purely electronic composition or composition for acoustic instruments accompanied by prearranged electronic sounds, the primary interest among contemporary electro-acoustic composers lies in the use of electronics as an extension of the analog acoustic. By putting the human performer in center stage, complete with the requisite human capacity for endless shades of expressive control over an acoustic instrument, composers have found new ways to seamlessly incorporate electronic sounds and real time processing to augment human musical expression. (Harlow 2011, 40–41)

A hyperorgan should be able to connect to computers through MIDI and Open Sound Control (OSC) interfacing protocols, Harlow writes.[18] It should have means to remotely control performance parameters such as touch. And finally, it should provide the possibilities for live electronics where the sounds of the physical organ are recorded to be transformed through electronic means. This requires that microphones and speakers can be placed inside the organ case.

The concept of the hyperorgan was picked up in 2014 by Hans Fidom in connection to the central design ideas behind the new baroque organ. In its digital mode, it was to combine quite a few of the innovations that can be found in the organs that Harlow described: regulation of wind pressure, control over individual pipes and pipe valves, control through MIDI and OSC protocols, and provisions for live electronics. To understand how the "seamless blend" between the acoustic new baroque organ and its electronic interfaces was conceived, we return to the Sauer organ in the Orgelpark and its two consoles.

INDIVIDUAL PIPE CONTROL

In 2011, standard action electromagnets from the catalogue of the German organ parts supplier Laukhuff were mounted to the pneumatic windchest of the Sauer. Thanks to these electromagnets, the keys of the new console could now be linked to the organ through electric cables instead of lead tubes. The reason for doing this was not only to have a moveable console, but, more importantly, to give the organist new possibilities to play the Sauer organ. The electronic console uses the MIDI protocol to communicate its information to the organ. In addition, it has three manuals instead of the two in the original pneumatic console and also additional knobs and switches (Fidom 2006a).

The software for the new console was developed by the Sinua company in Düsseldorf, Germany. This small firm specializes in designing and making innovative computer-controlled interfaces for pipe organs. The Sinua system and its artistic possibilities can be experienced in the St. Peter and Paul Church in the town of Ratingen near Düsseldorf. Here Benedikt Aufterbeck and Thomas Stöckl, who founded Sinua in 2007, developed a new console for the electropneumatic Romanus Seifert & Sohn organ from 1953, in close collaboration with the church organist Ansgar Wallenhorst. Aufterbeck and Stöckl made use of the fact that in the organ's stop channel windchest, every pipe has its own valve. By giving every pipe valve its own electromagnet, each pipe can be linked to the console in multiple ways. In addition, any pipe can be linked to any key on the four manuals of the console, which enables the organist to make stop combinations that are not possible on a normal organ.

The electronic console for the Sauer organ in the Orgelpark made in 2011 did not allow for individual pipe control, but all its registers can be used on all three manuals. To make this possible, Sinua developed the ideas and software behind the console, whereas Hans Fidom designed the appearance of the console itself. The seven windchests on the organ—four for the original two manuals and three for the pedal stops—have been distributed over six groups. Every group of stops has its own batch of colored switches on the digital console, allowing the organist to assign registers to the three manuals.

In addition, the cheeks next to the keyboards contained three knobs per color, so that the organist could use them in normal pitch, an octave lower, or an octave higher. These properties of the electronic console are a move away from traditional mechanical organ designs going back to the days of Peter Gerritsz where a fixed group of stops would always be linked to a specific manual. Next to the new ways of using the instruments stops, the console has a memory for three thousand stop combinations. It allows performances to be recorded in MIDI and played back without the performer being present. The organ can be played together with a laptop computer or by a laptop alone, allowing for live electronics, such as the real-time processing of the organ's sounds or sample sets of other organs.[19] Finally, the organ can be played from other places in the world (Fidom 2014a).

The experiences with the Sinua system in Ratingen and the new digital console for the Sauer organ were crucial inputs for the design of the windchest for the new baroque organ. Whereas the new console for the Sauer organ already offered ways of blending acoustic and electronic worlds, the new baroque organ should provide organists, improvisers, and computer musicians with even more space for experimentation. Hans Fidom had advised the core team to invite the Sinua company to develop the software for the new baroque organ and to come up with a new design for the existing electronic console in the Orgelpark, which would be linked to both the Sauer organ and the new baroque organ.

But how to realize individual pipe control, a main feature of the system in Ratingen? The organ in the St. Peter and Paul Church allows for individual pipe control because it has a cone chest where every individual pipe has its own pneumatic valve that can be operated by individual electromagnets. Most baroque organs, however, have slider chests where pipe combinations are limited by the fact that whole ranks of pipes are switched on and off by drawing a stop. The only historical windchest that could combine a baroque organ design with individual pipe control was the spring chest with its rows of small valves or pallets right below the pipes. By manipulating these little pallets with an electromagnet, it would be possible to use the Sinua system on the new baroque organ.

DESIGN AS DRAWING THINGS TOGETHER

We can now see the strange structure of the test organ built by Hans Elbertse differently. The small windchest is made from oak, as it would have been in the fifteenth and sixteenth centuries when organ builders used the spring chest for its greater reliability and stability. The little pipe valves closely resemble their counterparts from centuries ago, yet they are operated by the type of electromagnet that was used in the late-nineteenth-century electropneumatic key action mechanism of the *Kegellade* to operate cone valves. Even more so, the very idea of individual pipe control is closely connected to the pneumatic and electropneumatic organ technologies that were developed in this period. The gray DIN cables that connect the rows of magnets in the pallets to the small keyboard and its switches follow a twentieth-century industry standard. Experiences with developing twenty-first century software and digital interfacing in the context of the St. Peter and Paul Church in Ratingen by the Sinua company helped to find a way to more precisely control the opening speed of the spring valves in the new windchest. And when the members of the core group discussed the workings of the test organ, they were concerned by how its construction would affect the goal of creating a seventeenth-century sound. In the test organ, elements from different times coexisted in new interrelations.

The test organ, as a "quasi-object," exemplified not simply a movement back to the spring chest mechanism but rather a reworking of this mechanism through the principles of electropneumatic key action as well as digital software and interfacing. As Latour suggests, rather than going back to a baroque era, bypassing the innovations of modern organs, designing the windchest of the new baroque organ required bringing together heterogeneous artifacts, knowledge, skills, and technologies of different periods. In a conference paper on what he calls a philosophy of design, Latour contrasts the modernist view on making new things as a break with the past with his own definition of design as always being redesign (Latour 2008a). Following the argument he developed in *We Have Never Been Modern* (1993), if things such as the test organ always exist in the variable ways in which they are related to other entities, then designing a thing is the work of drawing together other things.

Latour gives five advantages of his redefinition of design. First, it implies modesty: Everything builds on what was already there. Second, design is not progressive and a radical break with the past but has an attentiveness to details, to skillfulness and craftsmanship (2008a, 3). Third, we should stop seeing modern artifacts as finished, but rather realize that they are "things" or as Latour formulates it, "complex assemblies of contradictory issues" (2008a, 3). Fourth, if we realize that design is never a process from scratch, that to design is always to *redesign*, than it can be "an antidote to hubris and to the search for absolute certainty, absolute beginnings, and radical departures" (2008a, 5). And finally, all design necessarily involves an ethical dimension. In the modernist version of design, objects are facts that stand in opposition to the values that allow to evaluate them. In Latour's view, designing things always involves a coalescence of materiality and morality.

In the final pages of his paper, Latour reflects on the challenges his definition of design offers to the practical skills for drawing objects, such as architectural drawings, mechanic blueprints, scale models, and prototypes. What is missing in these drawings, he argues, are the controversies and the many contradicting stakeholders that are drawn together in any design. From the age-old perspective drawing to the more recent computer-aided design software, he writes, "we are still utterly unable to draw together, to simulate, to materialize, to approximate, to fully model to scale, what a thing in all of its complexity, is" (2008a, 12). The collective thought experiment in the core group during the visit to Hans Elbertse's workshop exemplifies the contradictory and controversial nature of the test organ as a "matter of concern."[20] As a prototype of the windchest of the new baroque organ, it allowed the core group to put the central design ideas and assumptions in the project to the test.

TESTING THE WINDCHEST PROTOTYPES

At the start of the project, the need for a reference group of organ experts contributing to the design process had been acknowledged. As it turned out in the first year, this group was not formalized in terms of appointed members, regular meetings, or advisory status and mandate. Rather, the network of people that the Orgelpark had established in previous years, both through the programming of concerts and through the work of the research program, was mobilized depending on the topics or issues at hand. These could be related to the wind system of the new organ, its temperament, its disposition, or the integration of electronic elements.[21] Experts were invited to give lectures during the colloquia and symposia, and some of these lectures were published in the Orgelpark Research Reports series. In addition, they would be consulted ad hoc if this was thought necessary or fruitful. In some cases they gave concrete advice about aspects of the organ's design, or reflected on its future use based on experience in their musical and artistic practices. This flexible setup of the reference group enabled them to benefit from and incorporate relevant external knowledge and expertise, while at the same time centralizing the actual design decisions in the core group.

The staging of the colloquia and symposia vacillated between explanation and demonstration of the new organ's design ideas on the one hand and an invitation to critically reflect on these ideas on the other. The listening test during the symposium in June 2014 is an example. At the end of the second of the day of the symposium, the participants were invited to listen to the test organ and evaluate its sounds. It was placed on the north balcony in front of the Van Leeuwen organ. The setup of the experiment was simple: A sound would be played twice, and the respondents were asked to mark whether they observed a difference and, if so, to specify its characteristics.[22]

Three situations were compared in random order so the listeners did not know which situation they were listening to at any given time: a key channel as in the windchest of the Hus/Schnitger organ in Stade (chapter 2), a key channel that was higher than in Stade, and a key channel that was both higher and contained a Laukhuff action magnet. The goal of the experiment was to establish whether the size of the key channel and the presence of a magnet inside the channel had any perceptible effects on the sound. While Hans Elbertse operated the windchest setting and organist Andrei Ferrara played tones and chords on the test organ and changed the tone sliders, the audience below listened and wrote down their comments. These were summarized by Hans Fidom after the symposium. The results were ambiguous. The participants were never unanimous in hearing differences. In some cases they heard differences even if the same situation, such as Stade and Stade, was compared. From their comments it is clear that they struggled to describe the character of the differences. Examples of descriptions are: "1st a bit more brilliant, 2nd more body," "2nd time more 'stringy' quality," or "Tone and attack more lively 2nd time."[23]

In the evaluation round after the listening test, some participants questioned its setup. They argued that it was hard to assign the cause of possible differences in sound to the variables that were tested, because other factors, such as room reverberation and acoustic reflection as well as differences in the actual playing of the tones and chords, influenced the sound, or even the simple fact that one was tired after a day of lectures and discussions. A repeated test the next day rendered comparably diffuse results. In his report on the conference, Hans Fidom therefore concluded that the shape of the key channel and the presence of magnets inside it "does not have a significant influence on the sound."[24]

DRAWINGS ON A TABLECLOTH

Three months after the 2014 symposium, the issue of how the pipes will sound on an organ that has two modes of playing was back on the table. In September, the core team traveled to Germany to visit Hildebrandt organs in Sangerhausen, Hettstedt, Störmthal, and Naumburg with Munetaka Yokota. One of the goals of this new trip as Johan Luijmes formulated it earlier in the year was to listen to the sound of Hildebrandt's organs and to discuss it with Yokota to shape a shared understanding of the sound ideal for the new organ. Another reason to visit Hermann Eule Orgelbau in Bautzen was to see if and how the firm could contribute to the project. Three of the four organs that were on the agenda of this trip had been restored by Eule; the organ in Hettstedt has only a Hildebrandt facade. Hermann Eule Orgelbau thus acquired an extensive knowledge of the Hildebrandt organs and, more specifically, of the scalings of the pipes. This knowledge is personified in the firm's master organ builder Helmut Werner, who has been working for Hermann Eule Orgelbau for decades (Fidom 2023). Multiple outcomes of the meeting are foreseen. Hermann Eule Orgelbau might either provide information about the scalings of Hildebrandt's pipes so that they can be made by another pipe maker, or the firm might be invited to make the pipes itself.

The night before the visit to Eule, the team arrived at a hotel in Bautzen. During dinner, the conversation turns to the issue of pipe voicing—giving a pipe its sound character by manipulating its characteristics and, in the case of flue pipes in particular, the various parts of the pipe mouth (see chapter 5 and 7)—and Yokota brings up a problem that he anticipates. It is related to the core design principle of the windchest, the electronic manipulation of the pipe valves (see figure 3.3a–c).

The windchest of the first prototype has two modes. In the mechanical situation, the pallet box is under wind pressure and the pipe valves are open or closed depending on the stops that have been drawn. When the organist presses a key, the pallet opens and air flows into the key channel and from there into the pipe foot. This is characterized as indirect wind (see figure 3.4).

a

b

c

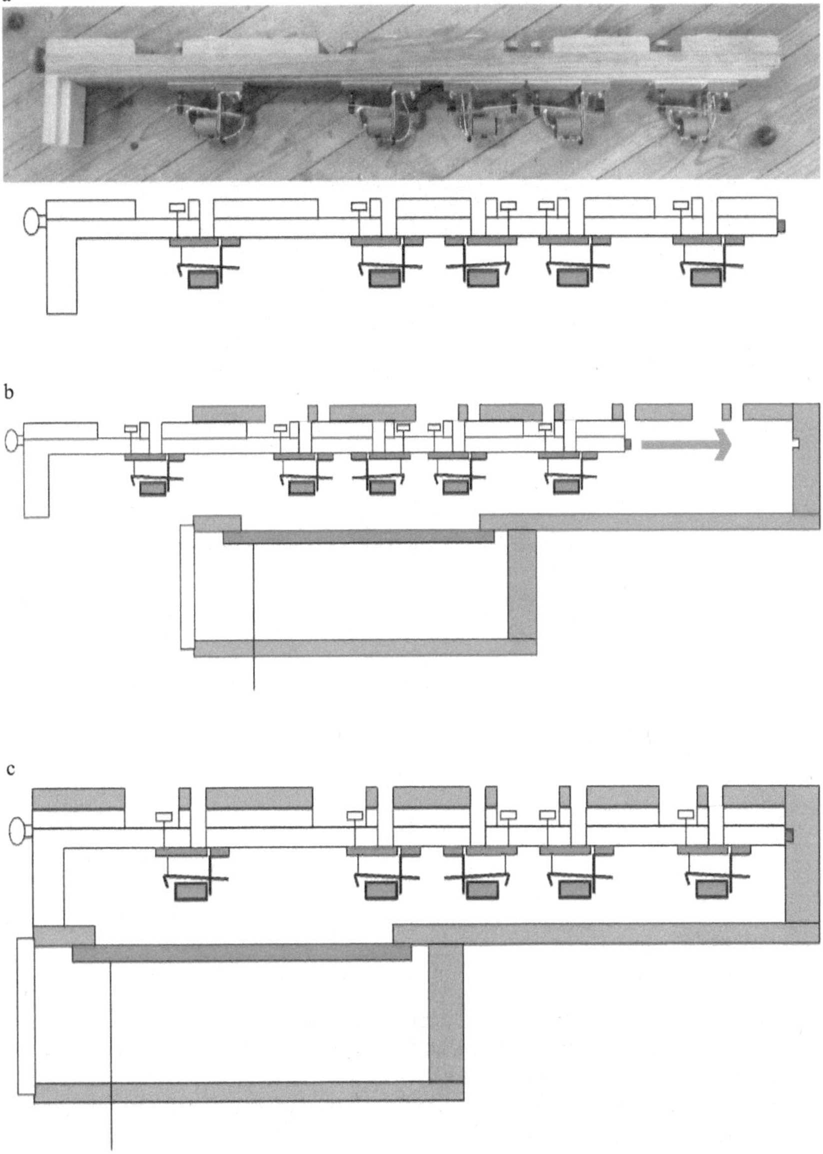

Figure 3.3
A picture of a tone slider from the first test organ with valves and electromagnets (3a). Schematically represented, it slides into a tone channel that corresponds to one key on the keyboard (3b) to connect to the electronic adapter inside the tone channel (3c) (drawing made by Hans Fidom, source Fidom [2018], reprinted with permission).

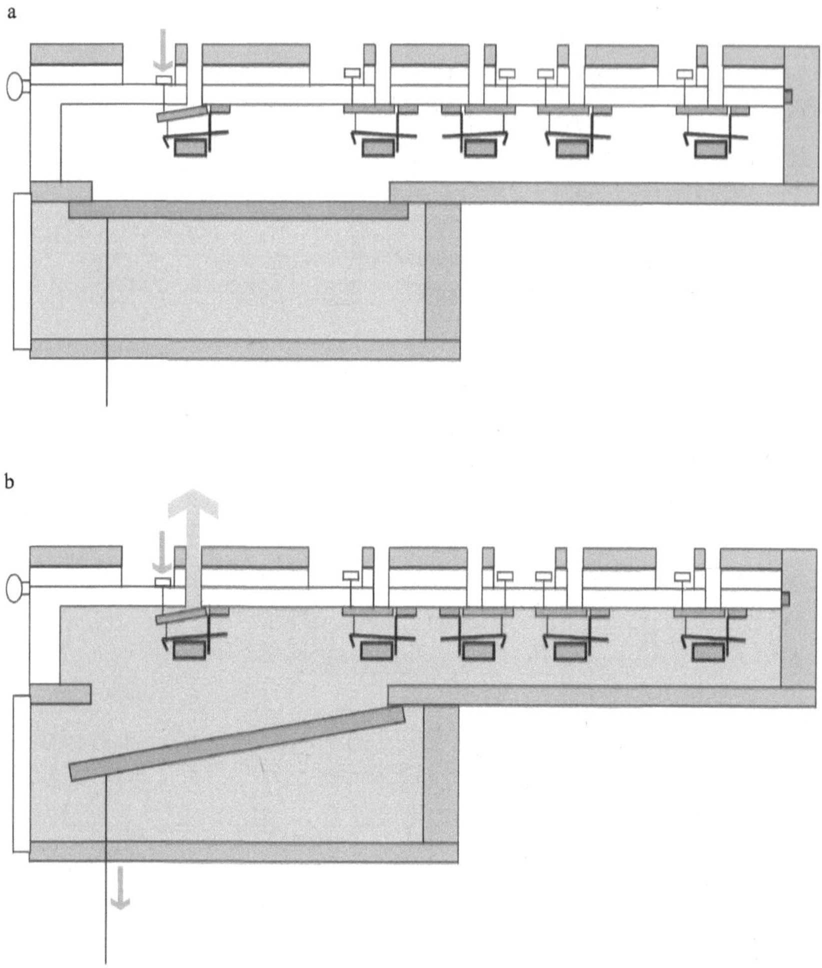

Figure 3.4
The hybrid spring chest in the first test organ in mechanical mode: The pipe valves for one pipe rank are opened (small arrow), and pressurized air is in the chamber below the closed large key pallet that is mechanically connected to the key (3.4a). If a key on the keyboard is pressed, wind will flow from the pallet box into the tone pallet and the pipe will sound. Note that the electromagnets are placed inside the tone channel, which reduces its volume (3.4b) (drawing made by Hans Fidom, source Fidom [2018], reprinted with permission).

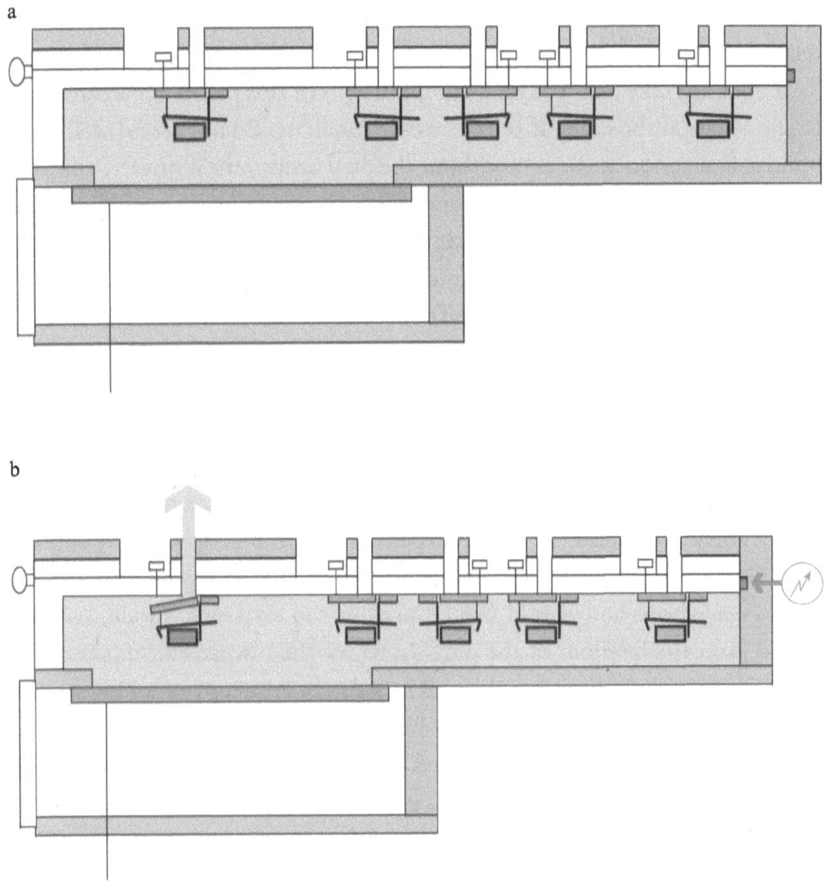

Figure 3.5
The hybrid spring chest in the first test organ in electronic mode: All key channels are filled with pressurized air. The small pipe valves are connected to the keyboard with an electric cable (3.5a). If a key is pressed, the corresponding electromagnet will be switched on, the pipe valve will open, and the pipe will sound (3.5b) (drawing made by Hans Fidom, source Fidom [2018], reprinted with permission).

In the electronic situation, all tone channels are under wind pressure. When a key is pressed, the pipe valve is opened by the electromagnet and the wind flows into the pipe: direct wind.

According to Yokota, especially for the lower pipes, the differences in wind behavior between the mechanical and the electronic mode might become so big that it would be very hard, if not impossible, to voice the pipes in such a way that they sound good in both modes. This is so because the manipulations of the pipe and its mouth to give it its "voice" are closely related to the amount of wind a pipe gets.

It does not take long before the implications of this problem become clear to the others at the table. If the electronic mode would not sound well, why build a dual action at all? But without the dual action, the innovative dimension of the organ would be lost and only a baroque organ would remain. Yokota had challenged the basic principle that underlies the decision to build the organ: "New is fine, but compromise on some parts is not good." Loek Dijkman seconds this: "If we don't succeed in the electronic system . . ." and Johan Luijmes completes his sentence: "We don't build the organ at all." Dijkman: "Yes."[25]

How to proceed? In the discussion that follows, more or less radical solutions are tried out. Dijkman explores the option to build two organs instead of one: a mechanical Bach organ and an organ with electronic possibilities, but he quickly disposes the idea as expensive and impractical. Johan Luijmes suggests a situation in which the lower pipes would be voiced only for the mechanical mode and would not be used in the electronic mode, but this would limit the freedom of the organist to use the instrument in new ways. Hans Elbertse reminds us that the goal is not to build a copy of a Hildebrandt organ but to create a new organ that has baroque sound qualities; therefore, both the mechnical and the electronic modes should sound good. Luijmes understands that for Yokota the voicing of the pipes according to the Hildebrandt is equally important in the baroque mode as in the electronic mode: "You don't want to compromise the Hildebrandt concept."[26] Yokota agrees. As a voicer, he can accommodate pipes so that they all sound more or less the same. This, however, means manipulating the pipes in a way that ignores the nature of each individual pipe. "Whatever it is, you have to get the best, the happiest type of sound. The conditional approach to the voicing of the pipe is very different from just trying to make it the same, to manipulate it."[27]

While the others are talking, Yokota draws on the paper tablecloth (see figure 3.6). He sketches a design for a windchest that combines a slider chest according to Hildebrandt's dimensions with a separate mechanism that allows one to play the same pipe through a valve and a magnet. There would be two key channels for every pipe, and the magnet would be not inside the key channel but on top of the toeboard. This design would combine the best of both worlds: a mechanical situation that is exactly according to Hildebrandt, and an electronic situation that solves the problem of

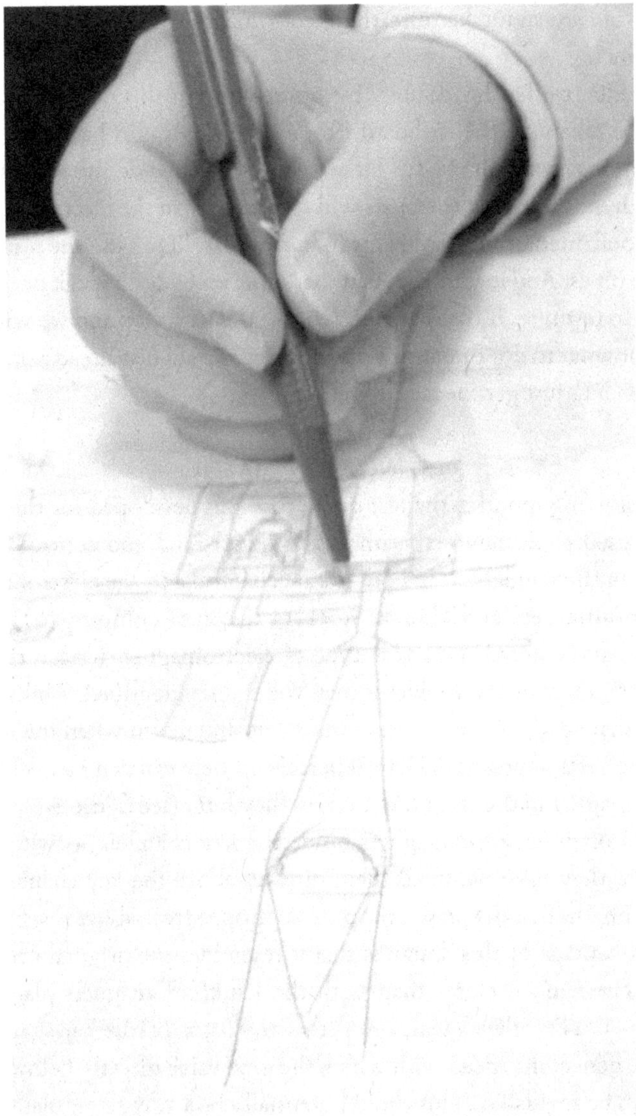

Figure 3.6
Munetaka Yokota sketching a new design for the windchest on the paper tablecloth during a dinner in September 2014 (photo taken by the author).

erratic wind behavior because the electromagnet is not placed inside the key channel.

Hans Elbertse has his doubts. He argues that in this design all the pipes have to be taken off the toeboard if one of the magnets has to be repaired. "It is a risky system," he says.[28] Elbertse suggests they do more tests to see if Yokota's hypothesis that the pipes will speak badly in the electronic situation will be confirmed or refuted. Hans Fidom agrees. "Let's use the test organ to voice the pipes. And it could happen that some pipes don't speak or speak bad. We have to test that. If that happens we take the next step and we will have to figure out what to do. But at this moment we should not already try to find a solution—let's just get the facts together."[29]

TESTING A NEW MAGNET

In the following months, the test organ that was developed for the listening test was used to do new experiments to understand and control the wind behavior in the windchest, in both the mechanical and the electronic modes. Benedikt Aufterbeck and Thomas Stöckl of the Sinua company in Düsseldorf developed and constructed a new type of electromagnet that has the potential to solve some of the problems that Yokota has identified. This magnet is cylinder shaped and has a core that moves up and down when the magnet is placed in a vertical position. Hans Elbertse and Bert van den Heuvel had used this new magnet in the second test organ they built (see figure 3.7).

Instead of putting the magnets inside the key channels, as with the first test organ, they have mounted them directly above the key channel. When switched on, the magnet presses the pipe valve open from above (see figure 3.8).

The advantage of this design is that it resembles the original principle of a spring chest much closer than with the Laukhuff magnets placed inside the channel. The other advantage is that the force of the bar-shaped magnets, and, hence, the speed with which the pipe valve directly below the pipe opens, can be regulated. This could potentially be a way to regulate the flow of wind into the pipe (see figure 3.8).

On January 15, 2015, a new listening test is organized at the workshop of Elbertse Orgelmakers in Soest. Next to the members of the core team and Bert van den Heuvel, Munetaka Yokota, Benedikt Aufterbeck, and Thomas Stöckl from Sinua and Dirk Eule and Helmut Werner from Hermann Eule Orgelbau are present. After the visit to this company in September 2014,

Figure 3.7
The second test organ that was used during the listening test in January 2015 (photo taken by the author).

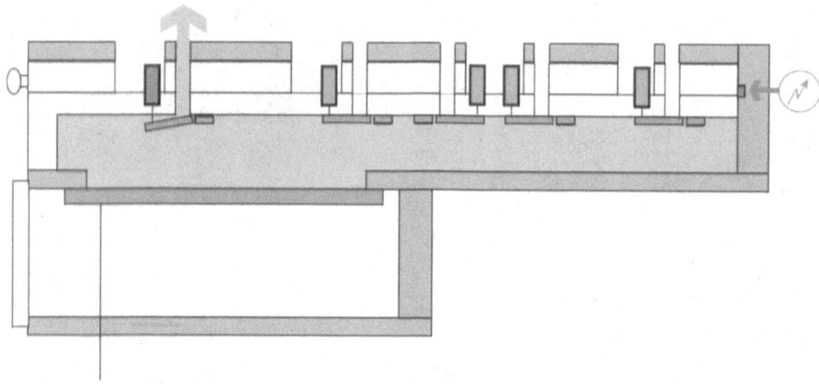

Figure 3.8
Design of the windchest in the second test organ. Contrary to the Laukhuff magnets in figures 3.3a-c, 3.4a-b, and 3.5a-b, the vertical Sinua magnets are positioned outside the key channel, which can have the same dimensions as a Hildebrandt key channel (drawing made by Hans Fidom, source Fidom [2018], reprinted with permission).

it has been decided that they will make the pipes according to Yokota's instructions and he will do the voicing of the pipes.

The main goal of the test is to compare two types of wind behavior in the mechanical and electronic modes. As Munetaka Yokota had argued, the difference between indirect and direct wind might cause differences in wind behavior that result in different pipe sounds. To minimize these differences, Sinua has developed software to regulate the opening speed of the individual electromagnets. An additional goal is, therefore, to see what effect the new magnets have on the sound. The pipes on the test organ have been newly made and have been voiced by Yokota. To also test the lower pipes that need more air to sound, two pipes from the Van Leeuwen organ have been connected to the test organ.

The setup of the test is similar to the listening tests during the symposium in 2014. Two sounds are played and compared.[30] Everyone present is invited to note whether the differences, if any, are acceptable. There is some discussion about the term "acceptable." Loek Dijkman says that he needs a point of reference, preferably the mechanical mode, which will be built anyway. "Suppose that the result is that we only want to hear magnets."[31] He is keenly aware of the importance of the test; if the majority of those present would say that the differences are unacceptable, it would mean the end of the project. Munetaka points out that in the mechanical situation, the speed with which the key is pressed influences the tone. Yet during the test it becomes clear that the new Sinua magnets function well.

The new magnet that Sinua has developed promises to solve the problems that Yokota identified in September 2014. Also, the magnet is no longer in the key channel but on top of it. This approaches the historical design of the spring chest. The dimensions of the key channels are no longer dependent on the size of the Laukhuff magnets and can now be built according to Hildebrandt. And finally, the opening and closing speed of the new magnet can be controlled, which means that for every individual pipe, the amount of air that flows in can be regulated. Even when the voicing has been done on indirect or "baroque wind," unwanted fluctuations in wind behavior in the electronic mode and with direct wind can thus be repaired electronically. In his report on the listening test, Hans Fidom compares the fine-tuning possibilities of the new magnet with a tool that enables the optimization of the sound of the organ in the electronic mode so that it sounds as if it is played mechanically.

PROJECTING PAST AND FUTURE

In his canonical article "'Testing—One, Two, Three . . . Testing!': Toward a Sociology of Testing," Trevor Pinch (1993) explores testing as a research site in the sociology of technology. He defines testing as "the attempt formally to specify and identify how the technology will perform, is performing, or has performed." (1993, 25) In testing, Pinch argues, something is at stake and many tests are performances that can be witnessed (1993, 27). The witnesses or audiences can have different interests in the outcome of the test. Prospective testing concerns the future performance of a technology and may be carried out on scale models or prototype models to determine "if a design is feasible, whether the technology works as specified in the design, whether different components can be integrated, and to monitor the performance for a range of putative uses." (1993, 27). According to Pinch, testing is always a process of projection, from the present to the future, from the present to the past, or from the small to the large. This projection depends on establishing a similarity relationship between the state of affairs in the test case and in the actual operation of a technology (1993, 27).

Pinch's sociology of testing is helpful to analyze how in the two prototypes of the baroque organ's windchest, the core design idea of combining a mechanical and an electronic mode of operating was put to the test (see also

Pinch [2020]). A central question was how the wind would behave under different conditions. In the first test organ, the key channels were dimensioned after Schnitger's design in Stade, and electromagnets were placed inside the channels to see if and how they influenced the sound. In the second, the magnets were placed outside the channels that were now dimensioned after Hildebrandt. Crucially, at least three types of similarity relationships were projected. The first and most obvious one concerned the relationship between the test organ and the actual organ: If the technology worked in the test organ, it could be expected to work in the future instrument. The second relationship was the projection from the test organ's design to its historical counterparts. Compared to the first test organ, the second more convincingly established this relationship. The test organs thus allowed the projection both to the past and the future and can be thought of as "past-making" and "future-making" devices.

The third, and most contested, similarity relationship projected the *sound* of the test organs, both in their mechanical and electronic mode, to the sound of historical organs. It was this relationship that caused debates during the first listening test in 2014 at the Orgelpark symposium, led Yokota to voice his concerns, and was the topic of the second listening test at Elbertse's workshop in 2015. Evaluating the different sounds of the test organs required the expert knowledge not only of the core group members but also of a forum of international organ experts whose sense of hearing was mobilized and tested at the same time (Tkaczyk, Mills, and Hui 2020). As witnesses to the performance of the test, they contributed to the design by drawing on their artistic knowledge on the quality of the speech and the sound of historical organ pipes to assess this third similarity relationship. The results of the listening test during the 2014 symposium were interpreted by the core team as a result that validated the design. If experts do not detect the same differences in different situations, and if they also hear differences when the same situation is repeated, it follows that any possible differences would not be significant enough to question the approach that had been chosen. During the second listening test, the design of the windchest model had evolved in a way that addressed the perceived shortcomings of the first test organ. The team was now confident that the sound of the test organ could be projected both to the past and the future.

One way of seeing the test organs is how they helped to explore and project similarity relationships. As Marres and Stark (2020, 437) have observed,

Pinch's approach of testing is concerned with the problem of reference, where the test stands for something in the real world.[32] Following Latour's ideas on the role of things in design, however, the test organs also allowed the testing of the ontological transformations that the various components would undergo after they had been drawn together. In this view "the test [organ] enables the performative specification of a new set of connections between technical elements, social actors, objects, interests, and so on" (2020, 428). Importantly, these elements are not just assembled to explore a certain outcome and its qualities, but their ontology changes and their various histories are made to matter in new ways.

FINALIZING THE PLAN

On March 10, 2015, two months after the listening test in Soest, the members of the core team meet at the office of the Utopa foundation in Leiden. On the agenda is a discussion of the final plan for the organ that Hans Fidom has written. The three-page document with eleven appendices summarizing the main design decisions reflects the outcome of many study trips, meetings, and discussion, as well as experiments and tests. Some elements of the plan are uncontested, such as the two goals that the organ will have: to give it a sound that refers to the organ that Zacharias Hildebrandt built in Naumburg in 1746 and to give the organ a windchest that fits the Hildebrandt concept while making it possible to combine any pipe with any other pipe. This dual design is reflected in the two "interfaces" that the organ will have: a mechanical console integrated into the main organ and a digital console integrated into the existing mobile console that was built in 2011 for the Sauer organ. Other issues are still undecided and require discussion, such as the pitch and temperament of the organ, as well as details in the design of the organ case and the wind system, and whether the mechanical action will have a sequencer system allowing organists to make presets of stop combinations.

The organ will have three divisions, each connected to its own keyboard: the Hauptwerk, Oberwerk, and Pedal. The pipes for these divisions will be built by Hermann Eule Orgelbau in Bautzen according to the scales of the Hildebrandt organ in Naumburg. The master organ builder, Helmut Werner, has indicated that these scalings are suitable for the Orgelpark. It was decided that Hermann Eule Orgelbau will provide a detailed plan for the scalings and measurements of the various stops in the disposition. Because the Naumburg organ is too big for the Orgelpark, the disposition of the new baroque organ is based on a smaller organ that Hildebrandt built in the St. Jacobi Church in Hettstedt, Germany. This church has almost the exact same volume as the Orgelpark. The voicing of the pipes will be done on the mechanical key action; adjusting the sound quality of the pipes in the digital action is possible by fine-tuning the opening speed of the pipe valves through software.

The issue of the pitch and the temperament of the organ was discussed extensively with experts such as Ibo Ortgies and Rudolf Rasch. Because the organ will be used to accompany early music performances on period instruments, a choice was made for a pitch of a1 = 415.3 Hz, which is commonly used on period instruments. The chosen pitch is a semitone lower than the other organs in the Orgelpark, which have a standard pitch of 442 Hz. In the choice for a temperament system, the argument that Bach's son-in-law, Johann Christoph Altnikol, who was organist of the Hildebrandt organ in Naumburg from 1748 until his death, remarked that Hildebrandt himself followed Neidhardt's temperament for bigger cities and courts. This system was rather close to equal temperament, which allows the use of all keys. It is decided, however, that the new organ will have an adjusted equal temperament to give the chords of C major and D major purer thirds than in equal temperament.

The construction and design of the organ case and its details was a topic of long discussions. Should the organ underline its historical reference to Hildebrandt by following the design of his organ facades? Or should it make clear that it is a twenty-first-century instrument by deviating from historical designs? The outcome of this debate, which mainly took place in the core group, was that Hildebrandt would be followed. His organs in Hettstedt and Sangerhausen were used as models. After studying these organs, it became clear that Hildebrandt himself freely modified his own design principles when the situation required it. Although the main contours of the organ were now set, decisions about details such as ornaments and the color scheme are expected to be twenty-first-century style. An important decision was to place the four wedge bellows inside the organ case. The organ will be placed against the wall, but the windows should remain accessible for maintenance and repair. Placing the organ against the wall also leaves more room on the balcony in front of the organ. Having space around the organ is something that Johan Luijmes had suggested because he wants to use the new instrument to accompany cantatas by Johann Sebastian Bach. To facilitate communication among the organist, singers, and ensemble musicians, Luijmes had suggested creating enough space around the organ so that the orchestra and singers can stand on the balcony.

Concerning the wind system, it had already been decided that the four wedge bellows will be operated by a blower as well as four beams that can be operated by foot, with dimensions derived from the Hildebrandt organ in Sangerhausen. The windchests will be built as spring chests with magnets in such a way that the magnets can easily be repaired and replaced.[33] They are the Sinua type and are placed above the key channels, in accordance with the historical spring chest design of Arp Schnitger in Stade. The blower will be placed in a box to insulate the motor noise. Its speed will be controllable from the digital console with a swell pedal.

The mechanical action will be simple. The keys will be linked to the windchests through trackers and rollerboards following the examples of Hildebrandt, and it will be possible to couple the two keyboards and the pedal. The construction of the stop action will follow the historical design of the spring chest, where wooden beams with pins press open the pipe valves when a stop is drawn. The stop drawknobs will be designed according to the Naumburg organ. The keys will be made out of ebony. The frequencies of two tremulants will be controllable.[34] The number of keys and their shape are similar to those of the Hildebrandt organ in Sangerhausen. Finally, the mechanical console will have a sequencer allowing the organist to preprogram stop combinations. The existing digital console will be given additional switches and knobs to control the organ, as well as a touch screen, which will allow the organist to assemble new stop combinations, to assign free combinations of pipes to certain keys, and to change the timing of the speech of stops and individual pipes. The touch screen will be built into the console to help the organist to create new sounds and offer presets of innovative stop combinations. To link the organ to computers and other electronic devices such as speakers or lights, it will have connection panels or patch bays inside the case, one on the organ floor, one on the first floor near the Hauptwerk, and one on the second floor near the Oberwerk.

Now the building of the new baroque organ can start. Based on the final plan, Elbertse Orgelmakers, Hermann Eule Orgelbau in Bautzen, and Sinua in Ratingen will write their tenders. Munetaka Yokota will be responsible for supervising the making of the pipes and for voicing of the organ.[35] It is expected that both the pipes and the hardware and software of the Sinua system will be ready in the summer of 2016. The organ is expected to be inaugurated in October 2017, the year in which Elbertse will celebrate the 100th anniversary of his firm and the Orgelpark will be in its tenth year. This means that the concert program in that year will have to be attuned to building and voicing activities in the Orgelpark.

In retrospect, the acceptance of the final plan symbolized a moment of closure in the design process that involved a cascade of interrelations: The mechanical parts of the organ had to be connected to the digital interface; the dimensions of the key channels in the windchest were related to the size of the magnets as well as to the channel volumes in the Hildebrandt slider chests in Sangerhausen; positioning the four bellows inside the organ case meant that there would be no room for a mechanical stop action, which therefore had to be built using electromagnets and motors; and the dimensions of the Orgelpark as well as the construction of the balcony limited the weight of the organ and its parts. These and many other tensions and dilemmas were solved on paper and in experimental trials. But the drawing together and experimenting would not end once the actual making of the organ had started. On the contrary, new problems would emerge.

4

MAKING THINGS FIT TOGETHER

ENTERING A CONSTRUCTION SITE

It is early September 2014. Munetaka Yokota has joined the core team for a study trip to Hildebrandt organs in Sangerhausen, Hettstedt, Störmthal, and Naumburg. After we listened again to the instrument that impressed us when we first heard it in December 2013, we go to the upper balcony of the Naumburg organ to look around inside.

I enter the organ through a narrow wooden door in the right side of the case. On my left, I see rows of large wooden pipes and, to my right, a ladder leads me to a wooden boardwalk along rows and rows of ascending and descending metal pipes. Below me, Hans Elbertse points out that some of the pipes have extensions at the top, probably to change their tuning. In the dimmed light, I can see that most objects are covered with a thin layer of dust. Woodwork is all around, some of it looking fairly old, some new. To my left, I can look through the feet of the front pipes. Far below is the church floor. Ahead of me are the white, sheep leather folds of one of the bellows. I meet Yokota, who carefully climbs onto one of the big wind channels, made from newer wood, to take pictures of pipes with his tablet.

The inside of an organ, John Shannon writes in his guide for organists and novices to organ building, "*is an exceedingly dangerous place*" (2009, 163, italics in the original).

> It has much in common with a construction site. Walk boards are often quite high and narrow. Ladders are often the means of moving from one place to another. Lighting is often poor. One often has to lean far over to fix a certain pipe. And, there are literally thousands of fragile and breakable parts, and a miscue can be devastating. (2009, 163)

Although I had not yet read this text when we explored the Naumburg organ, I was acutely aware of my responsibility to be very careful and move slowly. I could see the others doing the same.

Munetaka Yokota takes out a small pipe, looks at its mouth and foot, and puts it back. He takes another, lays it down in front of him on the boardwalk, and takes a picture of the mouth area. Invisible to us, Johan Luijmes is sitting below at the console, playing a Bach prelude. The sound is deafening. Unlike listening below in the church, inside the organ the sound comes from all directions. Pipe tones mingle with the noise of moving wooden beams and trackers. Gone is the auditory canvas of the church acoustic. We have entered the backstage of the organ and hear the machinery that is kept out of sight and ear by its beautifully crafted facade.

Climbing up a second ladder, I see how metal rods connect the wooden frame of the organ to the walls of the church. There are more bellows and wooden pipes. I am now in the upper parts of the instrument, right behind the wood carvings that crown the central tower. Not far above me are the embellishments on the ceiling. From here, I can look down into some of the larger wooden pipes below. As Johan Luijmes plays the last chord of the prelude, the sound reverberates through the church, making me suddenly aware of the size of the space and the height of where I am standing. Suppressing a slight sense of vertigo, I descend.

Nine months later, in May 2015, I visit the organ workshop of Hans Elbertse in Soest. I have an appointment with the master organ builder, Bert van den Heuvel, who, together with Hans Fidom and Hans Elbertse, is responsible for the overall design and building of the wooden parts of the new baroque organ: the framework, the bellows, the windchests, the wooden pipes, the key action, and the case. Van den Heuvel shows me the drawings he has made.[1] One drawing depicts the facade of the organ, another the construction behind the facade. Below are four large bellows inside the organ case, two on either side. Above the bellows, there are the four windchests for the manuals, two per manual, and two windchests for the pedal stops. The contours of some of the wooden and metal pipes have been drawn as well. I have seen earlier sketches and drawings, but this is my first view of what the new organ will actually look like.

Van den Heuvel explains how all the inner parts have to fit, and how making them fit means solving bigger and smaller puzzles. One of these puzzles is that all the parts have to be easily accessible when repairs have to be made. "There is nothing worse than having to disassemble everything to reach a place. So I made sketches first. I need a boardwalk, a small staircase here, a ladder there. Then this is what comes out."[2] In his imagination, Van den Heuvel has already walked inside the new baroque organ many times and climbed its ladders, even though he hasn't started building it.

In this chapter and the next, I follow Bert van den Heuvel, Hans Elbertse, Munetaka Yokota, and their colleagues in the actual construction of the new baroque organ in the two-year period between the acceptance of the final plan in 2015 and the summer of 2017, when the organ is transported from the workshop in Soest to the Orgelpark in Amsterdam. I focus on the crafts that they relied on, their experience from earlier projects, and the inquiries they had to make to solve problems. Often, these problems derived from the hybrid design of the organ. Whereas the previous chapter showed how this design drew together elements with various histories, here I describe the practical work to "fit" these heterogenous elements into a pattern that corresponded with the project's mission.

Analytically, I will use concepts that Richard Sennett discusses in his book *The Craftsman* (Sennett 2008). Against the background of the transformation of work and labor in industrialized societies, Sennett argues that craftsmanship involves an ethic—the desire to do a job well for its own sake—that helps us to rethink work practices. He studies crafts in a pragmatist register that draws on John Dewey's ideas about education, experience, and inquiry (Dewey 2008 [1916]), and highlights the intimate relations between problem-solving and problem-finding, the iterative process of making and understanding, of shaping objects and learning from how they behave under our hands. In following the making process of the organ, I explore how problems were partly given and partly made in situations that were, as Dewey writes, "doubtful ones." We encounter a problematic situation (so, the situation is given as problematic), but we need to articulate what exactly the problem is about (we make it or construct it).[3] In these problematic situations, the participants investigate, looking for "something that is not at hand" (2008 [1916], 218). This thinking and inquiry implies trying things out and taking risks.

To structure my story, I focus on the various parts of the organ that, together, lead the wind from the bellows to the pipes and control this flow of air. Following the technologies of transmission in the new baroque organ allows me to highlight problematic situations that emerged during the making process: transporting the wind from the bellows to the windchest, connecting the movement of the stop drawknob to the windchest, and transmitting the movement of the finger through the key to the pallet in the windchest.

Organ builders always had to find solutions to the problem of transmissions within the dimensions of the organ case, which was in turn related to the dimensions of the church. Traditionally, this problem-solving developed in the skilled crafting of various materials that organ builders used, such as wood, alloys, felt, leather, and copper. In the new baroque organ, however, these transmission mechanisms combine historical organ-building conventions with present-day organ-building skills, and the use of electronic components and software to control the movement of organ parts. The central question in this chapter is how the builders succeeded in making these many parts fit together. This started with making the drawings of the new organ's facade, its windchests, and most of its parts.

DRAWING AND MAKING

The drawings that Van den Heuvel showed me in the workshop in Soest were on paper and made by hand. Contrary to other organ builders, Elbertse's firm does not use computer-aided design (CAD) or drawing software. In *The Craftsman*, Richard Sennett presents the shift from paper to computer screen in design and drawing as an example of how hand and head have become increasingly divided because smart machinery has disrupted the mental understanding that results from the repeated hands-on learning associated with the work of the artisan (Sennett 2008, 39). Quoting the architect Renzo Piano, he explains that drawing by hand enables architects to imagine and understand a place in ways that are not possible through computer-generated visualizations:

> You start by sketching, then you do a drawing, then you make a model, and then you go to reality—you go to the site—and then you go back to the drawing. You build up a kind of circularity between drawing and making and then back again. . . . This is very typical of the craftsman's approach. You think and you do at the same time. (Renzo Piano quoted in Sennett 2008, 40)[4]

Drawing by hand matters, Sennett concludes, because it ingrains "the tactile, the relational, and the incomplete" as physical experiences in our plans (Sennett 2008, 44).

Sennett's analysis of CAD shows how ways of drawing matter in a making process, and the making of the new baroque organ was no exception. Drawing as improvising solutions was done in various stages of the design process.

Drawings finalized a design phase, solved its puzzles, and communicated the outcomes to those who would actually make the parts and the whole. A technical drawing also fixed the various relations between the parts: Making changes somewhere would come at the price of having to change many other parts as well. In this sense, the drawing could be seen as a plan that is simply carried out. In practice, however, the making process often developed differently from what the drawings showed. It was in these situations that problems were articulated because things no longer fitted together.

SKETCHING

When I saw Munetaka Yokota sketching on a tablecloth to communicate a solution to the problem of how to combine Hildebrandt's construction of the windchest with the use of magnets (see chapter 3), I recognized the visual language that he used. I had seen it before in books on organ building. The sketch also resembled the technical drawings I had seen in the workshop of Hans Elbertse and Hermann Eule Orgelbau, which showed an object from above or in cross section, displaying the various parts such as springs and valves, hollow spaces, and shaded contours. For Yokota, drawing was a way of visualizing what was difficult to explain in words. It allowed him to show details that would otherwise escape the attention of those at the table. And it functioned as evidence that the solution might be feasible.

Drawing by hand in situ happened at other times as well. During one of the study trips to Germany, the question of how to position the organ on the balcony was discussed. Loek Dijkman used a paper napkin to quickly draw his views on the matter. And when Munetaka Yokota traveled to Hermann Eule Orgelbau to discuss the construction of Hildebrandt pipes, he made sketches in his notebook to show the others present what he meant, helping him solve the problem of communicating in German and English.

Sketching was not only done by hand. In addition to making rough sketches, Hans Fidom used drawing software to create visualizations of the organ's main structure and facade. These were used during meetings of the core group to discuss how the organ should look. Was its appearance to follow the design of Hildebrandt's facades, or should it express that it was built in the twenty-first century? Similar drawings helped to show how various color combinations for the new organ would fit with the existing color scheme in the Orgelpark. Working with computer software had the advantage that the sketches could easily be adapted to the outcome of the discussions. Fidom also made many sketches of the layout of the new digital console, and he provided the drawings in the final plan that outlined the position of the main parts inside the organ, such as the bellows, the windchests, and the key and pedal action. Computer-assisted drawing also enabled Fidom to make diagrams and drawings for the slides he used for presentations at the symposia run by the Orgelpark's research program.

Finally, Bert van den Heuvel made many sketches in the early phases of the project to discuss possible solutions with Hans Elbertse, or to experiment with the composition of parts of the organ. These often simple drawings enabled him to play around with positioning parts in the organ without having to indicate their exact dimensions. Sometimes sketches were made at a location to analyze a problematic situation. This happened when Hans Elbertse investigated the structure of the north balcony in the Orgelpark to see what adjustments had to be made before the organ could stand there (see chapter 2). Sketches thus had multiple functions in the project. Some were used to facilitate the communication during discussions, others were a means of simulating various outcomes and possibilities in the design process of the organ and its facade, and yet others provided a low-tech means to try out solutions to technical and practical problems.

DRAWING

The drawings that Bert van den Heuvel showed me when I visited the organ workshop of Hans Elbertse in Soest were made on a sheet of white paper of about one by one meter. They are based on the specifications outlined in the final plan for the organ. The first showed the facade of the organ in schematic detail: the console and stopknobs, the front pipes in several fields, and the two pedal towers with their pipes and columns crowned with Corinthian wood carvings of acanthus leaves. The design of the base of the two pedal towers is slightly different. One looks solid and sturdy, the other somewhat slender.[5] At the bottom and on the side of the drawing, the height and width of the various elements in the facade is indicated in millimeters. I could see that the organ would be almost eight meters wide and almost seven meters high. A second drawing showed the inside of the organ in front and top view: its wooden framework; the metal and wooden pipes, some of them mitered[6]; windchests and bellows; the wooden beams to tread the bellows; and the rollerboard and trackers of the key and pedal action. This drawing also had the measurements of the various components in millimeters.

Van den Heuvel explains why he made the drawings on a drawing table instead of on a computer screen, which nowadays is common in organ-building workshops. Having to draw with a computer would mean learning new skills, which he does not desire. "This works very well and I like to do it."[7] Following the information about the scaling and measurements of the pipes of the organ, Van den Heuvel was able to calculate the dimensions of the windchests as well as the case and the framework as a whole. When doing so, he had to consider the thickness of the wood for the wooden pipes, the movement of the bellows, and countless other details. Making the drawings, Van den Heuvel says, fixed the various sizes, measurement, and proportions. "We will now have to stick to these measurements, because all other drawings are based on these two."[8]

The technical drawings that Van den Heuvel made indicated a point of no return in the making process. Given that all these drawings referred to each other and used the same measurements, even small changes would mean that all of them would have to be changed. The nearly three weeks spent on creating the drawings would be lost. In some cases, however, minor changes were indicated on the drawings by using sticky notes with texts such as "this part five millimeters to the right." Having a clear idea about the dimensions of the whole enabled the organ builders to actually make the parts. In the first phase of the building process, work started with making the wooden components of the key action, such as the rollerboards and the wooden arms that change the direction of the movement of a tracker connecting a key to the windchest, or even the little springs under the pipe valves.

MAKING

After having explained the two master drawings of the organ, Bert van den Heuvel took me to the drawings of the windchests of the Hauptwerk and Oberwerk.[9] These drawings had been made not on paper but on whiteboard. They showed the top view of the windchest in a 1:1 scale, the actual size of the windchest. Circles of different sizes indicated the exact location of all the pipes that were to stand on this chest (see figure 4.1). In addition, lines marked the places where the stop mechanism to open and close the pipe valves would be positioned, and green dots the place where trackers for the stop mechanism would go through the windchest. The dimensions of this drawing not only determined the location of the pipes but also the sizes of the tone channels and the tone sliders with their magnets inside (see chapter 3). In addition

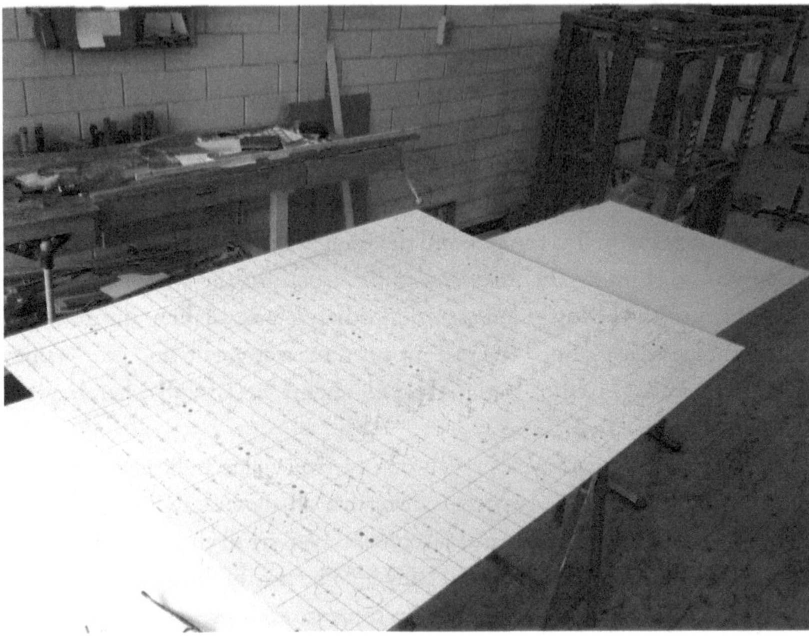

Figure 4.1
Whiteboard drawing of one of the windchests of the new baroque organ showing the exact location of the pipes and other elements (photo taken by the author).

to the drawings from above, Bert van den Heuvel had made 1:1 drawings of the front and side views of the wind channels that showed the mechanical parts in exact detail and dimensions. "You have to make these drawings all at the same time; otherwise you will make mistakes, so there should not be too much time between them."[10]

The drawing on whiteboard not only represents the actual windchest, it is also a template that helps to make it. Bert van den Heuvel explains that the drawing will be positioned on top of the wooden windchest and the actual holes for the pipe feet will be drilled through the template. Using the drawing as a template enables the organ builders to work very precisely. "It always fits," says Hans Elbertse. "You simply cannot make mistakes, unless you make them at the start, on the drawing."[11] Because the three divisions of the organ—Hauptwerk, Oberwerk, and Pedal—contain two windchests each, with exactly the same layout, the template has the advantage that it can be reused to make them identical.

Making the new baroque organ started with drawing. Sketching revealed possible problems resulting from limited space in the case and allowed solutions to be tried out. Making technical drawings was a way to fix the exact distances and dimensions inside the organ. And using a template helped to transfer these distances and dimensions very precisely to the actual material. Sennett analyzes transformations in architectural drawing practices as a shift from hand to computer that has risks. It disconnects simulation and reality, hinders relational understanding and leads to overdetermination. The practice of the builders of the new baroque organ offers a more nuanced picture. Hand drawing is valued, both as a means of communicating ideas and solutions in the situation, and in making the technical drawings of the organ. Yet computer aided drawing also is used, such as for simulating the colors of the new organ in the Orgelpark.

From the improvisatory sketch, via the technical drawing to the template, degrees of freedom decrease. Because almost all distances and measurements in the new organ are derived from the design of the windchest, its whiteboard template determined an almost absolute space within which the builders would have to make things fit. A large part of this space was to be filled with the four bellows, the new organ's lungs.

BREATHING WIND, MACHINE WIND

To make them sound, organ pipes need a constant and rather large flow of air or, as organ builders would say, wind. The pressure of this wind flow influences the tone. When it is too low, the pipe will not produce a steady sound; when it is too high, the tone will become harsh or the pipe will overblow, producing sounds at a higher pitch than intended. Given the crucial effect of wind pressure on the tone quality, organ builders have long used manometers to measure it precisely. A manometer is connected to the hole in the toeboard on which a pipe stands. The wind pressure pushes up a water column in a glass tube. The difference in height of the column in inches or millimeters indicates the wind pressure (Shannon 2009, 14–15).

A problem that organ builders have to solve is that continuous wind flow has to be supplied at a low pressure, but the wind demand varies. When more keys are pressed down or more stops are drawn, the pipes consume more wind. Since the early days of organ building, this problem has been remedied

Figure 4.2
Board for one of the bellows of the new baroque organ under construction in Elbertse's workshop on September 9, 2015. The rectangular holes will let the air in and out (photo taken by the author).

by constructing two or more wedge-shaped, or cuneiform, bellows. These bellows consist of two rectangular wooden boards that are connected in a hinge at one of the shorter ends. A reservoir is formed using a pliable material such as leather combined with thin wooden ribs, and by moving down the top board, the air is pressed out of the bellows. The bellows will empty themselves because of the weight of the top board, but normally extra weight is provided by using stone or metal to regulate the wind pressure precisely.[12] When more wind is needed, they will empty quicker. Bellows usually come in pairs. Their bottom boards are connected to the wooden framework and contain holes for letting the wind into the bellows and out into the main wind channel that leads the wind through the organ (see figure 4.2).[13]

In historical organs, the mechanism to raise the bellows to let wind in was usually operated by foot, but sometimes by hand (Shannon 2009, 17–18).[14] Until the late nineteenth century, playing an organ required at least one person, also called a calcant, to tread the bellows. Treading the bellows requires a certain amount of skill to raise them in a constant manner to avoid shocks that may negatively influence the sound (Salmen and Zepf 2007; Yearsley 2009). In the early twentieth century, manually operated bellows were increasingly replaced by electric centrifugal blowers.[15] These provide a steady flow of air into reservoir bellows, controlling the flow of air from the blower to the organ's wind channels through a system of curtains, valves, weights, and springs.

Because air will return to its original density after any form of compression stops, organ builders take great care in making their wind systems airtight. This also means that the system of wind channels that connects the bellows to the windchests has to be carefully designed. If wind is transported over longer distances through wide channels, a larger volume of air has to be displaced to achieve a pressure effect. A sudden demand for wind resulting from playing many notes at the same time may also lead to unwanted fluctuations in the wind pressure.[16] Ever since the advent of the reservoir bellows, it has become easier for organ builders to make a wind system that provides stable wind. Since the practice of historically informed organ building emerged in the 1960s (see chapter 2), however, it has been argued that a more flexible wind resulting from subtle changes in wind pressure adds to the liveliness of the organ sound. To produce this "breathing" wind, wedge bellows that can be operated manually have been added to the organs, both in restored historical instruments as well as in new organs.[17]

BALANCING THE WIND SYSTEM

From the start of the project, the idea was to give the new baroque organ a dual wind system. In his report on the core group meeting on February 5, 2014, Hans Fidom noted that the system would have to be flexible enough for playing baroque music, yet it should also supply enough wind for playing contemporary music.[18] This could be realized rather simply, he wrote, by using not only baroque wedge bellows but an additional reservoir bellows as well. An important feature of the wind system would be that the organ player would have control over it through valves, manipulated through software. As for the place of the bellows, Fidom thought of the space to the left of the organ on the balcony, or below the balcony, next to the Molzer organ. Later in 2014, however, Loek Dijkman argued that there was not enough room on the balcony or in the adjacent staircase. This meant that the four wedge bellows had to be placed inside the organ case, in contrast to the practice of

Hildebrandt, who would have used the space behind the organ in the church tower.[19] The final plan for the organ that was discussed and accepted in February 2015 specified that the organ would get a treading installation on the right side of the organ case and that the electric blower would be placed in a soundproof chest.[20]

The dual use of the organ and its wind system had its problems. To construct the bellows, master organ builder Bert van den Heuvel had to make calculations about their size as well as that of the wind system. Compared to organs from the nineteenth century, baroque organs tended to have smaller wind systems, as Van den Heuvel explains: "They did not use five fingers in their left hand. The bellows treader would get mad, because they would have to pump very hard and they did not like that."[21] Following the measurements of Hildebrandt's bellows and wind channels, however, was not possible. Played in its electronic mode, the wind demand in the organ could easily surpass the amount of wind needed for baroque playing. In the digital mode, it would be possible to link many pipes to one key, potentially even all the pipes of the organ. If this key were pressed, all pipes would demand wind at the same time, which would lead to a shock in the wind system big enough to damage parts of the organ.[22] A compromise had to be found between the historical dimensions of the wind channels and the potentially large demand for wind in the electronic mode.

To enable historically informed performance of music, the organ would offer the possibility of operating the bellows by foot. Below the four bellows inside the organ case space had to be created for four large wooden beams for the bellow treaders to tread.[23] The beams exit the organ on the right side, and the treaders will step on the beams from a low bench while holding onto a wooden railing with their hands. They have to develop the skill to fill the bellows gradually with wind and release it in a steady flow. The slow and regular movements of the heavy weighted bellow boards will press the wind through the channels to the windchests, resulting in "beautiful," as Hans Elbertse calls it, or breathing wind. Hans Fidom is hesitant to share the argument of some historically informed organ advisors that baroque organ builders had the explicit goal of making the wind flexible. "I think it is ventriloquising for the baroque to say that this flexible wind was intentional. It was more a result from the way they built their organs. Until deep in the nineteenth century, building practices resonated more with medieval and renaissance ideas about numbers, proportions, mathematics and mechanics than with expressivity as we came to understand it in the Romantic era."[24]

NEGOTIATING A PLACE FOR THE ELECTRIC BLOWER

If the organ was not to be tread by foot, it needed an electric blower. The final plan for the new baroque organ that Hans Fidom drew up in March 2015 specified that this blower would have to be closeted in a soundproof chest, but not where this chest was to be placed. This led to a discussion in the core group. Hans Elbertse claimed that there would not be enough room inside the organ case itself. Hans Fidom suggested placing the blower below the balcony, next to the Molzer organ. This meant that the main wind channel would have to pass through the balcony floor. Loek Dijkman was not happy with this. Making a hole in the concrete would lead to a loss of strength in the construction, he thought, because the balcony floor is not of the highest quality.[25] The Park Church was one of the first buildings in Amsterdam in which concrete was used, he said, and the builders economized on material, not on labor costs. Dijkman suggested placing the chest with the blower on the balcony to left of the organ.

The issue remained undecided. Over the course of 2015, Hans Elbertse and Hans Fidom developed the idea to place the blower in the left staircase and to lead the main wind channel through the wall of the hall. This would have the benefit that any remaining motor noise would not be audible during concerts. Elbertse bought a rather powerful blower, so having a longer main wind channel would actually be beneficial, he said. In January 2016, Loek Dijkman visited the Orgelpark for a listening test (see chapter 5) and also discussed the proposed location for the blower. He voiced his opinion that the staircase was problematic as the location for the blower. Not only did he object to making a hole in the supporting wall to let the main wind channel pass through, he also pointed out that the door to the staircase is an emergency exit and blocking it in any way would be illegal. Given that the Park Church is a listed monument, Dijkman felt responsible to leave it untouched as far as possible. He also thought that a long and winding wind channel would be detrimental to the quality of the wind.

Several other options were discussed, including putting the blower right next to the organ or below the organ.[26] Both options had their disadvantages. In the weeks that followed, Elbertse and Fidom explored the possibility placing the blower in the cellar, where parts of the Molzer organ reside. Hans Elbertse explained that the capacity of the blower is calculated to bridge a distance of seven to eight meters. Placing it too close to the organ would mean that its pressure would be too high. Given this situation, he would prefer to have the blower below the balcony, where it would also be out of sight.[27]

In November 2016, Hans Fidom wrote a memo in which he summarized the situation. The blower would be placed in the cellar of the Molzer organ, below the floor of the hall. This, together with the fact that cables for the live electronics facilities inside the organ would also have to lead to the cellar, meant that two holes would have to be made in the balcony floor.[28] Loek Dijkman disagreed, however, saying that these changes to the construction of the building would weaken the floor and affect the status of the church as a monument. A final decision was made to place the blower chest on the balcony, close to the wall of the hall and partly embedded in the balcony structure. Parts of the chest would thus be visible from the floor of the hall. Whereas a vertical wind channel from the cellar to the balcony would be made out of PVC tube, in the end, Hans Elbertse decided to make them from wood, which has the benefit of being more similar to Hildebrandt's design. In the spring of 2017, the main wind channels were constructed under the balcony floor to connect the blower to the bellows inside the organ case.

MAKING THE WIND FIT

The account of the making of the wind system of the new baroque organ shows how the organ builders had to solve problems that were partly of their own making. The dual design of the instrument resulted in a wind system that had to serve two uses. When played according to insights from historically informed performance practice, the goal was to have flexible, or "breathing," wind that would impart a lively character to the pipe sounds. This had consequences for the dimensions of the wind channels and also required the the treading of bellows. In the electronic mode, the wind system had to operate differently. Here the wind channels and the electric blower should have the capacity to respond to a large demand for wind when many pipes had to sound at the same time.

Making the wind system was not a straightforward process and led to problematic situations. Following Dewey, it is in these kinds of situations, where "things are uncertain or doubtful or problematic," that thinking occurs (Dewey 2008 [1916], 155):

> Only what is finished, completed, is wholly assured. Where there is reflection there is suspense. The object of thinking is to help reach a conclusion, to project a possible termination on the basis of what is already given. Certain other facts about thinking accompany this feature. Since the situation in which thinking occurs is a

> doubtful one, thinking is a process of inquiry, of looking into things, of investigating. Acquiring is always secondary, and instrumental to the act of inquiring. It is seeking, a quest, for something that is not at hand. (2008 [1916], 155)

Because the problematic situation has an uncertain outcome, Dewey argues, all thinking implies taking risks. Making the wind fit revealed that the risks were related both to the building in which the Orgelpark is housed and to the status of the new baroque organ as a meaningful instrument. The process also shows how solving a problem can lead to new problematic situations.[29] This became clear in the making of the stop action, the mechanism that allows the organist to open and close stops.

INVENTING THE STOP ACTION

When a stop is drawn on a mechanical organ as Hildebrandt built them, this movement is transferred to the windchest through a stop action that resembles the key action's trackers, stickers, rotating rods, and squares (Shannon 2009, 73). Because more energy is needed to manipulate the stop action than the key action, its components are bigger and require more space. This space was not available, however, in the new baroque organ because the bellows had been placed inside the organ case.

The lack of space caused a problem that becomes clear when we go back to the design of the spring chest as it was built by Hus and Schnitger in Stade (see chapter 2). Here, when a stopknob is drawn, the stop action transfers that movement to the chest where a slat is lowered that presses down a series of metal stickers that open the spring valves. When the slat is in the down position, the valves are open and the pipes of that stop will sound when a key is played. Because every one of the pipe valves has a little spring below it to keep it closed when the stop is not drawn, the stickers are pressed back, and this counterforce is transferred through the stop action back to the stopknob at the console. For this reason, stopknobs on a spring chest have to be hooked into the panel of the case to prevent them from retiring after they have been drawn. Not having the space for a mechanical stop action in the new baroque organ meant that the historical design to manipulate the stops could not be used. How was this problem solved?

For Hans Elbertse, it had been clear early on in the project that he would have to use an electrical system to operate the stops. On March 12, 2014, the issue was discussed in the core group. Loek Dijkman was surprised to hear that the stop action would be electrical because the idea was to build an organ that was "played, felt and thought as in the days of Hildebrandt," he said.[30] On learning that the stopknob sliders will, in fact, be only electrical switches, he wondered if anyone would want to play an organ with "fake knobs."[31] In the discussion that follows, it becomes clear that the electrical system will be necessary because of the space limitations inside the organ case. Without an electrical mechanism, it would also not be possible to give the organ registration aids and the capability of preprogramming stop combinations (called *Setzer* in German). In the final design of early 2015, the stop action was described as a hybrid construction: The windchests would have stop slats according to the construction of seventeenth-century spring chests. These slats will be operated by electrical magnets, one on each side of the slat. Because both the windchest for the Hoofdwerk and the Bovenwerk will be built as two separate half chests, four magnets are necessary for each stop. In addition, all stopknobs will have their own magnets so that their manipulation can be preprogrammed.

A SIMPLE MECHANISM

Opening a stop on a mechanical spring chest organ takes considerable force.[32] After the stop has been drawn, the knob is prevented from jumping back in by pushing it down slightly so that a notch in the beam of the knob catches the edge of the case. Hans Elbertse and his colleagues had to find a way to mimic this basic mechanical solution using electromagnets. The catalogue of the Laukhuff organ-building firm offered several types of drawstop solenoids—electromagnetic coils that can operate the stopknobs on a mechanical organ. It also offered electromagnets that could be used to manipulate the slat in the spring chest that pressed down the pipe valves.

The first problem to be solved was that these magnets did not have enough power to push down the slats. Another problem was that they are currentless in both the on and off positions. This meant that the spring valves would push the slat back to its original position as soon as the drawknob was pulled out and the current on the corresponding magnet switched off. This might be solved by using magnets that remained on all the time, but as Elbertse explained, with sixty-four electromagnets, this would require enormous amounts of current. At first, Elbertse thought he had found a solution by using a device that used a screw thread motor instead of an electromagnet.

When switched on, it would turn until the slat was pressed down and remain in that position until it was switched off and turned back. In the end, the device turned out to be too slow.

The lasting solution came during one of the many conversations that Elbertse had with Van den Heuvel. They thought up a simple mechanism that consisted of two vertical metal rods connected through a hinge. A horizontal sticker from the magnet was connected to the hinge. In the off position, the two vertical rods would be in a wedge shape and the slat would be up. When switched on, the horizontal sticker from the magnet would press against the hinge, resulting in a movement that pulled down the slat. By making sure that the hinge would be pushed passed its "dead point," the spring valves would not be able to push the slat back. When the electromagnet was switched off, Hans Elbertse explained, it would draw the horizontal sticker connected to the hinge back and the slat would move up again:

> Once you think of it and it works, it is actually very simple. What if we push it beyond its dead point? We made a small model and then it worked. And you wonder: why didn't we think of this before? This happened with so many things in the organ.[33]

Yet, another problem existed. If the movement of the electromagnet is abrupt, it might cause sudden shocks in the stop action when large numbers of stops were operated at the same time in the digital mode or when registration aids were used. This problem was solved by using electromagnets that could "brake" at the start and the end of their movement. When I visited the workshop in September 2016, Elbertse showed me a model of the mechanism. In all its simplicity, it looks as if it had been around for a long time (see figure 4.3).

IMAGINATIVE LEAPS

Where did the idea for the simple stop mechanism come from? Richard Sennett analyzes how intuitive leaps happen, such as the one that Elbertse and Van de Heuvel experienced. He claims that they may come from the frustration resulting from the limitations of a tool. "Intuition begins with the sense that what isn't yet could be" (Sennett 2008, 209). The electromagnets in the Laukhuff catalog were designed for standard situations in slider chest organs. Because no one builds new spring chest organs, there are no standardized parts available for this organ type. If we think of the Laukhuff drawstop

Figure 4.3
A model of the hinge mechanism in the stop action in the on position. The horizontal rod marked "25" is connected to the electromagnet and moves to the left when a stop is drawn. It pushes the two vertical parts of the hinge beyond the dead point (where they would be verically aligned). This ensures that the counterpressure of the pipe valves is not enough to push the horizontal rod back into the off position once the current for the electromagnet is switched off (photo taken by the author).

electromagnets not only as parts but also as tools, their improvised use in the new baroque organ make them an example of what Sennett calls "difficult tools" (2008, 195). These are tools that have certain purposes and flaws, and yet they challenge the imagination through their potential to be used in new ways.

Sennett argues that there are four ways in which a tool can organize untested possibilities. The first step is one in which "we break the mold of fit-for-purpose [which is] no more and no less than the willingness to see if a tool can be changed in use" (2008, 210). Instead of using the electromagnets to operate the slat directly, a detour was needed to compensate for its limitations. This was found in what Sennett sees as a second step in how imaginative leaps occur, by establishing adjacency of two unlike domains. In the case of the stop mechanism, these were the domains of electrically and mechanically operated actions. In a third step, the juxtaposition of domains can lead to surprise and wonder, Sennett argues. "Surprise is a way of telling yourself that something you know can be other than you assumed" (2008, 211). It is this sense of wonder and surprise that Hans Elbertse highlights in his account of the construction of the stop device. A final step in intuitive leaps is the insight that any technical solution will bring its own problems, which may in turn challenge the imagination. This new problem was the fact that the mechanism could cause shocks when many stops were manipulated at the same time.

Why do we call the intuitive leap "intuitive," Sennett asks, if we could also describe it as a form of reasoning (2008, 212)? His answer is that the kind of reasoning of craftsmen like Elbertse and Van den Heuvel does not start from universal propositions as in deductive or syllogistic reasoning. Instead, it starts from inductive understandings of situated problems. Theirs is a strategy of trials, or in the words of Sennett, of "what if?" (2008, 213). The intuition is triggered in the use of tools that have to be used even though their functionality is limited. To make them work requires tinkering with what is at hand. This was the case not only in the making of the stop mechanism but also for the technologies that transport the movement of a finger on the keyboard to the windchest.

CHALLENGED BY PRINTED CIRCUITS

The building of the organ started with making parts that would be assembled later on. Some of these parts were very small, such as the pipe valves, and some were bigger, like the rollerboard for the pedal. This board enables the transfer of movement from the pedal keys to other parts of the organ. When I visited the workshop in May 2015, the board had just been finished. During my visit, I tried to understand and visualize where all these smaller and larger objects would fit in the organ and what their role was. In doing so, I slowly began to understand the inner workings of this complex machine even before it was put together.

From the pedal rollerboard, I learned how the pressure of the organist's fingers and feet on the manual and pedal keyboards is mechanically transmitted through the organ. Like many other aspects of organ technology, the anatomy of the connective system between the keys and the pipes of organs, generally called the organ's "action," has been remarkably consistent over the centuries.[34] Because the distance between the key and the valve or pallet in the windchest is considerable and given the small force of a finger pressing a key, the transmission of this force rests on a combination of relatively simple mechanical machines such as levers, trackers, stickers, backfalls, and squares.[35] Because the distance across a keyboard is smaller than the space it takes to place the corresponding pipes on a windchest, an additional device is used to spread the action to the sides of the organ. This consists of a series of rotating wooden rods, one for each key, that are mounted on a rollerboard or frame. Each rod has two short wooden arms, one attached to the tracker leading to the key, and the other to a tracker leading to the valve in the windchest (Shannon 2009, 39–42). To make the mechanical action as efficient as possible, the distance between manuals and windchest is minimized, which means that the console of a mechanical organ is usually integrated in the case.

As Hans Elbertse explained, he was not able to use the exact layout of the mechanical action of Hildebrandt's organs because the new baroque organ has its bellows inside the case. Therefore, the action had to be placed around them. This meant that the trackers would make a longer route through the organ and that the transmission of the forces had to be carefully calculated to prevent the mechanism from becoming too heavy to play.[36] A heavy mechanism would hinder what is seen as one of the main advantages of a mechanical action: the tactile sensation of a direct, physical contact between the key and the valve in the windchest.

In the organ's digital mode, however, the movement of the fingers is transmitted through electrical wires connecting the digital console to the windchest. This route also aims at precision and control, but it is achieved

Figure 4.4
Pedal rollerboard for the new baroque organ (photo taken by the author).

through very different means, which I will explore in more detail in chapter 6. Here I focus on the last stage in this route: the hybrid windchest with its complicated tone sliders that combine wood, valves, magnets, connectors, and wiring. As became clear in the previous chapter, the possibility of regulating the opening and closing speed of the pipe valves by controlling the magnet was an essential element in the digital mode of the organ. Whereas the construction of the mechanical action required a fair amount of skill and experience, for an organ builder like Hans Elbertse, this is familiar territory. Building the hybrid windchest, however, offered challenges that Elbertse and his team did not fully anticipate at the start of the project.

THE HINGE

Of all the many parts of the new baroque organ, the tone sliders are the most complex. They are at the heart of the dual design of the new organ because they have to both emulate the workings of a historical spring chest and enable the electronic manipulation of its pipe valves. After the listening test in Soest in January 2015, the decision was made to use new magnets supplied by Sinua. These would be mounted on the tone slider, one for each pipe valve, thirteen per tone slider.[37] The wires for these magnets were housed in a plastic cable tray set in a groove milled out on one of the sides of the slider. Following the dimensions of the chest, the length of the tone slider was designed to be 130 cm. As was the case with the Hus/Schnitger organ in Stade, the model for the spring chest, the tone sliders could be taken out of the windchest without taking all the pipes off. The length of the tone sliders, however, turned out to be a problem.

From the start of the project, Johan Luijmes had made it clear that he would like to use the organ to accompany the performance of Bach cantatas.[38] In his view, the musicians and singers would ideally be situated on the balcony, in front of or next to the organ because the musicians and the conductor would be able to communicate better and the sound of the organ would blend with the ensemble. The issue was taken up during a core group meeting in February 2015. The outcome of the discussion was that there should be space for approximately twenty musicians and that the chairs on the balcony "would not be sacred." If necessary, the balcony would be rebuilt to acquire a stage.[39] For the same reason, it was decided to make the main case of the organ not too deep and to place it relatively close to the main wall.

For Hans Elbertse, the position of the organ on the balcony, close to the wall, meant that he had to find where and how to take out the tone sliders that contain the small valves for each pipe. One possibility was the front of the organ, but this solution would be cumbersome because panels and pipes would have to be removed first. To take out the 130 cm tone sliders and still place the organ 70 cm from the wall, Elbertse and Van den Heuvel made a test tone slider with a hinge. Now the slider could be taken out in two stages: After the first part is out, it can be bent upward to create space for the second part. From an engineering point of view, this solution was challenging. If not constructed properly, the hinge mechanism might cause wind leakage or weaken the structure of the slider. Also, the wiring in the cable tray had to be led along the hinge. And finally, the two parts of the slider had to be connected by a metal hook that also would have to fit. After some attempts, Elbertse and his employees were able to make the hinge work (see figure 4.5).

THE PRINTED CIRCUIT BOARD

During the listening test at Elbertse's workshop in January 2015 (see chapter 3), it became clear that the speed of the magnets could be regulated, which meant the opening and closing speed of the pipe valve and, hence, the amount of air and the speed of the airflow entering the pipe became controllable. On that occasion, Benedikt Aufterbeck and Thomas Stöckl from Sinua had used a laptop to run the software needed for the control. Shortly afterward, it turned out that regulating all individual magnets from one central computer would be impossible. Instead, every single magnet should have its own printed circuit placed very close to it.

For Hans Elbertse and Bert van den Heuvel, finding a place for the printed circuits in the tone slider was not easy. Because the dimensions for the windchest had been fixed once the final plan was accepted and the chest was drawn, the wood for the sliders had already been sawed. Holes had been drilled in the wood for the airways to the pipes. Space was made for the magnets, the cable tray, and the wires. Now for every magnet, a part of the wood had to be removed to make room for the printed circuit. After several experiments, Elbertse and Van den Heuvel found a solution: The circuit would be placed vertically, facing the magnet. It would be screwed in the wood using two screws, but care was to be taken not to drill the holes for the screws too

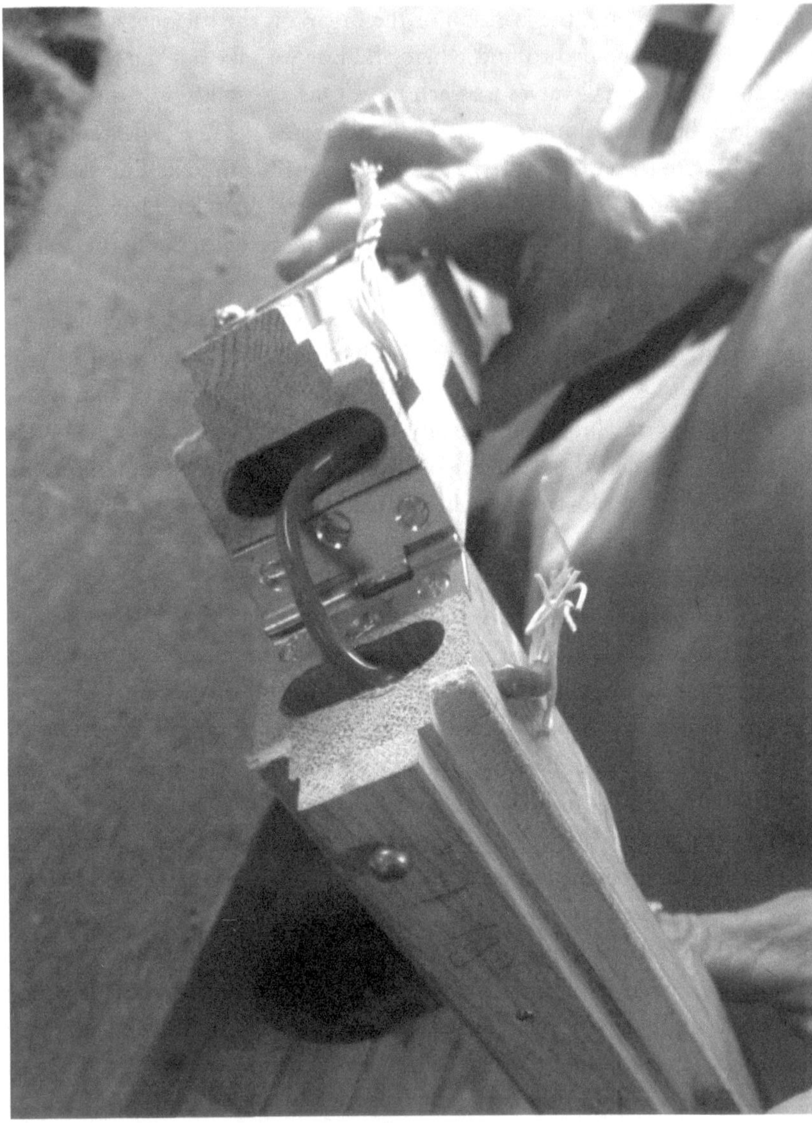

Figure 4.5
Prototype of a tone slider with a hinge (photo taken by the author).

deep so they would not enter the windway and cause leakages. Thus, the dimensions of the printed circuit followed directly from the dimensions and properties of the sliders. To communicate these dimensions to Sinua, in September 2015, Elbertse made a template that indicated where the two holes for the screws and the connectors for the wires had to be (see figure 4.6).

Sinua now had to find a way to fit the components they needed on the exact space that was given to them. Elbertse had argued that two screws in the printed circuit were necessary to prevent it from breaking when the slider was drawn out of the windchest. This meant even less space for the components. As Benedikt Aufterbeck said: "It is as if you have to build a house on a piece of land where there are already other objects. You have built around it, and that is more complex than when you can make what you want."[40] Contrary to what one might think, even the micro world of computer chips has its limits when it comes to reducing space, Aufterbeck points out. "We are using the smallest components that are available. They are so tiny. Sometimes we breathe and the components are away. And sometimes they do not fit the electrical needs. For a current running through a resistor or a wire, it needs to have a certain size."[41]

At Sinua, the design and prototyping of the printed circuit takes more time than the actual production of the devices. Software code has to be written. Electronic parts cannot be tailor-made but must be bought from specialized vendors who often do not sell individual parts. Sometimes finding a part with the specific properties necessary to fit in the design of a printed circuit takes a few minutes, sometimes weeks. Thomas Stöckl explains, "We have to use things that are meant for something completely different. The connectors that we use were made for LED stripes that are glued at the inside of cupboards. The driver IC, the main component we use to control the opening and closing speed of the pipe valves, is made for the automotive industry."[42] While Sinua worked on the software and the design of the printed circuits, the organ started to take shape in the workshop of Hans Elbertse.

THE DB19 CONNECTOR

Using thirteen printed circuits for every tone slider, each of them needing its own power, meant that more wires had to be placed in the cable tray. The thickness of these wires depends on the amount of current that has to run through them. Originally, the magnets were operating on a tension of 12 volts. According to Hans Elbertse, it would be better, however, to have 24

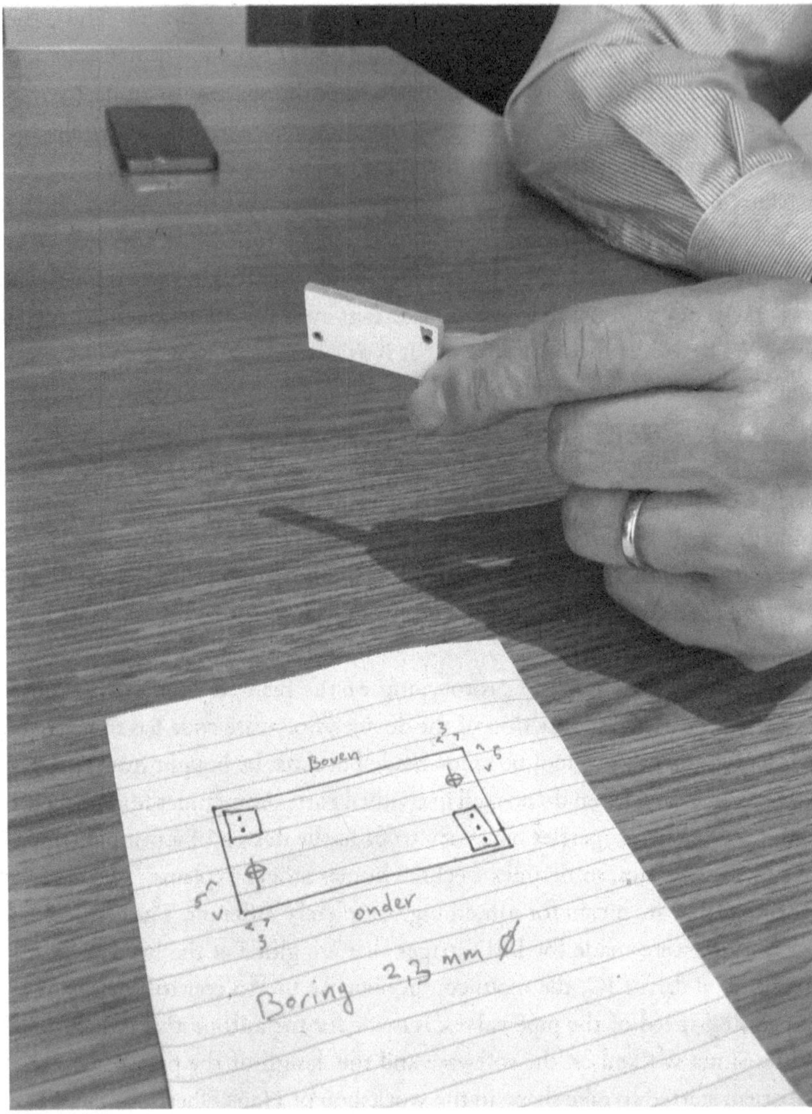

Figure 4.6
Template that Elbertse made for Sinua for the dimensions of the printed circuit to control the opening and closing speed of the pipe valves (photo taken by the author).

volt magnets to reduce the current running through the hundreds of thin wires inside the tone sliders. For Bert van den Heuvel, the new wiring design meant that he would have to find a cable tray with more space inside. After several attempts with more expensive versions, he decided that a standard cable tray that he found at a local hardware store worked best.

From the magnets and printed circuits, the wires were led past the hinge in the tone slider to the connector that was mounted at the end of the slider.[43] In the original design, the number of pins in the connector corresponded to the smaller number of wires. After the decision was made to use printed circuits, more wires were needed and, hence, more pins and a bigger connector. Because the wood for the tone sliders had already been sawed, space for the DB19 connector was limited. These connectors are produced in standardized dimensions. Using a larger connector would have consequences, as Hans Elbertse explained: "Suppose we take bigger connectors, then the tone sliders have to be six millimeters broader, which means that all tone grooves in the windchest would have to be broader and thus the whole windchest. The pipework would have to be redistributed and parts of the organ would have to be redesigned. Making the drawings on whiteboard of the windchest has cost 240 man hours that would be wasted."[44] According to Elbertse, it was pure luck that the standardized DB19 connector exactly fit the head end of the tone slider. Screwing the connector on the head end of tone slider, however, was difficult because there was hardly enough wood to securely tighten the connector screws.

While Sinua was designing the printed circuits, work on the windchest continued. In early 2016, Jos Elbertse worked on the chests for the Hauptwerk and Oberwerk. The bottom plate of a chest has long rounded holes that will be covered by the pallets in the pallet box. On his plate he glues the vertical oak beams that separate the tone grooves in which the tone sliders, one for every key on the two manuals, will have to fit. The beams will be covered by the toeboards with holes for every pipe that stands on the windchest. Jos uses clamps and chisels, glue, and a big wooden hammer, which he made himself during his years as a cabinet-making student (see figure 4.7).

In February 2016, I travel with Hans Fidom to Ratingen, where the layout of the digital console will be discussed with Sinua. On our way back we pick up the magnets for the pipe valves, 1,800 in total, one for almost every pipe in the organ.[45] Now the work on assembling the tone sliders can begin. When I visit the workshop in the summer, Jos Elbertse is busy soldering the wires to the connectors that connect the magnets and printed circuits to the

Figure 4.7
Jos Elbertse working on one of the windchests (photo taken by the author).

back walls of the windchest: fourteen wires per connector (see figure 4.8). From there, other wires connect every magnet to a switchboard behind two of the wooden panels at the front of the organ case. The soldering is work of great precision. One of the employees is busy screwing printed circuits onto the tone sliders, one circuit for every magnet, two small screws per circuit—another reminder that building the new baroque organ requires a lot of patience and attention to detail, even in endlessly repeated tasks.

RESISTANCE AND IMPROVISATION

Craftsmen learn from dealing with resistance and ambiguity, Sennett argues. He defines resistances as "those facts that stand in the way of the will," and he distinguishes two types: "found and made" (Sennett 2008, 215). The story of the design and building of the tone sliders can be retold using these terms. Both found and made resistance often took the form of limited space. Because room was needed for musicians near the organ, its case was placed close to the wall. From the three strategies that Sennett identifies to counter resistance—"reconfiguring the problem into other terms, readjusting one's behavior if the problem lasts longer than expected, and identifying with the problems most forgiving element" (2008, 222)—the first applies here. The tone slider was reformatted by using a hinge gave it the required length of 130 cm while permitting someone to take it out in the space between the windchest and the church's wall. Some of the resistance that the organ builders met was of their own making. Sinua's idea to regulate the speed of the magnet meant that printed circuits were necessary, which required extra parts to be added to the tone slider. In a way, things would have been easier if the team had accepted that the magnet would have just one speed. Complexity was added as a result of investigations by Sinua into new types of magnets and the means to control them. For Sennett this working with resistance can be positive, an insight that mirrors John Dewey's pragmatist embrace of problems as a key to positive learning (2008, 226).

For Dewey, resistance is never an isolated event but always has a context, Sennett remarks (2008, 277). And for Sennett, sites of resistance can often be exemplified by walls or membranes, structures that separate. In a metaphorical sense, the tone slider itself could be seen as a site of resistance that both separates and combines the worlds of historical organ building and twenty-first-century digital technology. It is at the active edge of this border that ingenuity is required. This ingenuity can be found both in the work of

Figure 4.8
Jos Elbertse soldering wires to a DB19 connector. As can be seen, the connector is exactly the same width as the wooden tone slider (photo taken by the author).

Hans Elbertse and his employees who had to find space for plastic and metal parts on, and in, the oak of the tone slider, the volume of which was limited by the overall design of the windchest and the organ as a whole. It can also be found in the time that Sinua spent to find the components that would both do the job and fit the space on their printed circuit, the size of which was fixed in the negotiations with Elbertse.

As Sennett points out, at these sites of resistance, there is often ambiguity when the separation is ill defined. Membranes are porous, which means that they facilitate both separation and interaction. Ambiguity can lead to improvisation, Sennett explains, when objects are used in a new, unexpected way.[46] Both Bert van den Heuvel and Sinua used materials and objects that were not originally meant to be used in an organ, as the examples of the cable tray and several of the components for the printed circuits show. A closer look at these printed circuits reveals how technologies of control from different time periods coexist in the hybrid windchest.

WOOD AND SOFTWARE

In May 2016, I received a phone call from Hans Fidom, who told me that the project would be delayed. When the final plan was accepted, it aimed for the completion of the organ in autumn 2017. Hans Elbertse had made it clear that this would not be possible and that he needed an extra half year because he is still waiting for the printed circuits that Sinua is making. He needs these to finish the tone sliders and, thus, the windchests. Mounting the windchests in the framework without the finished tone sliders might lead to problems, he thinks. Sinua, for their part, needed time to work on the design of the printed circuits and fit the required parts on the template that Hans Elbertse has provided.

A few weeks after the phone call, in early June, I met Benedikt Aufterbeck and Thomas Stöckl during the yearly symposium at the Orgelpark. Over dinner at the end of the first day Aufterbeck and Stöckl showed me a version of the finished printed circuit board. On a piece of green plastic approximately 20 × 10 × 2 mm., I read the words "SINUA Magnet Driver." In two of the corners are white connectors for three electrical wires each. Soldered on the green plastic base are five computer chips and a range of smaller components. In the other two corners are the screw holes, marked by white circles. Next to one of them I read "v0.9," which stands for version 9 of the prototype (see figure 4.9).[47]

Figure 4.9
Sinua Magnet Driver version 1.0 on a workbench in the workshop of Hans Elbertse. The dimensions are according to the template shown in figure 4.6 (photo taken by the author).

Aufterbeck and Stöckl explained how every printed circuit is a computer in itself:

> What this printed circuit does is actually switch on and off the power for the magnet at a very very very fast frequency, which is 30,000 times per second. It's the same principle as when you dim these LED lights. It switches on and off so quickly, that it has an analog feeling. You can adjust how long per period the magnet is on or off. If you start at one percent of the time, it still switches 30,000 times a second, but just for tiny fragment of the period the magnet is on, and most of the time it is off. The other way around, if you have 99 percent, most of the time the magnet is on. But this needs a lot of computing power. It would need a huge amount of computing power to do that for 1,700 magnets with a computer in one place. So we decided to have one controller for each magnet. The other problem is that when you switch the magnets on and off very quickly, the energy leaves the cable and creates electric interference, like a sender antenna. If you have the printed circuit close to the magnet, you reduce this effect.[48]

To control the force curve and, hence, the opening and closing speed of every individual magnet, Aufterbeck and Stöckl wrote software. The level of control that they were now able to achieve was key to the dual design of the organ. First, it was now possible to assign certain fixed-speed characteristics to a magnet and its valve. Thus, it would be possible to predefine the wind flow to a pipe—and, therefore, its sound—through electronic means.[49] Second, having control over the opening and closing speed of the spring valve was necessary to allow the organist to tinker with the sound characteristics of a pipe at the digital console, such as opening a pipe valve only halfway. Third, by using a touch-sensitive manual in the digital console, it would be possible to communicate the nuances by touch directly to the magnets and the valves. The organist could now, through electronic means, control the behavior of the spring valve from the keyboard.

Making the Sinua system for the new baroque organ involved work on both software and hardware. Asked how their workshop compares to that of Hans Elbertse, Aufterbeck answers that they do not really have one:

> You can imagine a small room with two desks, or three. On it are computers and a soldering station to prepare the prototypes for the experiments. And then you have a board with lots of wires, and an oscilloscope where we measure things and the behavior, and we have the magnets lying there, and some test situations. We try and test the printed circuit and evaluate the construction.[50]

Similar to the workshop of Hans Elbertse, the making process started with brainstorming, discussions, and drawing. Aufterbeck and Stöckl sketched force curves, circuits, and wiring. They also used CAD software to design certain components that were not available on the market. Like Elbertse's test organs, Sinua built prototypes of the printed circuit on matrix boards.

> This looks quite wild, but the result is the finished circuit. Then we make the circuit diagram, the official blueprint with all the logical connections. The next step is the design of the actual printed circuit, where you really say: I put this component here, so then we have space for the connections there, which means that that component has to be there. Thus we make the design of the printed circuit. When it is ready, we send it to the manufacturer to mass produce it, and it is out of our hands.[51]

Despite the resemblances, an important difference existed between Sinua and Elbertse in their making processes. In the case of Elbertse, the phase of drawing the organ and its parts was relatively short and the actual making took most of the time; for Sinua, this process was exactly the opposite: Drawing, designing, and prototyping took a long time, whereas actually mass-producing the printed circuits took only weeks.

TWO FORMS OF ADJUSTMENT

One way of conceptualizing an organ is to see it as a collection of devices that transfer, change, and control forces. In a fully mechanical organ, these forces originate in the human body—think of the movement of the bellow treaders' bodies and the organist's fingers and feet. In the nineteenth century, electro-pneumatic key and stop actions offered new ways to exercise and control the forces in an organ. The twenty-first-century digital technologies that Sinua mobilizes to control the forces needed to open and close the pipe valves can be seen as the next step in the manipulation of the physical components inside an organ. Yet, in the new baroque organ, given its dual design, digital control has to coexist with a time-tested means to regulate force, namely, the spring.

The basic idea behind the spring chest rests on the combination of a valve, a spring, and a sticker. When the sticker is pushed down by a slat, it opens the pipe valve; when the slat is lifted, the spring pushes the valve upward to close the toe holes. On one of the technical drawings that Bert van den Heuvel made, we see a side view of the mechanism in the new baroque organ (see figure 4.10).

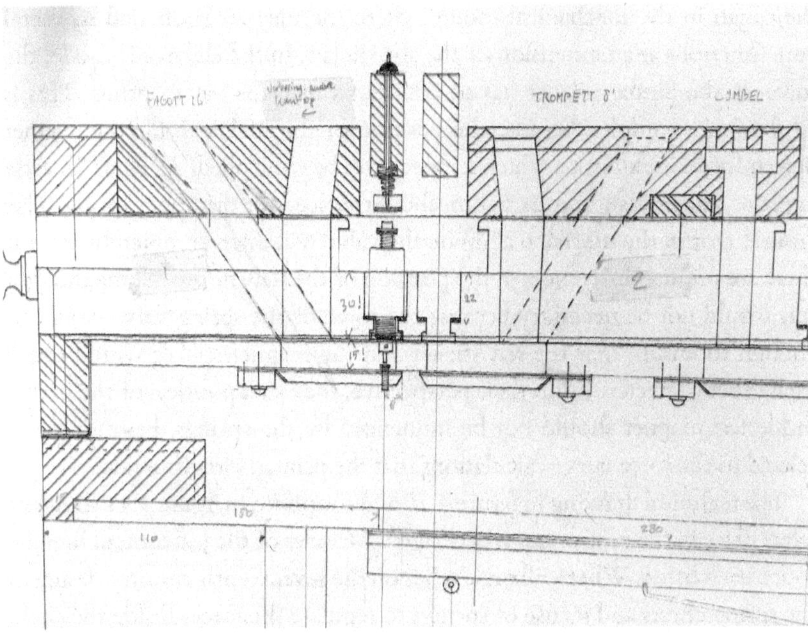

Figure 4.10
Side view of a tone slider. It shows in the upper-center part of the drawing the slat that presses down the pipe valve in the mechanical mode and the magnet that performs the same operation in the electronic mode (see also figure 4.11; photo taken by the author).

The lower part of the drawing shows the pallet box with the pallets, the valves that are directly connected to the manual and that open when a key is pressed. Because the pallet box is under wind pressure, wind then flows into the tone channel that we see in the center of the picture. Here we see two objects that represent, from top to bottom, the wooden slat, which can also be found in historical spring chest organs, that contains a sticker that is lowered when a stop is drawn, and the Sinua magnet, with its own sticker that touches the small valve directly under the windway to the toe hole on which the pipe stands. Invisible on this drawing is the small spring under the pipe valve that closes it when the slat is raised.

The two objects represent two forms of adjustment. The first is based on the workings of two springs: one below the pipe valve and one between the slat and the part of the slat sticker that presses on the metal rod inside the magnet.[52] Together, these two springs have to be carefully adjusted to use the organ in the mechanical mode, where the magnet is off and its metal core functions as an extension of the slat sticker. In the electronic mode, the force of the Sinua magnet has to counterbalance the valve spring. This is all the more complex because the position of the sticker inside the magnet determines the extent to which its speed can be controlled. As Hans Elbertse explains, there was some discussion about the need for the upper spring. This ensures that in the mechanical mode the valve will open completely, even if there are minute differences in the position of the slat stickers. Sinua thought this would not be necessary because the force of the spring valve would be enough to ensure that the slat sticker and the magnet sticker would touch each other correctly. From their perspective, the exact position of the sticker inside the magnet should not be influenced by the springs, because it was related to the force curve calculations that the printed circuits made.

The technical drawing in figure 4.10 and the photo in figure 4.11 show the exact place in the new organ where the two modes of the instrument literally touch each other. Where Elbertse relies on the seventeenth-century design of the spring chests and its use of springs to regulate the forces inside the chest, Sinua uses electromagnets, printed circuits, and software to fine-tune these forces. Computing capacity is placed next to each magnet to give its functioning an "analog feeling." The springs and the digital technology not only control the movements inside the windchest, they also allow the different material components to cooperate correctly with each other.[53]

DEALINGS WITH THE FUTURE

The collaboration between Elbertse and Sinua and the tensions that sometimes arose suggest two different worlds, one of working with wood and one of developing software and printed circuits. Bringing them together exemplified the main challenge in this phase of the project: How to actually *make* the dual design of the organ. Asked if Sinua could understand the problems of Elbertse's wood world, Benedikt Aufterbeck answered that he feels at home in both worlds as he is trained as a professional organ builder himself: "We are also wood people. I understand Hans very well because I know exactly what problems he has to fight. We can imagine what the wood people would like."[54]

And Thomas Stöckl added: "It sometimes turns out that somebody is only a wood person [laughs] and does not exactly understand what kind of

Figure 4.11
The slat sticker and its spring touching the magnet and its sticker. These are the components shown in the upper center of the technical drawing in figure 4.10 (photo taken by the author).

problems we have, or what is really a problem and what may not be a problem at all."[55]

Making things fit together related not only to the parts in the organ case and to the components on the printed circuit but also to different types of craftsmanship, each with their own problematic situations and ways of problem-finding and -solving.

According to Richard Sennett, "three basic abilities are the foundation of craftsmanship. These are the ability to localize, to question, and to open up" (Sennett 2008, 277–280). The first ability enables the craftsman to localize where something important is happening, or where a material, practice, or problem matters. For Elbertse, the springs and stickers were necessary to solve a problem that, according to Sinua, did not exist. But even though Elbertse and Sinua sometimes localized problems differently, they clearly share this ability. Questioning implies that one reflects on the characteristics and qualities of something, but it also demonstrates curiosity and the postponement of judgment. This ability is crucial to the trial-and-error process that characterizes craft practices, and the practices of Elbertse and Sinua are no exception. They both used prototypes and experimented with various solutions to a problem. And finally, opening up a problem depends on the ability to give new meanings to it, to make intuitive leaps while retaining the tacit knowledge of past experience. Elbertse and Sinua opened up problems in different ways, yet they both made intuitive leaps that brought them to use parts and components that were not originally meant to be used in the context of an organ.

In the windchest of the new baroque organ, two solutions to the problem of how to control the minute forces transmitted from the keyboard to the pipe valve coexist. Both Elbertse and Sinua worked in a situation that was, to use the words of John Dewey, "incomplete and hence indeterminate." And yet, as Dewey argues, "the perplexities of the situation suggest certain ways out." Reflecting on the situation required mobilizing past experience and knowledge to project the outcomes of trials and tests into the future functioning of the organ. What they knew from their past experience, the knowledge of their crafts, was of value to Elbertse and Sinua since it afforded their different "dealings with the future," as Dewey puts it—in other words, their imagination (see also chapter 6).[56]

FIRST SOUNDS

In the conclusion of *The Craftsman*, Richard Sennett argues that making does not precede thinking and judging. Time and again, he shows how craftspeople "pause in their work and reflect on what he or she is doing" (2008, 296). This view on craftsmanship, Sennett explains, can be associated with the pragmatism that William James and Dewey developed in response to idealist and pessimistic European philosophers such as Hegel and Nietzsche. Seeing himself as heir of this first wave of pragmatism, Sennett points out how the themes he discussed in his book appear in the work of James and Dewey: The attention for small details in daily practices to deal with large philosophical questions (James) and the "intimate relations between problem solving and problem finding, technique and expression, play and work" (Dewey; 2008, 287). At the heart of craftsmanship lies pride as the reward for commitment and skills that mature over time and evolve through the reflection, curiosity, and imagination that accompany working with resistance or dealing with the ambiguity that can be found at uncertain boundaries (2008, 289).

During my visits to the workshop of Hans Elbertse and in my conversations with Benedikt Aufterbeck and Thomas Stöckl, I could tell that they were proud of what they had made. Understood in the pragmatist vocabulary that Sennett offers, their work focused on solving problems and finding new ones in a variety of situations. Generally speaking, these problems followed from the dual design of the new baroque organ and the mechanisms of transmission that it needed. On a closer look, problems occurred when things had to fit: in the building that houses the Orgelpark, in the case of the new organ, in the tone sliders, or even on the printed circuit board. Each of these "problem spaces" set limits and more or less ambiguous boundaries that had to be dealt with.

The discussions during the making of the new organ show how some of these limits turned out to be more negotiable than others. When finding a place for the blower, the status of the building as a monument and the safety regulations limited the possibilities of placing it out of sight and hearing. In another case, the idea to have musicians around the organ determined the position of the organ on the balcony and required the construction of a hinge in the tone slider. As a concrete location, the building and its architectural characteristics determined to a large extent the conditions under which the organ could be built.

Organs have always been constructed in situations that at least partly determined their structure. Once decisions were taken and the organ and its parts had been drawn, certain aspects that had been flexible became fixed. Making adjustments then came at a price, either in terms of working hours and money, or as ingenuity and improvisation. Within the various types of limits, creating space was part of the builders' strategy. Sometimes this space was extremely limited, as in the case of placing the printed circuits, wires, and connectors in the tone slider. Sometimes it could only be created by taking more time, as in the case of the new design of the printed circuit. Degrees of freedom in the project were not only of an epistemological kind, where certain options were ruled out on the basis of what was known about historical organs and building practices. And they were not only of an esthetic kind, following from the artistic programs that underlie the project. As we have seen in this chapter, degrees of freedom also had to be negotiated in the decision-making process of the core team and in the actual making of the organ. Things had to be made to fit together, and in the work it took to do so, limits were set, boundaries challenged, and new spaces created. Something truly new was made, and this could only be done, as both Sennett and Dewey remind us, by combining work with reflection, curiosity, and imagination.

During my visits to the workshop in the early months of 2017, I saw how the organ grew. After the framework had been erected, the bellows were placed and the wind trunks connected. Work on the windchests and tone sliders continued. Parallel to this, the decorations and embellishments of the organ were made, as well as the panels for the case. Elbertse made most of the wooden flue pipes. When the windchests and tone sliders were finished, the heart of the organ was put in place, first the pedal chests, then the Hoofdwerk chests, and later the Bovenwerk chests. Now the key, pedal, and stop action could be connected. Trackers spanned the distance between the keys and the windchests. They had not been made in advance because their exact length had to be measured in the actual situation. Bert van den Heuvel showed me how he strengthened the head ends of the trackers in the style of Hildebrandt, not with the usual thin copper wire that was wrapped around the wood but with parchment glued on it. The various electronic parts of the organ were connected by gray cables mounted on the wooden frame. In June, Hans Elbertse celebrated the 100th anniversary of his firm. The guests, including Loek Dijkman and Sylvia de Munck, could see the organ in an almost finished state. Some of the embellishments had already been guilded. The instrument looked brand-new.

Building up the organ in the workshop was a way to test it. One of the things tested was the wind system. When put under wind pressure, the amount of leakage in the channels or even in the windchest should be at an acceptable level. One day in February 2017, I visited the workshop when a big wooden pipe had just been connected to the wind system. This pedal pipe produced the first sound of the organ. When Hans Elbertse pressed on the lever that would be connected to the pedal, this movement was transferred from the pedal to the windchest, and wind flowed from the bellows through the wind channels to the tone channel of this particular pipe. A soft, very low rumbling noise filled the workshop. Now that the organ was ready, it was time to put the pipes on their windchests where they would produce sound. How this sound was crafted will become clear in the next chapter.

5

CRAFTING THE SOUND OF HILDEBRANDT'S PIPES

A HISTORICAL SENSATION IN STÖRMTHAL

With its pastel, pale yellow tints, sturdy tower, and gently sloped planes of red roof tiles, the small Church of the Cross in Störmthal, close to the city of Leipzig, Germany, exemplifies the architecture of eighteenth-century Saxony. Its left wall overlooks a well-tended cemetery. The interior of the church consists of two rows of benches, which have rather broad balconies above them. Facing the pulpit on the opposite end of the room is a small organ. Its keyboard and pedal are hidden behind the balustrade. The marbled colors of its facade return in the panels of the church benches and balcony sides.

Even though I know that both the church and the organ have been restored recently, entering through the wooden front door feels like stepping back in time. Adding to this experience is what I know about the church and its organ after having read the biography of Zacharias Hildebrandt written by the German musical historian Ulrich Dähnert. Published in 1962, this is the only book-length description of Hildebrandt's life and the organs that he built. Dähnert recounts how Hildebrandt was born as the son of a cartwright master in 1688, went to school, and worked as a carpenter. After he traveled around as a journeyman, he started to work with the renowned organ builder Gottfried Silbermann in 1713. His years of training in the workshop resulted, in 1721, in a new organ in the St. Nicolai Church of Langhennersdorf, near Silbermann's workshop in Freiberg. Keenly aware of the talents and skills of the young organ builder, Silbermann feared him as a competitor. Tensions between the two men arose, and eventually they both signed a contract that stated that Hildebrandt could accept a commission for a new organ only after Silbermann had rejected it. In 1722 or 1723, Hildebrandt decided to move to Leipzig.

Work on the new organ in Störmthal started soon after. Although its design and disposition were heavily influenced by Silbermann, it contained were new elements, and Dähnert conjectures that these may have been inspired by Johann Sebastian Bach, who had taken up his position as cantor in the Thomas Church in nearby Leipzig in May 1723 (Dähnert 1962, 31). That the two men were acquainted is certain. Bach approved the organ before it was inaugurated on November 2, 1723, during a festive service that celebrated the renovation of the church. Bach had composed a new cantata for the occasion, *Höchst erwünschtes Freudenfest* (BWV 194), performed with his choir from the Thomas Church and his wife, Anna Magdalena Bach, singing the solo soprano part. Bach directed the musicians and most likely played the new organ.

The study trip of the core team to Saxony in early September 2014 brought me to the space where Bach's cantata first sounded. Looking around in the relatively small church, I cannot help wondering if the organ balcony and the two side balconies would have been big enough for the orchestra, the choir, and the soloists. When I walk around the organ, I see an old organ bench. A copper plate on one of its sides tells me that this is the organ's original bench, made by Hildebrandt in 1723. Its simple but graceful design, which seems to embody the needs of the organist sitting at the console, evokes a historical sensation.[1] Between the sides are two bars on which the organist's feet can rest when the pedal is not played. The seat is long so that the organist can move sideways, and right below is a wooden box that probably contained music sheets, or could it have been candles, matches, or pencils? The blueish-gray paint shows signs of use in the many years after Bach first sat on this bench.

The organ in Störmthal is the only instrument made by Hildebrandt that has been preserved almost completely. In the nineteenth century, one of its pedal stops was replaced, and during the First World War, the original front pipes were used by the army. The organ builders of Hermann Eule Orgelbau restored the organ in 1934, and again in 2008.[2] On the left side of the organ is a narrow door. Behind it are two wooden beams to tread the bellows of the organ, the upper sides of which are worn out by the feet of generations of calcants.[3] When we look around inside the case, it is clear that most parts of the key and stop action as well as the wind channels are old. The rough wooden structures filled with written and carved inscriptions point to the carpentry work that shaped them. I realize that this place is full of what Valeriani (2011) has called "travelling facts," facts that are recorded and transmitted in material objects. Drawing on examples from architecture, Valeriani shows how these objects carry knowledge over time and space, far beyond the contexts in which they were originally produced and used.

Skilled organ builders like Hans Elbertse and Munetaka Yokota can read and interpret the material traces and facts in the Störmthal organ. They study them to learn more about the way Hildebrandt and his assistants worked and then use that understanding in their own practice. Organ pipes are of particular interest to today's organ builders (Peters 2019). The traveling facts they embody tell a story about how they were made. What happened to the pipes during their life can often be reconstructed in some detail. Doing so is a way to have an idea of how the pipes sounded when they were built. But contrary to Hildebrandt's organ bench and the historical sensation it evoked, we cannot encounter historical sound. Sound's fleeting character has not changed since the days of Hildebrandt, but the ways in which sound can be known and reproduced have changed dramatically. In fact, our notion of sound itself is very different from in Hildebrandt's time. This means that our understanding of what organs sounded like in the past requires not only questioning the materials and operations of historical builders but also ways of hearing and listening (Kursell and Peters 2020, 116). This, in turn, affects how we can evaluate the quality of the sound of an organ pipe.

In this chapter, I focus on the work that was to be done to recreate or re-enact the sound of Hildebrandt's pipes. I follow Yokota and the organ builders at Hermann Eule Orgelbau in Bautzen, Germany, in their attempts to craft the sound of Hildebrandt's pipes. Having restored many of Hildebrandt's organs, including the ones in Naumburg and Störmthal, Eule's chief restorer, Helmut Werner, has experiential knowledge of both the measurements of Hildebrandt's pipes and how they had been made. For that reason, it was decided that Hermann Eule Orgelbau would make the pipes for the baroque organ. Yokota was to be responsible for determining their final sound character—the voicing of the pipes. After an organ pipe has been constructed, it takes the hands and the ears of a skilled voicer to give it its final voice or sound character through manipulating all the parameters of a pipe that determine its sound character. This was done in two stages: the first in the workshop of Elbertse in Soest in the autumn of 2016, and the second in the Orgelpark in the months before the inauguration of the new instrument. In this chapter, I describe the first phase or prevoicing. The work started by determining the design of the pipes of the various registers of the new organ. Measurements of the Hildebrandt pipes were used as a point of departure but also ways of pipe making in his time. The ideas of Yokota about the importance of following traditional pipe-making practices as closely as possible played an important role in how the pipes for the new organ were actually made.

Whereas the previous chapter analyzed how practical problems and puzzles were found and solved to make things fit, here I focus on how the making of the pipes coevolved with the ability to value their sound. In doing so, I will look at the role of materials such as alloys, wood, leather, and cloth, as well as the use of machines and tools in manipulating these materials. I draw on the work of the anthropologist Tim Ingold and the sociologist of music Antoine Hennion. Ingold has extensively written about the role of materiality and skill in making (Ingold 2013). He thinks of making as "a process of growth" in which the maker is placed in a "world of active materials" (Ingold 2000, 3). His ecological approach is informed by James Gibson's view on perception as an achievement not of a Cartesian mind in a body but of an organism exploring its environment (2000, 3). Hennion, together with Bruno Latour (2013), uses the concept of instauration to express the idea that things in the making are calling us: We feel a responsibility to bring them to "more" existence through our actions. Instauration is an active and reciprocal process, defining at the same moment what we make and what makes us (Hennion 2017a, 78). I will use the concept of instauration in relation not only to the pipes but also to their sound.

To understand how Munetaka Yokota and the people at Hermann Eule Orgelbau were able to build pipes in the style of Zacharias Hildebrandt, I start the chapter by introducing the anatomy of organ pipes in some detail. This anatomy can to a certain extent be measured and visually represented, which allowed the builders to determine the sound character of Hildebrandt's pipes. I then follow the preparations for the actual making of the pipes and focus on the casting of the metal sheets. Here, Yokota's previous experience with constructing the pipes for the North German Organ Research Project (see chapter 2) played an important role. To make a pipe sound good, it has to be voiced. The final part of the chapter describes how Yokota worked with the pipes to instaurate their sound according to Hildebrandt's preferences. Before I go into detail about the work on the pipes of the new baroque organ, it will be helpful to explain their anatomy in some detail.

HOW ORGAN PIPES SOUND

Organ pipes sound because a column of air inside the pipe is set into vibration.[4] There are two types of organ pipes, flue pipes and reed pipes, each generating the vibration in their own way. The air column inside flue pipes vibrates because the wind blown into the pipe periodically exits the pipe mouth, causing the air column to take on a standing wave, which we perceive as a tone. Reed pipes produce their tone in a way that resembles the clarinet and the oboe, where a reed is made to vibrate. Organ reeds are made of a strip of brass mounted on a shallot inside a boot that is connected to a resonator.

The length of a flue pipe determines its pitch; a long pipe has a longer air column, a longer wave with a slower vibration, and, hence, a lower tone. The length of the reed determines the pitch of reed pipes. Flue pipes can be open, partially closed, or closed, where a closed flue pipe will sound an octave lower than an open pipe of the same length.[5] The sound character or timbre of a pipe is determined by the relative strengths of the harmonics in the sound. These variations depend on the pipe's diameter in relation to its length, the shape of a pipe, and the construction of its mouth or, in the case of reed pipes, the form and material of the reed. The material of the pipe can also play a role, as a smooth inside keeps the higher harmonics alive, whereas a rougher inside, like in wooden pipes, quenches them, making the sound less bright. Given the fact that on an average organ four out of five stops are flue pipes, I will focus on this type in the remainder of the chapter.

All flue or labial pipes share the same principle of sound production. A pipe has a body, a languid, and a foot. The air enters the pipe through the toe hole in the foot, which is placed on the toeboard of the windchest (see figure 5.1).

The pipe's mouth is positioned in between its upper and lower lips. The height of the mouth is called the cut-up. The foot of the pipe is covered by the disc-shaped languid, which leaves open a narrow slit in the pipe's mouth, called the windway or flueway. The edge of the languid has a bevel angle determined by the pipe builder, called the languid phase (see figure 5.2).

When a pipe is blown, a sheet of air formed between the languid and the lower lip is directed toward the upper lip. As the sheet of air is led inside the pipe, the pressure inside the mouth increases. As a result, the pressure outside of the mouth drops, which then pulls the sheet of air in that direction, causing an underpressure on the inside. The vibration that this back-and-forth movement causes excites the column of air in the pipe, creating a standing wave, which can be heard as a tone (see figure 5.3).

The pitch of the tone depends on the wavelength inside the pipe. Increasing the wind pressure at the toe hole will increase the pitch; soon, however, this will cause the pipe to overblow, which means that the second harmonic

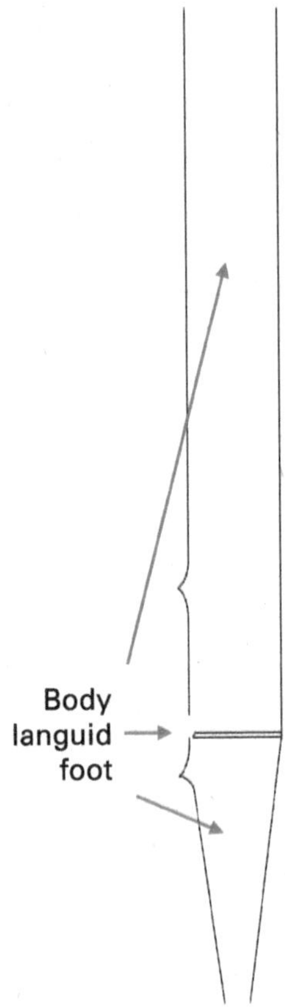

Figure 5.1
Schematic representation of the position of the pipe body, languid, and foot (drawing made by Hans Fidom, reproduced with permission).

becomes the new fundamental, which, in turn, means that it will sound an octave higher.[6] Adjusting the various parameters of the toe hole, the pipe mouth, and pipe body enable a pipe maker and a voicer to give the pipe its timbre or sound color, and to determine the way it starts "speaking," as organ builders say, the way it stops doing so, and to adjust its loudness.

Flue pipes come in three families: the principal pipes with normal body width; the flutes, which are wider; and the strings, which are narrower. Pipes in these families share certain general sound characteristics but will still sound different from other members. Until the end of the fifteenth century, organs contained only principal pipes. These are made in different

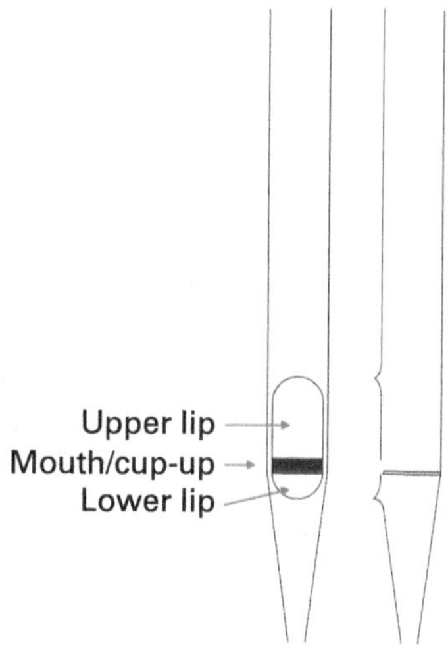

Figure 5.2
Schematic representation of the position of the upper lip, the mouth and its cut-up, and the lower lip (drawing made by Hans Fidom, reproduced with permission).

pitches.[7] Principal pipes are open pipes of medium scale, an organ builder's term that refers both to the relation between the length and the diameter of a pipe and to the point in a series of pipes where the diameter halves.[8] The principal family is the main one in organs. In most organs, a Principal 8' is the basis, 8' indicating the length of the longest pipe, which makes this stop sound on the pitch one would expect also on a harpsichord or piano keyboard. The higher stops include pipes that sound an octave higher (4'), a duodecime (3'), a superoctave (2'), and so on. The family is usually crowned with Mixtures, stops that sound a combination of even smaller pipes, adding high to very high frequencies.

Flute pipes have wider scales then principal pipes, are less bright in tone, and bring out the fundamental tone rather than its harmonics. Whereas all principal pipes have the same form and are usually made of metal, flutes are made in various forms, and they can be made of metal or wood. Flute pipes may be open, stopped (i.e., covered at the top), or partially stopped; these options are tools to reduce the number of harmonics in the pipe's sound. In the latter case, the pipe is covered but has a small pipe on or in the cap.

The third family of pipes are the strings. They have narrower scales than principal pipes, which accentuates the higher harmonics. Because of these

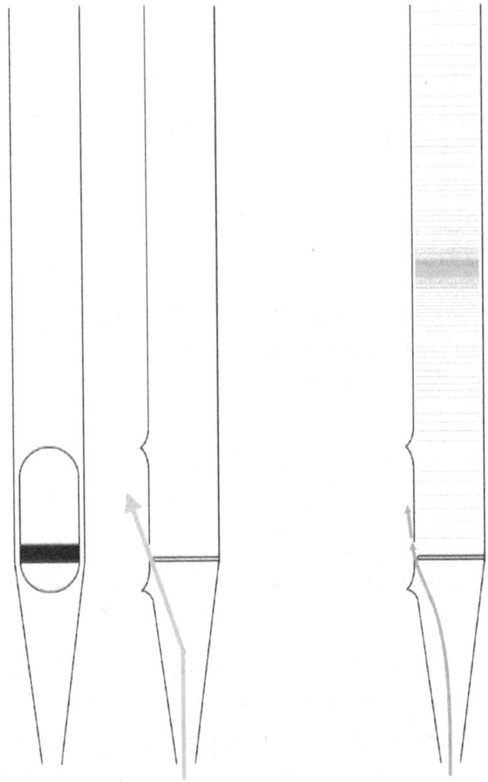

Figure 5.3
Schematic representation of a pipe that is blown (drawing made by Hans Fidom, reproduced with permission).

characteristics, string pipes resemble the strings in an orchestra, hence their name. They can be open or stopped. In baroque organs, they form a minority; it was only after 1850 that they became a major factor in organ sound. Combined with the reed pipes, which can also have a wide variety of shapes and scales, the different families of flue pipes enable an organist to create many different sounds by using the various colors in combination.

NORMALIZING PIPE SCALES

Organ pipes are made according to fixed proportions and ratios called scalings. Since the frequency of the vibrations inside a pipe is related to the length of the air column, the lowest-sounding pipe in a big organ resembles a modest tree standing thirty-two feet tall, its foot not even counted, and the highest is 1/32 of a foot, which is aproximately one centimeter. The familiar ascending and descending rows of pipes in an organ facade enable us to hear all these pipes with the same timbre because their diameter changes in relation to their height. To achieve this equality in tone color, organ builders use scaling diagrams where the diameters of octave pipes increase by a ratio as they descend into the bass. In contemporary organ building, this ratio is fixed as the fourth root of 8 and called the normal scale or "Normalmensur," developed by J. G. Töpfer in the nineteenth century (McNeil 2012, 11).[9] This normal scale not only provided a standard scaling method, it also expresses the ratio needed to give the pipes equal timbre.

In the normal or standard scales, as they were developed in the nineteenth century, the length and diameter of a pipe were related to each other according to fixed proportions. The mouth width and height, the latter also called the cut-up, and depth of the windway would also be fixed.[10] The goal of this system was to evenly balance the sound character in every stop and in the organ as a whole. For any stop, these parameters and characteristics can be plotted on a sheet of paper, called a scaling diagram. In the diagram, the vertical lines represent the pipes in a stop, starting from the lowest (see figure 5.4 for an example). A pipe's parameter is represented by a line that crosses the vertical lines. For any pipe, the distance between the intersection of the two lines and the bottom line gives the exact value in centimeters or millimeters for that parameter.

Scaling diagrams combine all the relevant information about an organ pipe, and an expert like Yokota can have an idea of the sound of a pipe based on this information. For his work on the new baroque organ, Yokota used a scaling diagram with Pythagorean ratios of 2:1 instead of Töpfer's "Normalmensur" because, according to him, this type of geometrical configuration was used during the time that Hildebrandt built his organs. Pythagorean scaling leads to subtle differences in the tone color and perceived loudness of pipes within a stop. This results in a liveliness of the sound that modern organs often lack, Yokota argues, but that was common in the days of Bach and Hildebrandt.

SOUND THAT HAS A SENSE OF LIFE

How did organ builders work in the past, and why would it be relevant to know? Finding answers to this question summarizes much of the work that Munetaka Yokota did over the past forty-five years. Yokota developed his idea to work according to eighteenth-century principles when he studied to

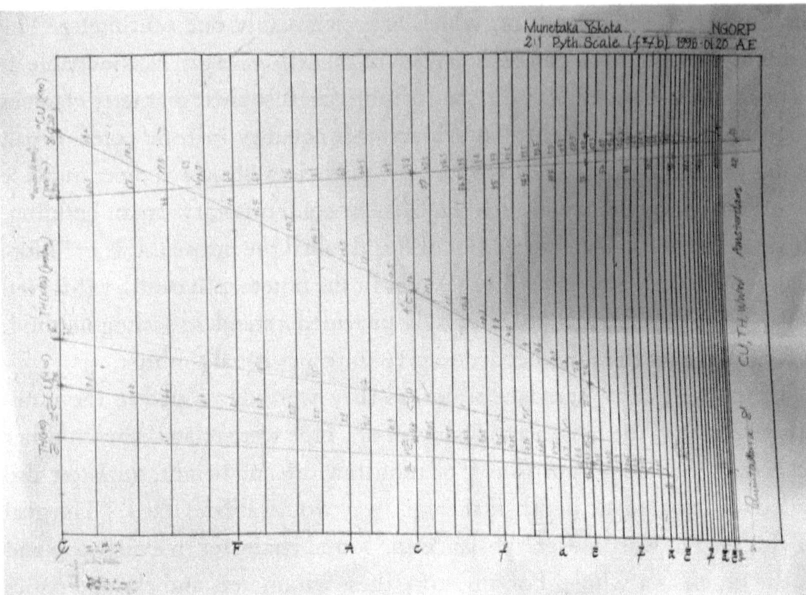

Figure 5.4
Scaling diagram made by Munetaka Yokota for the Quintadena 8' stop of the new baroque organ, showing the exact measures of the toe hole, cut-up, and windway width from the intersection to the top or bottom line of the diagram. Yokota used a template with the 2:1 Pythagorean scale that he also used during the North German Baroque Organ project in the 1990s (see chapter 2; photo taken by the author).

become an organ builder. The sound of historical organs had fascinated him ever since he was a fourteen-year-old schoolboy. He bought an LP of a recording made by the German organist Fritz Heitmann in 1938 on the Schnitger organ in the chapel of the Charlottenburg Castle in Berlin. Because the organ was destroyed in the Second World War, the recording is the only way we can still know how this instrument sounded. In addition to this document, there are photographs and measurements of the pipes. Having listened many times to this and other recordings of historical European organs, Yokota concluded that these "old organs sounded better than new ones" (Yearsley 2010, 5).

After studying economics at Tokyo's Gakushuin University, Yokota became an assistant to Japanese organ builder Hiroshi Tsuji who was inspired by European historical organs (Yearsley 2010, 5). Another strong influence in the early 1970s were his encounters with Harald Vogel, who organized the German Organ Academy and was one of the leading voices in historically informed organ building and performance practice (see chapter 2). In 1978, Yokota moved to the United States where he started working with John Brombaugh, who took historic European organs as a source of inspiration. In 1984, California State University, Chico, invited him to become an "artist in residence." The university celebrated its centennial anniversary, and the university's organ professor had asked Yokota to build an organ.

In the six years that followed, Yokota recruited volunteers from the university's students and from the Chico community. By himself, he taught the knowledge and the skills to build a big organ in the eighteenth-century style of Silbermann and Hildebrandt. This approach, he explained, mirrored the practice at the time in Thüringen and Saxony. Organ builders would come to a village or a town to work on site with the local craftspeople and use local materials: "It may have been a practical necessity, because transportation was expensive and road conditions not very good. As a result, each organ has its very distinctive character even though the designer and builder may have been the same person."[11] His organ building in Chico enabled Yokota to develop his ideas about the differences between historical and contemporary organ building practices.

Yokota tried to understand how eighteenth-century craftsmen worked with what was at hand, not only the materials they used but also for how they processed them. Not having machines to make a thick piece of wood thinner and throw away a lot of the material, they would look at the required strength and the qualities of the wood that was available. To make an irregular wooden part fit the other parts in the organ required extensive technical and intellectual knowledge, Yokota says. To explain the differences with later organ-building practices, he draws a parallel with performance practice in baroque music.

> This older music is a kind of outline. As a performer, you have to do lots of things that are not written on paper. Today's composers, depending on the style, write down their music in more detail. This also happens in construction work today, or in car making or architecture. Architects make very precise drawings, and pretty much everything is determined in design rooms. In order to do that, you have to have calculable and predictable materials. But the old method was not like that. Design had to be some kind of rough outline. The reason for that was not that designers were lazy, but they knew that the actual material that they had to work with could be different. So they could not plan things into small details. That was left to the craftsman who actually worked with it. They could do a lot

of judgment by just tapping on the material [he taps on the table with his fingers] and listen to it. In this way, they could choose whatever was best for the job based on qualities that cannot be written down in a design sheet. So the technical and artistic judgment was made on site, in the workshop, and this required a lot of empirical knowledge from the craftsman. Today, craftspeople are very good at making things very precise, and to precisely follow the design. In that sense the old way to design an organ resembles old written music. It is just the outline. The craftsman is expected to act like a good performer. If you play one note a bit too long, then in order to make sense, you adjust the length of the following notes within the bar.[12]

Using local materials thus introduced a contingency that required adaptation to the circumstances, Yokota argued. Understanding how historical organs were built means learning about these circumstances.

Yokota was able to further develop his ideas and experience when he was invited to make the pipes for the North German Organ Research Project in Göteborg, Sweden. The goal of this project was to have a north German baroque organ that sounded as good as the organs that were built by Arp Schnitger (see chapter 2). In his contribution to the extensive documentation of the GOArt project (Speerstra 2003b), Yokota describes the sound of historical pipes:

These old pipes have a beautiful balance between the fundamental pitch and overtones, a balance between "musical" sound and "noise," as well as a good sense of balance between the strength, length, and character of the speech and the sustaining tone. . . . Was the old sound partly a product of the aging of the materials, or could we reach this level of quality again in a modern instrument? Essentially I define "good sound" as sound that has a sense of life. The listener senses the life from the sound of each pipe, each stop, and several stops together interacting symbiotically just like in human society. (Yokota and Ruiter Feenstra 2003, 165)

To give the pipes a good, lively sound, the researchers and builders, including Yokota, had to learn the "pattern language" of Schnitger. They took this term from Christopher Alexander's books *A Pattern Language: Towns, Buildings, Construction* (1977) and *The Timeless Way of Building* (1979). These were an important early philosophical reference for the project, as Joel Speerstra writes in his introduction to the documentation volume:

So, using the most coherent collection of pipework to survive from any Schnitger organ, we tried to learn about the craft processes that produced the original

> object, in order to *perform* them well enough to build a new object in the same language as the original. . . . A living pattern language is one that is shared by a community. Everybody is empowered to make a new object in this language because everybody already knows the words in it. . . . Alexander suggests that historical building traditions took the same improvisational approach to shared community patterns. . . . Like Alexander's farmers, Schnitger and his fellow organ builders knew what an organ looked like and sounded like, because they knew the patterns of the language that made up all of the things that it would have to do. . . . The historical instruments themselves have to be read just like any other primary source, to coax out and define what the pattern language really was for the Arp Schnitger school of organ building. Then the ways of working had to be understood, copied, tried out, and finally performed, in a continuous dialectic interplay between theory and practice. (Speerstra 2003a, 17–19)

Understanding the pattern language of historical organ builders became a guiding principle in the GOArt project and in the later work of Yokota. In their years of research, the project members tried to do this not only by reading written sources but also by doing extensive experiments with reconstructing historical making processes. This process reconstruction does not aim at imitating but at learning. Imitating, Yokota argues, would be senseless precisely because one would only copy site-specific solutions that were developed under unique circumstances with local materials and tools. Learning why historical builders developed certain solutions is thus key to making new pipes that sound as good as the old. For the New Baroque Organ project, it was, therefore, important to know more about Hildebrandt's pipes and how he made them.

DISCUSSING HILDEBRANDT'S MATERIALS AND SCALINGS

Hermann Eule Orgelbau in Bautzen has built organs for almost one and a half centuries.[13] Nowadays, the firm combines the building of new organs with the restoration of historical instruments. After the first visit of the core team to the workshop in Bautzen in September 2014, the decision was made that the company would make the pipes for the new baroque organ and that Munetaka Yokota would be responsible for the voicing. This approach required deliberation on the characteristics of the pipes and how to make them.

In June 2015, Hans Fidom and I traveled to Bautzen and picked up Munetaka Yokota, who arrived from Japan, at Leipzig airport. The next day, we were welcomed by master organ builders Helmut Werner and Konrad Dänhardt in the office of managing director Dirk Eule. On the agenda was a discussion of the proposals that Werner and Dänhardt had made for the measurements, scalings, and construction method of the pipes. The proposals were based on Werner's long years of research on original Hildebrandt pipes, of which those in the organs of Störmthal and Naumburg were particularly relevant because these were used as a sound reference. In addition, the measurements listed in the book published in 1962 by Ulrich Dähnert on Hildebrandt and his organs was an additional source.

The discussion starts with the general building principles of Hildebrandt's pipes and the materials he used to make them. Yokota raises the issue of how the Hildebrandt scalings, many from the Naumburg organ, will work in the Orgelpark. The loudness of an organ is determined not only by the wind pressure but also by its scalings. Following Naumburg, which is a much bigger space than the Orgelpark's hall, might have the effect that the new organ would become too loud. On the other hand, making adjustments in the original scaling might disrupt the overall sound character of the organ. Everyone agrees that the original scalings should be followed. An important argument for this, Helmut Werner claims, is that Hildebrandt himself often used the same scalings in different churches:

> The big secret of the scalings does not exist. Hildebrandt simply had his own scalings. We found that the Quintadena in Störmthal, which he built at the start of his career, and in Naumburg are identical. The spaces are not, but the scalings work both in Störmthal and in the huge church in Naumburg.[14]

According to Werner, there was no reason to be afraid that the scalings would not work in the Orgelpark.

Depending on the stop, the flue pipes are made of an alloy of tin and lead in different proportions: The principal and string pipes have 83 percent tin and the flutes between 34 and 29 percent. The metal sheets from which the pipes were made were cast on a casting bench covered with linen. This has been concluded from cloth marks on the inside of some of the pipes. It has been established that the wall thickness of the Hildebrandt pipes decreases from the bottom to the top. The shape of the upper and lower lips on the front pipes varies; the larger pipes have lips soldered onto the pipe; for the smaller ones, Hildebrandt used a pointed shape. For the languids, Werner found that they were made from lead and that their bevel angle varied between forty and seventy degrees. After close inspection of the languid bevels, Werner observed that Hildebrandt sometimes made small cuts into the metal. These nicks were made to reduce unwanted noise or speech in the pipe sound.[15] Werner also found very tiny nicks in the languids of Silbermann, Hildebrandt's teacher:

> These languid bevels are sometimes so corroded or worn out because of the wind blowing along them that one cannot see the nicks at all. You look at the languid: no nicks. But if you clean them after two hundred years, then they emerge again: very fine, very regularly set thin nicks.[16]

It is agreed that Yokota, as the voicer of the new pipes, will decide if and how to make nicks in the new pipes' languid bevels.

Now and then the discussion would address how the new pipes should be made. When they discussed the measurements of Hildebrandt's cut-ups, the distance between the lower and upper lip of a pipe (see figure 5.2), Yokota suggested that these should not be made in a modern way. At Eule, the pipe makers make the cut-up only after the body has been soldered to the pipe foot. This prevents solder from leaking into the corners of the pipe mouth, which influences the pipe sound negatively. Yokota suggests that the pipe makers at Hermann Eule Orgelbau experiment with a procedure where the cut-up is made in the sheet before the body is soldered onto the foot. This also enables the pipe maker to adjust the position of the upper lip and the lower lip more precisely.

The scaling of the pipes as Werner measured them indicates that Hildebrand used scaling diagrams with divisions based on meantone temperament, Yokota observed.[17] He is surprised and claims that organ builders such as Schnitger often used diagrams with a Pythagorean division, and he also found this division in the important eighteenth-century source on organ building, *L'art du facteur d'orgues* by Dom Bedos de Celles.[18] It is agreed that Yokota will work with the octave divisions as they have been measured by Eule. The manufacture of the wooden pipes, the way the languids of these pipes are to be made, and the historical composition of the paint that is used to cover the inside of the wooden pipes are discussed. Details matter here. Some ingredients present risks, such as lead oxide; others might cause mold inside the pipes in humid environments.

After the general characteristics of Hildebrandt's pipes have been discussed, Werner gives his measurements of every individual stop. Sometimes the measured values of parameters such as mouth width (see figure 5.2) differ from what could be expected based on the scaling system that was used. In these cases it is decided to follow Hildebrandt's practice, even though explanations for the anomaly cannot be found. In rare cases, following the Hildebrandt scalings might lead to problems, as in the case of the Principal 4' of the Oberwerk of the Naumburg organ. This stop sounds rather loud and brilliant, which is not a problem in the large space of the St. Wenzel Church, with its high ceiling, but might not work in the Orgelpark where the Oberwerk is closer to the concrete and concave ceiling. It is decided that, should this be the case, adjustments will be made later, such as building a "roof" above the organ to slightly mute the Oberwerk stops.

The sound of the pipes of the new baroque organ is modeled after the remaining original pipes of Hildebrandt. Werner's work on measuring these pipes, and studying their scales and the shapes of their components and the materials that were used, will help the builders to bring their sound to the Orgelpark. After a long day of minutely reviewing what is known about the Hildebrandt pipes, a plan is made. Hermann Eule Orgelbau will make test pipes for a selection of the stops, three to five pipes per stop. These pipes will be made according to the specifications that Yokota will give. In November, the Orgelpark team will revisit the workshop of Hermann Eule Orgelbau to discuss the test pipes and their making. In January 2016, Yokota will come to Amsterdam to voice the test pipes in the Orgelpark. This listening test will provide further directions for the fabrication of the pipes.

CASTING THE PIPE METAL

Making a metal flue pipe starts with the casting of metal sheets on a casting bench.[19] The metal is usually an alloy of tin and lead, although historical pipes also show traces of other elements such as iron and copper.[20] After the sheet has been cast, its surfaces are scraped to make them flat. Front pipes are often polished as well.[21] The dimensions of the pipe bodies and feet for a particular stop are determined by the scaling diagram. Subsequently, pipe bodies and feet are laid out on the metal sheets with dividers and a straight edge and cut out with a sharp knife. The pipe body and the pipe foot are rounded by hammering them around a cylindrical or conical mandrel with a piece of wood. Soldering the various parts of the pipe requires great care and is considered to be the first step in the process of voicing the pipe (Blanton 2003).[22]

One of the processes that was reconstructed at GOArt in the 1990s was the casting of the metal for the pipes. In general, sheets are cast by pouring melted lead–tin alloy into a wooden casting box, which has a narrow opening at the bottom, and sliding this box over a casting bench. Thus the melt will run out and spread into a thin layer, covering the bench. The melt cools down and solidifies into a metal sheet. Important parameters in the historical casting process are the speed with which the metal cools down and the thickness of the sheet. Munetaka Yokota, responsible for the pipes in this project as well, used what he calls a tap-testing method to analyze the material from which a historical pipe was made and to know its scaling and wall thickness. Tapping on the metal with his fingernail gave him two tones: The metal tone produced by the pipe body and the air tone produced by the air column inside the pipe. A higher metal tone indicates a harder metal. It turned out that even though the chemical composition of a modern alloy would be almost similar to its historical counterpart, it would be much softer than old pipes. This led Yokota to conclude that it was necessary to know more about how the metal was cast and what the relationship was among the materials, tools, and construction method.

Historical documents such as organ contracts, organ building treatises, and secondary sources were studied to gain information about the composition of historical pipe metal.[23] Metal samples of remaining organ pipes were analyzed by using spectrometrical methods. Using this knowledge, the alloys of the pipe metal for the new organ were established (Yokota 2003, 165–167). Once the alloys for the pipes had been determined, the next step was to reconstruct a seventeenth-century practice of casting the metal sheets. Relearning the skills of Arp Schnitger and the people in his workshop proved to be arduous. Hundreds of casting experiments were carried out before sheets could be cast that resembled the metal from Schnitger's pipe in great detail.[24]

In modern organ-building traditions, the casting bench is made of stone or wood and is covered with a fire-retardant cloth. In historical traditions, however, the casting bench was covered with a layer of sand or with linen cloth on stone. An important question was whether the material properties of the pipe metal were dependent on using sand or cloth when casting. Yokota and his team established that casting on sand gives the metal a completely different quality than when cast in more recent ways. Due to the sand bed beneath the molten metal, the metal cools quickly, causing the pipe metal to become harder than present-day pipe metal. The type of sand, the proportions of impurities or trace materials in the metal, and the casting temperature all proved to be of vital importance for the end result. During the casting experiments that were carried out between 1995 and 1997, practical questions arose: Did Schnitger cast on sand or cloth? If he cast on sand, what thickness of sand did he use? What type of sand did he use? Did he mix any oil or water into the sand, and, if so, what type of liquid and how much did he use? Finding the answers required patience, Yokota writes:

> Many times, it is tempting to take modern shortcuts or to find a series of immediate solutions to specific problems; such solutions, however, often overlap and create a domino effect of new problems. Seventeenth-century builders must themselves have sought immediate solutions to problems, but they had many more technical limitations. Therefore they had to solve these problems in terms determined by their practical conditions. In addition, wood and metal have more complicated properties than modern materials such as plastic and steel. Our goal remained that of discovering the so-called simple techniques of the earlier organ builders, where each aspect of the process positively affected the others without creating unnecessary complications. (Yokota 2003, 168)

It turned out that a thick sand bed mixed with olive oil to make it flexible worked best to control the cooling speed. The GOArt team used a horizontal bench that enabled them to control the thickness of the metal sheets. This was important because historical pipes often are thinner at the top than at the bottom. This not only makes them more stable, it also improves the resonance of the pipe sound.

OLD-STYLE LINEN

In November 2015, I accompany Munetaka Yokota, Hans Fidom, and Johan Luijmes when they visited Hermann Eule Orgelbau to inspect and discuss the test pipes that had been made. After a brief discussion in the office of Dirk Eule, we walk to the main workshop building. On the first floor are the rooms where the firm makes and voices its flue and reed pipes. All the test flue pipes are ready and have been put on display, including the Rohrflöte, a partially stopped flute register that has strong fundamental tones. Helmut Werner decided it would also be good to make these.

Yokota and Werner enter into a dialogue about the pipes. Werner explains how they have been made, and Yokota asks questions. His main concern is whether the pipes have been made in the style of Hildebrandt in all their details. An example is the way the lower lip of one of the pipes is burnished. He argues that in the Hildebrandt style, the surface should be perfectly flat instead of very slightly rounded. Werner shows black-and-white pictures of Hildebrandt pipes from the Naumburg organ, made by the German organ builder Gerald Woehl in 1992, during an extensive study preceding the restoration of the instrument. Whereas the pictures seem to support Yokota's argument, Werner explains that it is very difficult to make lower lips that are completely flat. To do so, one would have to scribe the metal so deep that it might break when the lip is formed. Also, when transforming the round surface to a flat plane, the surplus material has to go somewhere.

The test reed pipes are not all ready yet, which gives Yokota the chance to see how they are made. He is happy with the construction of the wooden resonators of the Posaune, an 8' pedal stop. No nails have been used to put the sides together. Werner explains that the direction of tree rings determines the stability of the resonator and its sound. Yokota asks what the finish will be on the inside of the resonator. He says that during his extensive research on the Casparini organ for the Eastman School of Music in Rochester, he determined that the inside of the pipe was painted to make them airtight: "Casparini did it before gluing. I can share our experience how to do that. It is easier, I think. Eugen Casparini explained that to the organist in Boxberg. We have the recipe."[25] Werner: "But it works this way as well."[26] Hans Fidom, who has the role of interpreter between Yokota and Werner, remarks that "it would be nice" if the resonators could be made according to Yokota's understandings of the historical production style. The exchange of insights between Yokota and Werner about how to make the organ pipes takes place with the utmost respect for each other's expertise. Both have extensive knowledge of historical traditions, which they use to back up their arguments. A difference between the two is that Werner and the employees at Hermann Eule Orgelbau adhere to present-day ways of making, even when they closely follow the measurements and scalings of the Hildebrandt pipes. Yokota, on the other hand, brings in his long experience with experiments in relearning historical pipe making skills.

After the making of the test pipes have been discussed, we enter the room where the pipe metal is cast.[27] On the floor are piles of unprocessed metal sheets in different alloys. Next to the window is the casting bench. To the right of the bench is a small room with the melting pot and a pile of tin bars. Above the melting pot is a rail that holds a pan in which the melted alloy is transported from the pot to the bench. At the other end of the casting bench is a scraping machine. A rough metal sheet can be placed on a cylinder. By moving a chisel along the metal surface while the cylinder spins, the surface of the sheet can be scraped to the desired thickness (see figure 5.5).

Figure 5.5
Scraping machine at the Hermann Eule Orgelbau workshop in Bautzen (photo taken by the author).

The casting bench is covered by a Nomex fire-retardant cloth with a layer of flannel beneath it. "That should not be there," Yokota says. "I said: just linen, old-style linen. One layer, that's it, nothing more. On stone directly. I really want them to try this out and if they have a problem, I can help them with the technical details."[28]

He realizes that he asks the pipe makers to change their way of working, but he thinks it is crucial to the quality of the metal sheets to be cast according to the Hildebrandt tradition. What linen did he use? This can be decided from cloth traces on the inside of original pipes. Werner confirms that some of the pipes from the Naumburg organ show that Hildebrandt did not cast on sand but on linen.

To know the right linen, Yokota says that he has to see it. It turns out that there is a linen-weaving mill in the nearby village of Oberlausitz.[29] An appointment is made, and after lunch the group drives to the firm. The owner shows us several linen samples. Werner points out that if the linen is woven too coarsely, the cooling metal might stick into it, making it difficult to remove the sheet from the bench. After some consultation, two types of linen are chosen, one with a finer weave than the other. On the way back, Yokota explains the principle of linen on stone. The stone ensures that the metal cools down quickly, which gives it a hard surface. The linen is necessary to make room for small gas bubbles that would otherwise cause a bad surface structure. After the fire-retardant cloth has been removed and the granite slab has been put on the casting bench, the rougher of the newly bought linens is fixed on top of it. The next morning at 7 a.m., the casting of the test sheets will begin.

CASTING AND SCRAPING

It is still dark when we walk from the hotel to the workshop of Hermann Eule Orgelbau. In the casting room, the melting pot has been heating the organ metal throughout the night. Two of Eule's employees, Andreas Baum and Ingolf Huth, are wearing their protective overalls and gloves. Baum is stirring in a big black pan. Inside I see the swirling surface of the molten metal. Its movements and appearance remind me of mercury. The new linen has been tightly strung over the stone beneath it. It is fixed with two wooden slats that also serve as rails to guide the casting box. The casting box that will be used this morning has two compartments. The metal will be poured in the first, and when the casting begins, a valve will be opened so the metal can flow into the second compartment. From there it flows onto the bench when the casting box is slid from one end of the bench to the other. Attached to the lower end of the bench is a rectangular bin to recollect the remaining metal.

Baum keeps on stirring the pan, while Huth measures the temperature of the alloy in short intervals with an electronic thermometer. The atmosphere in the room is one of concentration and expectation. Suddenly the two men tilt the pan toward the casting box and pour the metal in it. The sound of flowing fluid fills the room. The empty pan swings back. They quickly grab the casting box. Their bodies slightly bent, they walk a few steps along the casting bench, steadying the movement of the box, which makes a whizzing sound over the linen (see figure 5.6). Behind the box, I see the shiny surface of the metal. At the end of the table, the remaining metal flows into the collecting bin and spats of it land on the floor where they immediately solidify. Huth and Baum raise their protective head shields and look at the metal. Where it cools down its surface becomes dull. They start discussing the quality of the metal sheet with Yokota, Werner, and Thomas Habelock, head of Hermann Eule Orgelbau's metal workshop. According to Werner, the metal has cooled down too slowly because it was too hot, and Yokota agrees. He suggests to cast at a lower temperature. Huth and Baum separate the metal from the linen by sliding a wooden slat under the sheet. The surface of the bottom side shows clear cloth marks.

Several sheets are cast before the right temperature is established. I ask Yokota how eighteenth-century craftsmen were able to measure the temperature of the molten metal.

> They did not know the absolute temperature, but they did know the relative temperature. If the procedure is always the same, they knew how long they had to wait for the right temperature. At the same time, the type of alloy allows us to judge the temperature by its stickiness. The alloy first has a completely liquid form, but when the temperature goes down, there is a kind of pasty, sticky form, and only then it becomes solid. So when they are stirring the pot, they can pretty much judge the temperature. When it has this pasty form, the temperature stays there for some time. At this point of metamorphosis, they should stop waiting and just pour and go.[30]

Controlling the temperature of the cast metal is a complex affair, all the more so because the stone gets warm after several sheets have been cast. This has to be compensated to get the right cooling coefficient, which is necessary for a good structure of the metal.

Figure 5.6
Andreas Baum (left) and Ingolf Huth casting pipe metal on linen at the Hermann Eule Orgelbau workshop in Bautzen, November 2015 (photo taken by the author).

When the thickness of the newly cast metal sheets is measured by Thomas Habelock, it turns out to be more or less equal. Yokota would rather see that the sheets were tapered: thicker at the start of the sheet, and thinner at the end. From his previous experiments he knows that this makes it easier to follow the style of historical pipe makers. How to produce this tapered shape? With the two-chamber casting box this will not be possible, Yokota argues. He points at the scraping machine at the foot end of the casting bench (see figure 5.5):

> That kind of scraping machine was invented in the nineteenth century. It can produce a constant quality. The drum spins, and the operator moves the cutter sideways over the metal to make it thinner. A machine like this requires a metal plate of constant thickness from the beginning to the end. So the double box casting procedure is connected to that machine.[31]

To cast a tapered sheet, however, a single box is necessary. This type of box was used in eighteenth-century casting operations, as Yokota knows from historical organ-building treatises. Nowadays, this casting box is sometimes used to cast pure lead. It turns out that Hermann Eule Orgelbau has one for that purpose (see figure 5.7).

After morning coffee, the second, finer woven linen is put on the table and the casting bench is prepared for casting with the single-chambered box. When operated in the right way, it will produce a tapered sheet. Because of the weight of the metal in the box, more will flow out at the start than at the end of the sliding movement when the box is nearly empty, resulting in a sheet that is thicker at the start and thinner at the end. The metal can thus be cast much closer to the final thickness. Yokota explains why this is important. In the eighteenth century, scraping had to be done by hand, which was time and labor intensive. Hence, it was rational to cast metal in a way that required less scraping: thin and tapered. With the coming of the scraping machine, sheets with one thickness became the norm. This led to two problems, Yokota explains:

> This kind of casting [on linen and stone] has the best part not right at the upside or downside of the sheet, but close to that. It's very good quality. If you cast very thick, the inside part cools down very slowly, so it is a weak part. So if worst comes to worst, if you cast very thick, which itself slows down the cooling, then the inside of the sheet is not so good. But if you then use the scraping machine and scrape off both sides of the sheet, you end up with the weakest part of the sheet, and maybe the poorest part in terms of quality.[32]

Figure 5.7
Andreas Baum at the Hermann Eule Orgelbau workshop preparing the single-chamber casting box, November 2015 (photo taken by the author).

Operating the single casting box turns out to be a challenge for the employees of Eule. They have to adjust their routine ways of working, but after several attempts they succeed in casting a sheet that meets the parameters set by Yokota.

During the two-day visit, many aspects of pipe making were discussed. Even when closely following the measurements and scalings of Hildebrandt, actually making the pipes in his style involved a tension between present-day habits of working and the historical skills that Yokota reconstructed and relearned during his own projects. Experimenting takes time, and for a commercial firm, it is a balancing act to weigh the value of historical skills against the economic benefits of machine-supported making practices. As one of the pipe makers said to Yokota after he had suggested a different way to solder the languid to the foot: "If I have time, I can make anything."[33] In the conversations during the two-day visit, Hans Fidom had the role of the interpreter, not only between German and English but also between the present-day organ-building practice at Hermann Eule Orgelbau and Yokota's historical experiments. Conscious of what it meant to ask the experienced workers at Hermann Eule Orgelbau to change their habits, he made notes of what was agreed upon. He later sent these to both Hermann Eule Orgelbau and Yokota to make sure that no misunderstandings would arise later on in the project.

HOW MACHINES AND MATERIALS MATTER

When I visited the workshop of Hermann Eule Orgelbau, I was struck by the immense number of tools and materials. In fact, an organ builder's workshop is basically a place where materials such as wood, metal, leather, and paint are stored and manipulated. The lead and tin bars for the new baroque organ were stored in the cellar; wood for the wooden pipes was stored outside under a roof, where it could dry. Work benches were covered with tools such as rulers, scribes, hammers, and many others whose function I did not know. There were machines, such as the scraping machine, but also a very large cast-iron machine used to cut large pieces of metal. On the walls were pictures from historical sources on organ building, such as Dom Bedos's eighteenth-century treatise *L'art du facteur d'orgues*, which can be found in virtually every organ builders workshop today (see figure 5.8).[34] In these workshops, artisans have worked for years to acquire the experiential knowledge to use tools and to shape materials. On the first day of our visit, a trainee showed me how

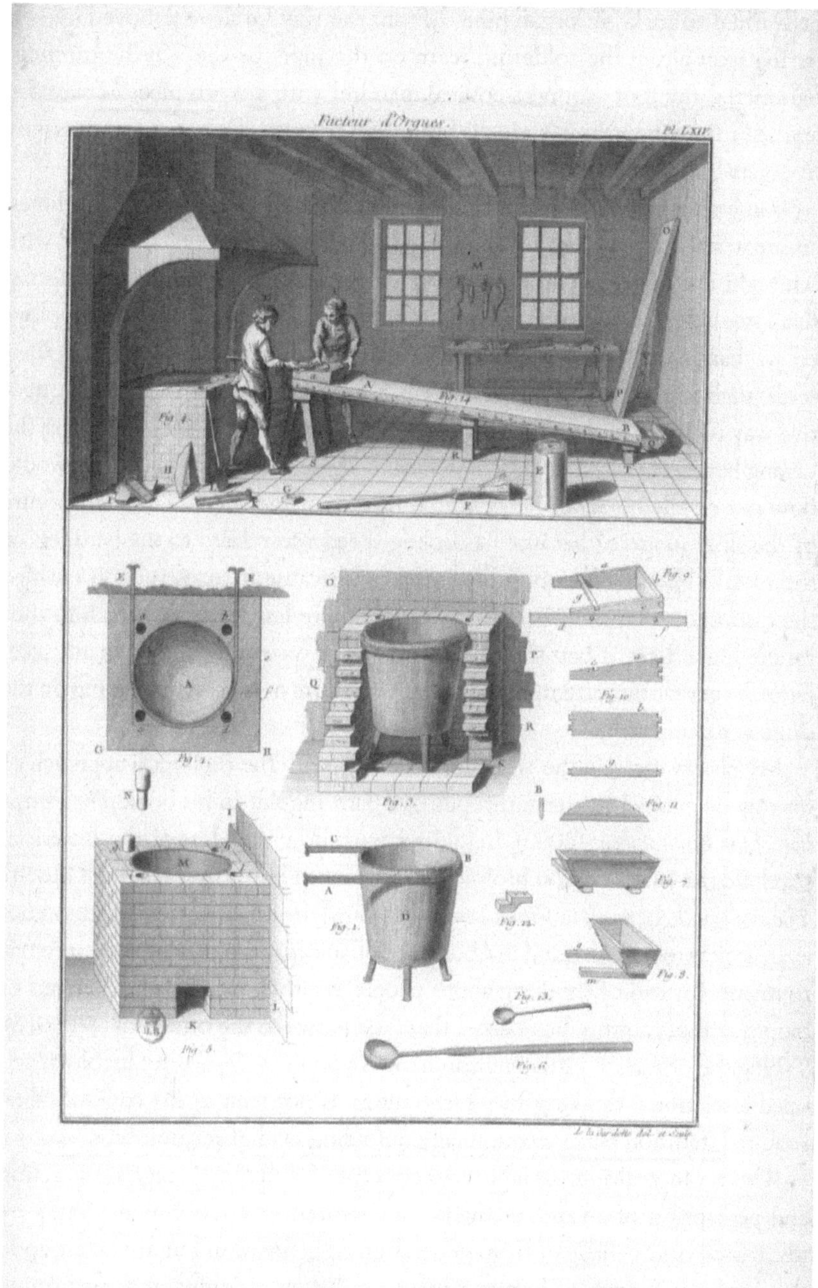

Figure 5.8
Plate LXIV from Dom Bedos de Celles's *L'Art du facteur d'orgues* (Paris: 1770) showing an eighteenth-century pipe-casting operation and the tools used.

he learned to make an organ pipe.[35] From the way he slowly moved his soldering iron along the soldering seam on the pipe, or the way he carefully hammered the foot around a conical mandrel with a worn piece of wood, I learned more about how a pipe is built than from reading about its anatomy in organ-building handbooks.

Immersing myself in this world of knowledge, experience, tools, machines, and materials, I struggled to understand what was going on. It is only with hindsight and after watching the many videos that I made of the pipe casting that I was able to reconstruct Yokota's arguments when he asked Eule's employees to change their work habits and their tools.[36] This required changing their tools, such as the casting bench and the casting box. Using these in an effective way required multiple casting attempts. Their speed of walking along the casting bench mattered because it determined the amount of alloy that would flow out of the box and, hence, the thickness of the sheet. The temperature of the alloy mattered because its cooling speed was related to the hardness of the metal. The single casting box mattered because its construction enabled the casting of a tapered sheet that would require less scraping, which in turn resulted in a sheet of better quality. These steps were not planned in advance; rather, every casting attempt might result in failure or success depending on the singular casting event.

My observations in the workshop resonate with the ecological approach of making developed by the anthropologist Tim Ingold. In his book *The Perception of the Environment* (2000), Ingold recounts how the ideas of the American psychologist James Gibson profoundly influenced him. After reading Gibson's *The Ecological Approach to Visual Perception* (1979), he came to see perception not as an achievement of a mind in a body but of an organism that explores its environment. Instead of thinking about people as either sociocultural persons or biophysical organisms, he realized that "the person is the organism" (2000, 3). Taking the organism-in-its-environment as a point of departure, Ingold developed a relational thinking in which culture is not seen as the content of an acquired tradition but as a continually unfolding field of relationships.

This ecological approach puts the concept of skill center stage as the action and perception of an organic being in a structured environment. Skills are not passed on unchanged from generation to generation but are taken up in the ways of working of human organisms through training and performing specific tasks. Studying skilled practice, Ingold argues, "demands a perspective which situates the practitioner right from the start, in the context of an active engagement with the constituents of his or her surroundings" (Ingold 2000, 7). This focus on continuous engagement challenges the opposition between intellectual design and mechanical execution. Instead of being prior properties, intentionality and functionality are immanent in a practice itself.

Skills do not execute a pre-existing design but generate the form of artifacts (Ingold 2000, 291). In historicizing these claims, Ingold points out that skills and, more specifically, tool using, changed as a consequence of the growth of industrial capitalism. In the relations among workers, tools, and raw material, subject-centered skills became replaced by what Ingold calls objective principles of mechanical functioning. "Thus whereas in the artisan's handling of his tools, the movements of their working points are guided by his own perception, the motions of the machine, and any tools attached to it, are predetermined" (2000, 289). For Ingold, the history of human technicity can be understood as a "history of externalization" in which the artisan is moved from the center to the periphery of production (2000, 289).[37]

In his book *Making* (2013), Ingold revisits the theme of the relation between mental image and material object, now asking the question what it means to make things. Whereas he gives many examples of making practices, varying from hand axes and watches to ceramic pottery and houses, his conceptual approach is straightforward. He starts from the opposition between a flow of consciousness with its experiences of light, sound, and feeling and a flow of materials that circulate, mix, and meld. In what he calls lateral thinking, these flows are related at a point in time where image and object connect. This line of thinking underlies the idea that making can be thought of as unifying materials provided by nature with conceptual representations springing from a cultural tradition. In this *hylomorphist* approach, Ingold argues, "practitioners impose forms internal to the mind upon a material world 'out there'" (Ingold 2013, 21). He contrasts this to a *morphogenetic* view on making as a process of growth:

> This is to place the maker from the outset as a participant in amongst a world of active materials. These materials are what he has to work with, and in the process of making he "joins forces" with them, bringing them together, or splitting them apart, synthesizing and distilling, in anticipation of what might emerge. The maker's ambitions, in this understanding, are altogether more humble than those implied by the hylomorphic model. Far from standing aloof, imposing his designs on a world that is ready and waiting to receive them, the most he can do is to intervene in worldly processes that are already going on and which give rise to

> the forms of the living world that we see all around us, in plants and animals, in waves of water, snow and sand, in rocks and clouds—adding his own impetus to the forces and energies in play. (2013, 21)

Ingold proposes thinking of making not as the lateral transposition from image to object but longitudinally as a confluence of forces and materials (2013, 22). Does this mean that the concept of design no longer has a meaning? Objects can have forms and a maker may have a form in her mind, but as Ingold underlines, this form does not create the work but the engagement with materials.

How do practitioners engage with the material world? As Ingold has argued, this engagement cannot be thought of as the imposition of ideas on a material world. Neither, he argues, can we think of it by affording material objects with some sort of agency thus enabling maker and material to interact. Rather we should think of making as "correspondence," Ingold claims. To explain this concept, he gives the example of activities such as flying a kite, playing a cello, or using a potter's wheel. In the case of the latter, the maker is not imposing a mental form on matter, but the making emerges from the working together of gesturing hands, the rotations of the wheel, and the behavior of the wet clay. The potter needs the wheel to correspond with the clay, just like the player needs the cello to correspond with the sound, or a flyer needs a kite to correspond with the wind. As Ingold puts it more abstractly: The kinesthetic awareness of the maker (the potter) is converted by way of a transducer (the wheel) into a corresponding flow of material (clay). This process should not be understood laterally as the interaction between three agents at a given moment in time, but as "points [that] are set in motion to describe lines that wrap around on another like melodies in counterpoint" (2013, 107). For Ingold, making something requires corresponding with the world. Rather than describing of representing this world, it means *answering to it* (2013, 108).

Ingold's ecological perspective can help to understand how Munetaka Yokota, Helmut Werner, and the craftsmen at Hermann Eule Orgelbau engaged with their materials and tools, drawing on experiential knowledge from making practices in different historical periods. Reading Ingold, one cannot help to see an irony in the attempts of Yokota and others to relearn the skills of organ builders in the eighteenth century. In an approach that underlines the importance of understanding the "organism-in-its-environment," it is relevant that Hildebrandt and his coworkers made their organs in a very different world than ours. To what extent can this world be emulated or recreated in the attempt of relearning a skill? This problem became obvious at several moments during the casting of the pipes. The employees of Hermann Eule Orgelbau used a thermometer to measure the temperature of the molten metal instead of their tactile experience while stirring. Their understanding of the casting was closely linked to the use of the scraping machine. They had to engage in a second-order type of learning that required to first know what they had to unlearn. To put it in Ingold's terms: Yokota asked them to correspond with the metal in a different way.

For Munetaka Yokota, the process that Ingold describes as the "history of externalization" of artisanal skills was crucial in understanding the value of historical organs and their sound. Precisely because we have tools such as the scraping machine, the fire-retardant cloth on the casting bench, and the double casting box, organ builders have unlearned the type of correspondence that was familiar to builders in the eighteenth century. We have to understand this type of correspondence while knowing that the skills that enabled it cannot be passed on unchanged but have to be relearned in our twenty-first-century environment. For Yokota this relearning aimed at understanding why the pipes of old organs sound so much better. What struck me during the two days in Bautzen was that making the pipes in the style of Hildebrandt was not a goal in itself. He judged details such as the quality of the metal or the construction of the pipe mouth not only as a pipe maker but also as a voicer who knows what it takes to make the pipe sound right.

VOICING THE PIPES

Newly assembled organ pipes do not sound very good. Even if they have been built precisely according to specified measurements and scalings, the work of a voicer is needed to give them their final sound character or voice. This work presumes a combination of knowledge, skill, manual dexterity, and a trained sense of hearing. To give a pipe its voice, the voicer manipulates all the parameters of a pipe that determine its sound character. In the case of a flue pipe, these are the size of the toe hole, the shape of the windway, the position of the upper and lower lips, the position of the languid, the cut-up, and the structure and phase of the languid edge or bevel. Changing one or more of these parameters can result in a marked change in tone characteristics.[38]

How a pipe sounds is determined not only by its physical characteristics but also by its spatial relation to other pipes and its position in the organ case.

Another important determinant of the sound is the wind pressure that is set for the organ as a whole. Finally, the voice of a pipe interacts with the room acoustics of the church or hall in which it sounds. A distinction is usually made between prevoicing and final voicing. Prevoicing is done in the organ workshop. It entails checking if the pipe's components have been built and positioned according to right measures, cleaning and repairing imperfections that arose during the production, and making sure that the pipe behaves properly when producing its tone. The final voicing and the tuning of the pipes is done when the new organ is placed in the room and all the pipes are placed in the organ case. Here the goal is to adjust the sound of the pipes to the other pipes of the same stop and in the organ, and to adjust the sound to the acoustical characteristics of the room.[39]

By determining the sound character of an organ, a voicer provides musicians with the sounds to make their music. This relation among sound, expression, and music leads to the idea, commonly shared in organ worlds, that voicing involves esthetic judgments. As Segurado (2015) remarks, this judgment is more complex than distinguishing between a beautiful and an unpleasant tone. It relates to all the elements in the organ sound. When a pipe is blown, it will first produce a very short noise that is usually called speech, or transients. This sound can be shorter or longer, depending on the way the pipe is built and voiced. Speech can have different characters, such as "pff," "tss," or "sch." The steady sound of the vowel type can have characteristics that resemble open and closed vowels in human speech such as "aa," "ee," and "ii." These vowels are part of the color of a tone depending on the relation among the fundamental and the harmonics, the loudness and strength of the tone, and the types of noise in the speech and the steady tone (Segurado 2015, 16–18).

Balancing and judging the various elements in the sound of a pipe is an arduous and time-consuming task that requires tacit knowledge, skills, and judgments. The difficulty in documenting the process of voicing may be the reason it is under-represented in historical treatises on organ building.[40] As Segurado points out, there is also very little research on the topic (2015, 11–12).[41] This does not mean that organ voicing is simply a subjective affair. Precisely because voicing involves esthetic judgments and reflects musical tastes, its history can be traced by studying remaining historical pipes and organs in relation to the esthetic ideals of a period. Baroque organ builders favored pipes that had a clear and pronounced speech, and they compared their sounds to human speech. In the Romantic era, many organ builders and organists preferred a cleaner, less noisy organ pipe tone that would blend easily with pipe tones from other stops. Understanding these historical conventions was an important aspect of the voicing work that Munetaka Yokota did on the new baroque organ.

SETTING THE WIND PRESSURE

In the first week of January 2016, all the test pipes were in the Orgelpark. We took some of the smaller pipes in the car after our visit to Bautzen in November. The larger pipes and some additional pipes were sent later. I meet Munetaka Yokota and Mariya Kato on the north balcony, at the console of the Van Leeuwen organ. Kato studied French literature. Her father is a physicist, and her mother a pianist. She decided that she would like to combine music with working with her hands. As an assistant to Yokota, she learns the craft of organ building. "The organ has ever more secrets the more you know of it," she says.[42] One of these secrets might be the relations among the wind pressure of an organ, the way the pipes have been built, and the room acoustics. The aim of Yokota and Kato this week is to check the test pipes, to set the wind pressure for the new organ, and to voice the test pipes. These test pipes will later be used as reference pipes when all the pipes will be prevoiced in the organ workshop of Hans Elbertse later in the year. They have set up a small workshop on the north balcony, next to the Van Leeuwen organ. On two tables I see a range of voicing tools, some books and folders, and many scaling sheets. Some of the pipes in the Van Leeuwen organ have been taken out to make room for the new test pipes.

I observe Yokota and Kato as they voice five pipes from the Praestant 4' of the Hauptwerk, all Cs in various pitches. Yokota measures the various parameters of the test pipes: the height of the cut-up, the width of the windway, and the diameter of the toe hole. He compares his findings to the scaling sheets. Sometimes he adjusts the pipes. Using a small knife he cuts millimeters of metal from the upper lip. He hammers the lower lip with an small hammer or pulls it out to widen the windway. With a toe hole cone, he is able to close the toe hole of a pipe or makes it wider with yet another tool. He cuts away small drops of soldering metal in the corners of the pipe's windway. His subtle and careful actions alternate with moments when he looks at the pipe mouth area or blows on the pipe—first soft, then hard, to see if and when they overblow. When a pipe is voiced, it is put back on the

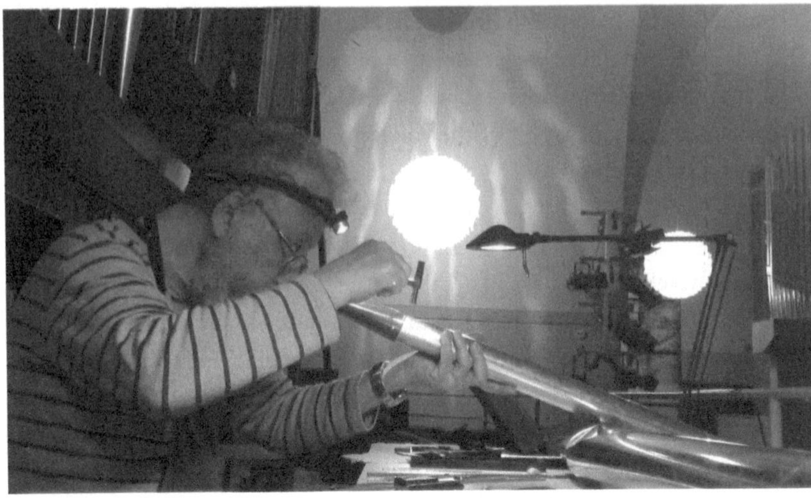

Figure 5.9
Munetaka Yokota voicing a principal pipe in the Orgelpark, January 2016 (photo taken by the author).

windchest. Kato presses the corresponding key and Yokota goes downstairs to listen how it sounds in the hall. To tune the pipe, Yokota cuts metal from the top of the pipe while Kato presses a key (see figure 5.9). This raises the pitch.[43] I notice the characteristics of the organ metal. It is flexible and soft enough to manipulate it, yet hard enough to retain its shape. It is this delicate balance that affords the minute changes that result from the movements of Yokota's and Kato's hands and tools. Fractions of millimeters matter to give a test pipe its voice.

How a pipe sounds is determined by how it is voiced but also by the room acoustics in the hall and the wind pressure that is set at the bellows. As we saw earlier, in the eighteenth century, organ builders such as Hildebrandt used the scalings they knew for various organs, independent of the size of the space. To adapt an organ to its space and to prevent it from being too loud, organ builders chose a disposition that fits the room and raised or lowered the wind pressure.[44] For the new baroque organ, Yokota and Fidom decided to voice the pipes on a wind pressure of 63 mm water column. This is lower than the wind pressure historically documented for some of Hildebrandt's organs and also lower than the 72–78 mm that Hans Elbertse had in mind when he ordered the blower for the new organ.[45] By setting the relatively low wind pressure, Yokota aimed to prevent the organ from becoming too loud.

Since it is going to be used as a concert instrument before a silent audience instead of a singing congregation, sounding too loud would be a problem. But Yokota realizes that when the hall is full, "the organ sound gets kind of thin. So it is not easy for the voicer where to set the wind pressure. . . . It is impossible to make it perfect for every situation."[46]

After four days of voicing the test pipes, Yokota and Kato present the results to the members of the core group. Dirk Eule and Helmut Werner from Hermann Eule Orgelbau, who have traveled to Amsterdam to discuss the test pipes, are also present. When everyone is seated in the hall, Yokota introduces the work that was done and the choices he made. He remarks that the choice for the Hildebrandt sound concept was partly made because it was expected to fit the dimensions of the Orgelpark. The pipe dimensions that Hildebrandt used in Naumburg and in Störmthal, two spaces very different in size, are nearly the same. The question was whether pipes with the historical Hildebrandt scalings would actually work in the hall of the Orgelpark. To learn more about the acoustic characteristics of the space, Yokota listened to Hans Fidom playing some of the organs. Acknowledging that he is not a scientific acoustician, he observed some "peculiar tendencies":

> One is that the bass is not that strong. The reverberation is there, but it has a quite strong mid-range. Higher organ pipes in the 2' or 1' range are very prominent, especially down in the center of the hall. Some very high hissing noises are very weak near the pipe, but very strong down here.[47]

Yokota explains that these acoustic characteristics complicate the voicing. Pipe tones contain a mix of the fundamental tone, overtones, and interesting airy noises that make a very complex and colorful sound "with some sort of spice, or salt or pepper in it. If you clean up too much, it is not interesting. On the other hand, if you have too much pepper it is too spicy."[48] In determining the sound character, Yokota takes not only the room into account but also the city of Amsterdam. He sees it as one of the most highly developed cultural centers and the Orgelpark as one of the pinnacles of its organ culture, where the organ will be used during concerts with very different musical styles. In determining the level of refinement he considered all these elements:

> Over-refining is not interesting, but too rough is not appropriate for this setting, I think. As a voicer, I can set just about any level, technically speaking. But in order to voice the pipes in a good way, their volume has to be in the right range, the tone colors have to be within the Saxonian style of the early eighteenth century, and the

> amount of speech has to follow the practice of that time. The sound of the pipes has to have a very lively character. Lively or dead is a very metaphysical way of saying, and too much poetic is not scientific, but this lively feeling is something that I have to maintain. In order to achieve all this, I used the scalings of Hildebrandt, but I used a wind pressure much lower than his. Some scholars might argue about this, but this is the only way to make a sound that is correct for the Hildebrandt esthetics.[49]

After Yokota's introduction, the group listens to Kato sounding the test pipes. Loek Dijkman suggests comparing them to pipes from the same stop in the Sauer organ and is happy with the sound of the new pipes. "They have much more difference and fullness. As a bunch of flowers, it is a big bouquet."[50] Johan Luijmes thinks that in the principal pipes the lower notes are much softer than the higher notes. Hans Elbertse is struck by the amount of speech in the pipe sound. When we walk to a nearby restaurant in the Vondelpark to have lunch, I talk to Helmut Werner. He is happy with how the pipes sound. They do what they have to do and the idea to use Hildebrandt's scalings in the Orgelpark works well. He uses the German metaphor of a "*Punktlandung*," or landing precisely in the right spot.[51]

HAMMERING WIND

Based on the results of the voicing of the test pipes and further specifications that Yokota has given, Hermann Eule Orgelbau begins making the actual pipes for the new baroque organ in the early months of 2016. In September, Munetaka Yokota is back in the Netherlands to work for three months on the prevoicing. Together with Kato and two other assistants, he set up a workshop in the storage space at Elbertse Orgelmakers in Soest. Here, the team will manipulate all the metal and wooden pipes that are ready. When I visit them, Mariya Kato is busy comparing the pipes, large and small, with the scaling sheets and all other information that is available. She wears a headlight and takes pipes from the rows in front of her on the table, inspecting them attentively. In some cases she removes surplus soldering metal from the corners of the windway of the pipes with a metal tool. After working on the geometry of the pipe mouth with careful movements, she blows on it, sometimes overblowing, to check the sound.

Yokota is standing at the second test organ that was used in the listening tests of 2014 and 2015, which now functions as a voicing bench. To listen to their speech and their steady tone, he places pipes on the windchest that has been set at a wind pressure of 63 mm water column. He uses the four or five test pipes that he voiced in January in the Orgelpark as a sound reference for the new pipes:

> Those are the pipes that I adjusted in the acoustics. Volume, timbre, speech, amount of noise: I decided that this is the way to go. Now, I can listen to these pipes in this room on that windchest and some of them sound exactly the same, some of them only partly. I have to find a way to make them sound best and also what could be the Hildebrandt style or not.[52]

Yokota tells me that he tries to respect what Hildebrandt did as much as possible. A problem is that the information he has is not always consistent. In almost all of his organs, Hildebrandt used a Quintadena stop, but they are not exactly the same from organ to organ. This depends on where the Quintadena was used, in the Hauptwerk or Oberwerk, or on the room acoustics. To Yokota, the historical pipes represent a grammar. Even though he follows Hildebrandt measurements, there are many different parameters he can change. "There is a certain amount of freedom. But if we deviate from that too much, then it is really out of Hildebrandt's range. I try to make a good theoretical guess, to recreate his language."[53]

Taking a Mixture pipe as an example, Yokota shows to me how he voices it. He blows on the pipe:

> Do you hear this, like scratching sound? I call this a "cough." This is created by a little too wide windway. [Blows again] And also wind is going outward too much. So in this case, I am going to close the windway just a bit [uses a small hammer to give short, soft taps on the lower lip]. This reduces the cough. [Blows again, and overblows] Maybe a bit more [gives very small taps]. [Looks at the pipe, and overblows it] Now I have to check the overbite.[54] And eh, the languid position. I could make the languid position a bit lower. [Takes a thin metal rod that he sticks into the pipe from above until it reaches the languid, puts the pipe against his body and hammers three times on the rod, then blows over again] So when I blow very hard, overblow, that means that air went into the pipe a bit more. So this is just the right timing. [Hammers on the corners of the lower lip] Now it is a little too early, so let's do the overbite to bring the air outward back a little bit. [Hammers on the upper lip] I already checked this amount of overbite. This is important because that and the width of the windway and the height of the languid, those three elements, determine how the wind hits the upper lip.[55]

There is still some "cough" left. Yokota observes that the speech is now very fast and has a little accent at the beginning. "But it is of a good kind. So it can function as a Mixture pipe, and the sound is not too soft."[56]

To voice the pipes, Yokota uses many tools: small knives, hammers, metal hooks and rods, and files. He has made or adapted most of these tools himself "improvisationally," as he himself says. "I have to make tools first to do the job that I want to do. [Laughs] And after thirty, forty years, I have a tool box full of home-made tools."[57] He shows a small knife with a wooden handle and a triangular blade. The flatness of the blade is important, he explains, because it enables him to make very straight cuts in the pipe metal when adjusting the cut-up of the mouth. Originally, it was a shaving knife. One of his small hammers is normally used by clock makers. Yokota adapted the surface of the hammerhead. It is polished and shaped slightly round, and its edges are strongly rounded off. He has used this hammer for forty years and knows exactly where to hold the handle to give the kind of taps that he needs. Another instrument is made out of a file. The surface is flattened and its tip bent into a hook. It can also be used to pull out the lower lip a bit to widen the windway. Because it is very thin, it can be used for narrow windways. Another small knife is used for making nicks, small cuts in the languid bevel that enable the voicer to reduce unwanted noise. Yokota sharpened the knife himself in a very sharp angle and made it narrow at the tip. "In order to put eighteenth-century style nicking you have to have some such narrow blade, thin and narrow."[58]

When I observe Yokota during the prevoicing of the pipes, it strikes me that judging the sound of a pipe is only part of what he does. Examining the pipe's geometry and comparing it to the scaling sheets that give the proportions of original pipes is just as important as an idea of how the pipe should sound. Some stops need a more "crisp" sound than others, which he knows from his studies of historical sources and also from listening to historical organs. He checks historical pipes "microscopically," and if it creates this crispness and looks close to the original, "then I can safely assume that it could be the original way."[59] Watching him work with the pipes, I ask Yokota if he imagines the sound of Hildebrandt's pipes in his head, as a goal that he wants to achieve.

> Yes and no. I do have some image of Hildebrandt organs in my head. On the other hand, the Hildebrandt organ sound we hear today may not be the original sound. I rely partially on what I hear, but partially on these geometrical relations. If we make the geometry right for the Hildebrandt sound and I use other knowledge, like what kind of treatment has to be done to the languid and so on, then the theoretically correct method is applied. Whatever the sound that comes out, I have

> to accept it as: this must be the original. . . . On the other hand, if it is obviously wrong or strange, first I doubt: my sense might be wrong. So it is a very complicated process to make this judgment. I hope to use my experience as a voicer or a researcher who has seen many old pipes, and some of these observations in the past inform this judgment. . . . Judging means including some acceptance of what sounds very strange, but what maybe should sound like that. Sometimes it is very painful and challenging. You see, because Hildebrandt is not myself. Hildebrandt might have liked the steak rare, I might like it more well done. But I have to accept his way. [Laughs][60]

Listening to the pipe sounds that emerge under Yokota's and Kato's hands at the voicing bench, I realize the ambiguity of the situation. Even if it would be possible to make and voice pipes that sound as the ones that Hildebrandt himself manufactured, our understanding of what sound is, and our ways of listening are very different from his.[61] Yet one of the central goals of the project was to offer future musicians sound material that has the same quality as historical Hildebrandt organs.

MAKING AND VALUING AS INSTAURATION

In the organ workshops of Hermann Eule Orgelbau and Elbertse, normative judgments were made based on the measurements and scalings of the pipes, the way the pipe metal was to be cast, and the voicing manipulations that determine their sound. All of these judgments derived from the commitment to bring the sound of Hildebrandt's pipes to life. In this chapter, the complexity of the task became clear. Given the fact that it is impossible to listen to the pipes of the organ in Störmthal in 1723, the pipe makers and voicers had to trace and interpret the many remaining facts that traveled across space and time (Valeriani 2011). Using various representations of pipe characteristics, such as measurements and scaling diagrams, they explored their anatomy, acknowledging that these representations themselves have histories: Using the Pythagorean scale, the meantone scale, or Töpfer's normal scale leads to small but relevant differences when used to build a pipe.

A similar historical hybridity could be seen in the casting of the pipe metal. Machines, tools, materials, and embodied skills from different time periods all mattered in shaping the sheets from which the pipes would be made. When Yokota explained to me how using a double-chamber casting box was related to the scraping machine, and why casting with a one-chamber box would render sheets that do not need much scraping would preserve the hardest—and, therefore, best—parts of the material, I gradually began to understand the interrelations between entities that made each other do certain things: the metal alloy, the melting pot and its temperature, the scraping machine, the linen that covered the casting bench, the casting chamber, the speed of moving it over the bench, and the cooling time of the metal. All of this mattered to make the pipes in such a way that they could be voiced in Hildebrandt's style. In the voicing of the pipes, historical knowledge, as well as listening and dextral skills, were needed to give a pipe its sound. How to conceptualize all of this work to make Hildebrandt's sound present?

We might say that in the organ-building workshop of Hermann Eule Orgelbau and during the voicing in the Orgelpark, sound was constructed by putting together and manipulating all the elements in the right way. But how to know what is the right way if we simply cannot know how Hildebrandt's pipes sounded in the small Störmthal church when Bach and his singers inaugurated its organ in 1723? According to Bruno Latour, this would be the wrong question. As we saw in chapter 3, posing it in this way assumes a practice of purification that presents the sound of 1723 as a singular essence that we can return to. Describing the making of Hildebrandt's pipes, as I have done in this chapter, suggests a practice of translation in which sound exists in the myriad of mediators that are mobilized in a collective process of making and judging through time.

In his book *An Inquiry into Modes of Existence* (2013), Bruno Latour returns to his inquiry into the Moderns that he started in *We Have Never Been Modern* (1993). He shows how the term "construction" creates a distinction between real and fake for the Moderns, for example, when it comes to scientific facts. These are either true or constructed, which then equals fabricated. Analogously, constructing a historical sound would reveal it as a fake past even though we are not naïve and understand that hearing the "real" sound of 1723 is not possible.

To circumvent the dichotomy of real and fake, Latour would argue, we need to look more closely at how we use the term "construction." If we say something is constructed, we say three things that are lost in the Modern use of the term. First, we ignore that the source of action is lost. If someone acts, others pass into action as well. "This is what the French expression *faire faire*—to make (something) happen, to make (someone) do (something)—preserves so preciously" (2013, 158). Second, to say of something that it is constructed obscures the direction of the vector. In the example that Latour gives: Balzac as the author of his novels created his characters, but we could as well say that they created him in their force to put them on paper. And third, when we say that something is constructed, we introduce a value judgment "not only on the origin of the action—double trouble, as we have just seen—but on the quality of the construction: it is not enough for Balzac to be carried away by his characters, he still has to be well carried away; it is not enough for the experimenter to construct facts through artifices; the facts still have to make him a good experimenter, well situated, at the right moment, and so on." In the Modern question "Is it true or is it actually a construction?," the three aspects are lost, Latour claims.

To find a concept to replace construction, Latour turns to the French estheticist Étienne Souriau (1892–1979). Souriau takes the example of the sculptor to argue that he does not create a work, but instaures it:

> If there is something that a sculptor never asks himself, it is this critical question: "Am I the author of the statue, or is the statue its own author?" We recognize here the doubling of the action on the one hand, the oscillation of the vector on the other. But what interests Souriau above all is the third aspect, the one that has to do with the quality, the excellence of the work produced: if the sculptor wakes up in the middle of the night, it is because he still has to let himself do what needs doing, so as to finish the work or fail. . . . The responsibility of the masterpiece to come—the expression is also Souriau's—hangs all the heavier on the shoulders of an artist who has no model, because in such cases you don't simply pass from power to action. Everything depends on what you are going to do next, and you alone have the competence to do it, and you don't know how. (2013, 160)

The notion of instauration, Latour writes, assembles the double movement of *faire faire*, the uncertainty about who is the author of an action, and finally the idea that the search for excellence is risky and without a pre-existing model (2013, 160).

Using the concept of instauration, it becomes possible to understand the quest for Hildebrandt's sound in a way that moves beyond the Modern dichotomy of true or fake. Yokota, Werner, and the employees at Hermann Eule Orgelbau instaured this sound. As something whose ontological status was open, that had to be made present, it was capable of worrying them, of making them do something, of testing and trying out. They felt a responsibility to bring the sound to life, yet without being able to work according to a pre-existing model. But how then do we value this sound?

Reasoning along the lines of Latour, music sociologist Antoine Hennion (2017a) suggests that valuation depends on a pluralist and relational ontology; in order to bring something into "more" existence, there is work to be done: "The word instauration expresses more strongly this idea of incomplete worlds, made of realities that are then 'calling us' because they need to be sustained to get 'more' existence" (2017a, 78; 2019). Valuing the sound of Hildebrandt's pipes should not be understood as assessing their qualities from the outside, as of inert things in a Euclidean space as Hennion writes, "but as multiple 'additive' relations, experimentations that help sustain those very things" (2017a, 79). The pipe measurements and scaling diagrams, the casting bench with its related equipment, the tools with which Yokota shapes the mouth area of the pipes during voicing are all means to value what the pipe makers make while they are making it. The work to be done when bringing Hildebrandt's sound into existence, assumed constant trials in which action and the criteria for assessing the results of the action could develop in tandem.

When I visited Elbertse's workshop in the fall of 2016, it was full of sounds. In the storage room, I heard repeated blowing and overblowing of high-pitched pipes and the growling of the lower wooden pipes connected to the blower of the test organ. In other rooms there was the sound of the sawing machine and the mortise drill that Elbertse had bought to make joints in the wooden framework. The work followed its routines and schedules; at fixed moments, a bell would chime to indicate that it was time for coffee. In one room the new organ slowly took shape; in the other, prevoiced pipes were lined up against the wall. Assembled in the new organ they would enable musicians to play old music in new ways, and new music on an old instrument.

6

IMAGINING AN ORGAN WE DO NOT KNOW

NO HANDS

The American organist Virgil Fox (1912–1980) became famous for the concert tours he gave in the 1970s.[1] Online videos of his performances show him playing Bach on a large console with dozens and dozens of stops, knobs, pistons, and pedals. The sumptuous electronic sounds of the organ's loudspeakers are accompanied by light effects and the rhythmic clapping of the audience. Trained as a church organist, Fox had well-known teachers such as Marcel Dupré and Louis Vierne, and in the postwar years, he was organist at the Riverside Church in New York City.[2] In 1965, Fox decided to perform only on stage. He started touring around the United States, giving performances in non-organ venues such as auditoriums and pop music halls. One of his famous instruments was a Rodgers touring organ, known as "Black Beauty." With this electronic organ, he gave "Heavy Organ" concerts where he performed music by Johann Sebastian Bach.[3] Fox hated the musical "purists," as he called them, who claimed to know how to perform Bach's music correctly. He himself played the music exactly as Bach wrote it, he said, and he "gave it color and depth."[4] To do so, he relied on the sound possibilities of the Rodgers, and later Allen, touring organs.

Virgil Fox not only played Bach's music, he also wrote adaptations of it. One of these, using the chorale melody *Komm, süsser Tod* (BWV 478), became particularly well-known. It was played by Hans-Ola Ericsson on the Sauer organ at a concert in the Orgelpark during the June 2014 symposium "Electrifying Baroque." In Fox's rendering of the chorale, the melody starts by slowly building the chord that supports the first note. This note almost imperceptibly changes into the second and third notes such that all seem stretched into eternity. From where I was in the audience, I saw Ericsson sitting behind the console and admired the calmness of his gestures. A few moments later, I realized that Ericsson's hands were no longer on the keys but on his knees. He looked at the score before him and listened attentively to Fox's Bach flowing from the Sauer organ. After a few minutes, the organist rose from the organ bench, walked to the first row of seats and sat down, listening to his own playing. After the last note of the piece he seemed to hesitate before joining the applause.

Ericsson's theatrical move could be interpreted as an ironic comment on the exuberant playing style of Fox. Technically, it was made possible because the Orgelpark's digital console records a performance in MIDI format and replays it. Ericsson recorded the Fox piece during a break earlier that day and had only to press the play button to start his performance during the concert. For me, the strange image of the organist leaving his console while the organ continued to sound raised more fundamental questions about the interactions between the player and the instrument. Probably without realizing it, Ericsson had performed a "breaching experiment" in the style of the ethnomethodologist Howard Garfinkel (1967). In these experiments, codes and conventions are breached to demonstrate that social situations are possible only because of the shared unspoken background knowledge and expectancies of the participants. In the case of a concert, we expect to listen to and watch a performer who physically interacts with a musical instrument (Pinch and Bijsterveld 2003). The possibility of separating the act of playing and the sounding of the organ that the digital console offers disrupts conventions deeply ingrained in our understanding of what it means to perform music.

This chapter concerns the interactions between musicians and their instrument that the new baroque organ hopes to afford. Combining a traditional and a digital console, the sound of the pipes that had been crafted by Hermann Eule Orgelbau and Munetaka Yokota would be made accessible in novel ways. As the performance by Ericsson shows, this opens a complex of interrelated questions on the gestures of the human performer and their relation to musical meaning, on the theatricality of the live performance as an effort aiming at excellence—in short, on the new interactions between the corporeal performing musician and the musical instrument. At both the mechanical and digital console an organist can follow baroque conventions in touch, articulation, and registration. The digital console also enables performers to link the organ to computers and other electronic devices so it can serve new musical purposes. By extension, it raises questions about what a musical instrument is or even what music can be. Whereas the traditional console is grounded in a conception of music as tones, intervals, and harmonies, the digital console connects the organ to the world of electroacoustically "organized" sound (Landy 2007).

To answer these questions, I start with an exploration of the mechanical console of the new baroque organ and the various stops that were selected for its specification. Accessing these stops from the digital console and expanding their sound by enabling various forms of real-time processing of the organ sound through live electronics were topics extensively discussed by musicians and composers during the Orgelpark symposiums between 2014 and 2017. After briefly outlining the tradition of electro-acoustic music, I show how their ideas provided an important benchmark in the design of the digital console. Concerts offered opportunities to experiment and demonstrate ways of interacting with the organ. For me as a researcher, these concerts were an important way to understand how the new baroque organ could be positioned in the landscape of twenty-first-century music. I then focus on how the existing digital console at the Orgelpark was redesigned to make full use of the possibilities that the electronic spring chest and its magnets offer.

Whereas any experienced organist will be able to play the mechanical console after some practicing time, interacting with the redesigned digital console requires new skills and imaginations. I will argue that the philosophy of art that John Dewey (1859–1952) developed in his book *Art as Experience* (2005 [1934]) offers a valuable framework for understanding how and why the digital console was designed the way it was because it gives a central role to the concept of interaction.

THE MECHANICAL CONSOLE

Historically, new organ designs always coevolved with the means to interact with the sounds of the instruments. The organs in the hall of the Orgelpark exemplify this coevolution. The short, sturdy keys of the Van Straten organ and its four robust stop handles, common in medieval and Renaissance organs, give access to three sounds. As a replica of the Peter Gerritsz organ built in 1479, the Van Straten is conceived as a *blokwerk* organ that has multiple pipe ranks sounding at the same time when a key is pressed (see chapter 1). As Barbara Owen writes in her study of registration in baroque organ music, around the year 1500, organ builders started to separate pipe ranks from the rest of the plenum, thus enabling organists to make more sound combinations.[5] This also increased the number of stop drawknobs. Next to this development, organ builders started to combine the larger organ with a smaller, or positive, organ. The practice of combining divisions or works with various sizes in one instrument would dominate organ building in the sixteenth and seventeenth centuries.[6] It is reflected in organ terminology: Great, *Grand orgue*, or Hauptwerk for the larger division, and *Chair* (Choir), *Positif du dos*, or Rückpositiv for the smaller. From 1500 onward, the construction of new types of windchests and actions coevolved with the development of new pipes and sounds. These sounds were often modeled after orchestra instruments at the time (Owen 1997, vii).

Larger baroque organs had multiple divisions. In north German organs, these divisions had their own sound character, based on their specifications.[7] The Hauptwerk division would contain the stronger principal and flute pipes, as well as reeds such as the Trumpet. The Rückpositiv, or back positive, a small organ behind the organist's back, also had principals, flutes, and reeds. The Brustwerk division (literally "chest division," but there is no fitting English translation) contained lighter and higher-pitched principal, flute, and reed stops. The Pedal division would contain lower principal pipes as well as reeds. Usually higher-pitched pipes were added to the Pedal specification. The list of pipe ranks in the various divisions of an organ is called the specification. Divisions each have their own windchests and manuals; to change between divisions, an organist can play on the corresponding manuals. In most baroque organs, manuals can be coupled. This means that when one presses a key on the coupled Hauptwerk manual, the Brustwerk or Rückpositiv keys will also play. Organists can now make combinations between the pipes in the various divisions, such as using principals from the Hauptwerk and adding flutes or reeds from the Brustwerk or Rückpositiv. This practice of choosing combinations of pipes by drawing stops and coupling manuals is called registration.

Organ structures as they developed around 1700 were by no means universal. Differences occurred not only between countries but also between regions. Whereas larger organs in north Germany at the end of the sixteenth century would normally have a Rückpositiv, this was often not the case in Saxony and Thuringia; both Silbermann and Hildebrandt did not build Rückpositivs.[8] Their organs would have a Hauptwerk, an Oberwerk, and a Pedal, depending on the size of the organ and the room. The new baroque organ's design followed this structure: It would have two manuals and a pedal, but what pipes were to be placed in its three divisions?

SPECIFICATION OF THE NEW BAROQUE ORGAN

The specification of an organ determines which pipes and pipe combinations an organist can use. In the eighteenth century, it was determined by the organ builder in consultation with the authorities who commissioned a new organ and in relation to the size and character of the room. The specification would be a part of the contract, which means that even when an organ no longer has its original pipework, this can still be found in historical documents.

As became clear in chapter 2, the specification of the new baroque organ was determined in the final plan of February 2015. During the early months of 2014, the core group discussed several options. Whereas the idea to copy the specification of the fifty-three-stop Naumburg organ was dismissed after the decision was made to place the bellows in the organ case, the initial suggestion by Hans Fidom was to use the stop list of the Hildebrandt organ in the *Dreikönigskirche* (Three Kings Church) in Dresden. This had thirty-eight stops on two manuals and a pedal.[9] This specification proved impossible because of the total weight that Hans Elbertse calculated exceeded the limitations on the north balcony. The Hildebrandt organs in Sangerhausen and Hettstedt, both with around thirty stops, offered an alternative reference. Of these two, the organ in Hettstedt was preferred because it provided slightly more possibilities and because the volume of the St. Jacobi Church in Hettstedt more or less matches the Orgelpark's hall. Of this organ, however, only the case remains, yet Hildebrandt's original specification can be found in historical documents in the town archives of Hettstedt (see table 6.1).

This specification was used as the main model for the new baroque organ because, as Hans Fidom writes:

> [i]t seems as if Hildebrandt went further than Gottfried Silbermann, for whom he worked during the first three years of his career, in establishing the three characteristics which, according to the latter, a good organ should possess: *Brilliance* is achieved thanks to the four compound stops on the Hauptwerk, even without considering the high-pitched stops on the Oberwerk; *Poesie* can be achieved in all manner of gradations thanks to the many 8' and 4' stops on the manuals, both in combination and also as solo colors; *Gravität*, meanwhile, is provided by the Pedal which, despite the organ's relatively small size, is equipped with one 32' and three 16' stops. (2018, 36–37)

Yet the Hettstedt specification was not simply copied. Hildebrandt might have used it for a space like the Orgelpark, but he would have made small changes

Table 6.1
Specification of the Hildebrandt organ in Hettstedt

Hauptwerk	Oberwerk	Pedal
Bordun 16 Fuβ	Gedact 8 Fuβ	Subbass 32 Fuβ
Principal 8 Fuβ	Viola di Gambe 8 Fuβ	Violon 16 Fuβ
Rohrflöte 8 Fuβ	Principal 4 Fuβ	Principal 16 Fuβ
Quintatön 8 Fuβ	Rohrflöte 4 Fuβ	Possaunenbass 16 Fuβ
Prästant 4 Fuβ	Nassat 3 Fuβ	Possaunenbass 8 Fuβ
Gemshorn 4 Fuβ	Octave 2 Fuβ	Octavbass 8 Fuβ
Weitpfeife 2 Fuβ	Waldflöte 2 Fuβ	Nachthornbass 2 Fuβ
Cornett 4 fach	Tertia	
Cymbel 3 fach	Quinte 1 1/2 Fuβ	
Sexquialter	Sifflöte 1 Fuβ	
Mixtur 5 fach	Scharf 4 fach	
Trompete 8 Fuβ	Vox humana 8 Fuβ	

Source: Fidom (2018, 36).

himself, as Fidom argued (2018, 37). Fidom suggested adding a Fagott 16' to the Hauptwerk, as could be found on the organ in the Dreikönigskirche in Dresden and the Rückpositiv of the Naumburg organ. To the stops of the Oberwerk of the new baroque organ, an Unda Maris stop was added.[10] The Hettstedt pedal specification was changed as well. The 32' Subbas, a very low and, therefore, large stop, was replaced to save space with a combination of a more conventional Subbas 16' and a Quintbass 12'. When played together, both stops give the impression of a 32' sound. The string stop Violin 16' and the Nachthorn 2' were eliminated, the first because its function could be taken over by the Principal 16', which according to Fidom would sound rather mild as a result of the low wind pressure of 63 mm water column, the second because a 2' stop would be available in the Pedal by coupling it to the Hauptwerk. Instead of the Nachthorn, a Clarin 4' reed stop was added to the Pedal that could be used in solo voices, just like in Dresden.

One of the strategies baroque organ builders used to make the organ sound more lively and expressive was to use a tremulant. This is a mechanism in the wind channel causing the wind supply to a division to fluctuate, which results in a tremolo and vibrato effect. The new baroque organ was to have a tremulant both on the Oberwerk, where it would mainly be used for the Vox humana, and the Hauptwerk. To make the speed of the tremolo effect adjustable, Bert van den Heuvel developed a new version of this mechanism (2018, 40). He also designed the shove coupler with which the two manuals can be coupled manually, while playing, because the organ would get a digital registration memory system. Normally, shove couplers can be engaged only while not playing and thus require a musical break. Like the tremulant, this coupling mechanism can also be operated electrically. Finally, the specification of the new baroque organ included two more playful stops, a Cymbelstern and a Nachtigall. The Cymbelstern is a series of eight little bells that sound when the stop is drawn; at the same time, a star crowning the middle of the facade starts to rotate. The Nachtigall is a bird call stop that functions by blowing wind through a small, high-pitched pipe positioned upside down, partly submersed in water or oil.[11]

AFFORDING ORGANISTS

The console of the new baroque organ follows the design of historical Hildebrandt consoles. The manuals were modeled after the Hildebrandt organ in Sangerhausen, down to the smallest details such as the semicircular embellishments in the front of every key and the white top layer on the black upper keys. The space above the manuals and around the stopknobs was painted in the exact color of the Naumburg console (see chapter 2 and figure 6.1). This organ was also the model for the design of the stop drawknobs. The Naumburg organ has printed stop name tabs in a combination of Latin and gothic letters. These fonts have been identified by the printing firm of Philip Elchers in Groningen, the Netherlands, and were printed on paper that matches the color of the original parchment labels in Naumburg. Whereas the original stopknobs as measured by Hermann Eule Orgelbau have a diameter of 27 millimeters, the knobs of the new baroque organ were made slightly bigger to improve the readability of the stop names (Fidom 2018, 52).

Although the historical reference to the Sangerhausen and Naumburg organs determined the overall look and feel of the new baroque organ, some details in the console are unmistakably from our time: two lamps that illuminate the manual and pedal keys, but even more so, the small display above the upper manual. When the electricity is turned on, the organist will see green letters and numbers. These show the position of the sequencer system. Here

Figure 6.1
Original console of the 1747 Hildebrandt organ in the St. Wenzel Church in Naumburg, Germany. It was played by Johann Sebastian Bach (photo taken by the author).

the baroque organ borrows a technology that was introduced in the twentieth century, the "Setzer," or sequencer (see chapter 3). Using switches hidden in a small drawer below the stopknobs on the left of the console, the organist can program a sequence of stop combinations and store it in the organ's computer memory. Stop combinations can be retrieved by pressing on the two iron foot levers above the pedal. These allow the organist to move forward or backward in the sequence of stop combinations while playing. In the baroque era, the work of changing stop combinations was done by the organist or with the help of stop pullers (registrants).[12] The use of a sequencer in the mechanical console of the new baroque organ resulted from the stop action being controlled by magnets. This was necessary because the organ's bellows were placed inside the organ's case and there was no room for a mechanical stop action (see chapter 4).

Knowing the exact specification of the new baroque organ did not answer the question of how the stop drawknobs should be ordered on the console. Because there were no mechanical limitations due to the electric stop action, any stopknob could be connected to any stop in the windchest. This made the ordering arbitrary. During the car trip to Hermann Eule Orgelbau in November 2015, Hans Fidom showed a proposal for the ordering. He tried to ensure that stop combinations that were important for Hildebrandt would be easy to reach for the organist. As a reference, he had considered the current situation in Sangerhausen and the ordering in Hettstedt as described by Ulrich Dähnert (1962). In the discussion between Hans Fidom and Johan Luijmes that followed, the pros and cons of the various stop orderings were explored. Because the orderings in Sangerhausen and Hettstedt were very different, it could be assumed that Hildebrandt did not work with a fixed ordering himself. It was also not clear whether mechanical aspects, such as the layout of the windchest or the amount of space inside the case played a role. And finally, little is known about how registration conventions affected the stop ordering on the console.[13] This meant that a historicizing approach based on Hildebrandt's style would not be possible. The issue was not decided until later in the process; in the end, the stop ordering of Hettstedt was used as a reference.

Choosing a stop ordering that feels logical for the player was all the more important because in the Orgelpark, most, if not all, organists are guests. They have limited time to get to know the organ. Knowing an organ well means not only that an organist has a clear idea about how stops sound but also that stops can be found without much time and effort, almost blindly. This awareness of how the sound can be changed by drawing or closing stops is part of the habitus of organists. It is quite common to play on two different manuals at the same time, one hand on each, while both feet are playing the pedal. In the musical flow, short moments have to be found to reach sideways, sometimes quite far, to close or open a stop. With the feet playing, the body of the organist balances on the bench. For experienced organists, finding this balance has become a tacit skill. Because no organ is the same, however, playing on an unknown organ requires time to calibrate the embodied repertoire of movements to the new instrument and the new situation.

Playing a medium-sized mechanical organ such as the new baroque organ demands a range of bodily movements. In most of these movements, the organist is constantly aware of the tactility of the instrument. Pressing a key when the windchest is under wind pressure enables the organist to feel the exact moment when the pallet in the pallet box opens and wind flows into the tone channel. The same is true for releasing the key and closing the valve; if this is done too quickly, the sound will end abruptly. It is the organ builder's task to construct key and stop actions that are well playable and afford the musicianship of the organist. On the other hand, it is precisely the subtle resistance in an instrument that results in the unique touch and feel of a mechanical organ. Learning to work with this resistance is part of what it takes to play and understand an organ as a musical instrument.

The mechanical and digital console of the new baroque organ would be another example of how innovations in organ building coevolve with the means to interact with the instrument. Using the archaic technology of the spring chest to combine a mechanical baroque action with twenty-first-century digital technology implied that the mechanical console, as the traditional interface between the organ and its performer(s), had to be rethought and redesigned. In addition, to use all the possibilities that the electromagnets in the spring chest offered, the already-existing digital console in the Orgelpark had to be refurbished, and features for live electronics had to be added to the organ. From the start of the project, musicians and composers who work in the tradition of electro-acoustic music have been consulted and their works were performed during concerts. Analogous to the historically informed practices of performance and organ building, recent developments in electro-acoustic music played an important role in the design of the organ and its digital console.

A LANDSCAPE OF (IM)POSSIBLE SOUNDS

Electricity entered the organ in the late nineteenth century when it was used to control electromagnets in electropneumatic stop and key actions (see chapter 3). The sound in these organs, however, was produced by blowing wind through pipes. The first organ that used electricity to generate its sound was the massive Telharmonium, developed between 1897 and 1902 by the American lawyer and inventor, Thadeus Cahill. It not only produced its tones electrically through a system of tone wheels—rotating cylinders with bumps that caused an alternating current in magnetic coils close to the wheels[14]—it also could transfer these tones via telephone lines. In the decades that followed, a whole range of new musical instruments emerged, many of them using electricity to generate their sound like the theremin (1919) and the ondes martenot (1928). Patteson (2015) quotes Leo Kerstenberg, a pupil of Federico Busoni, who wrote in his anthology *Kunst und Technik* (*Art and*

Technology) published in the mid-1930s that the meaning of the word "*Technik*" had changed in relation to music.

> While previously, it referred to a musician's cultivated skill in performing or composing—that is, technique—increasingly the term was used to describe the material means of tone generation and transmission: technology. This new sense of *Technik*, argued Kerstenberg, stands at the center of contemporary thought: not the discipline of the performer, but the capabilities of the musical machines. (Patteson 2015, 152–153)

Only a few years after the publication of Kerstenberg's anthology, in 1937, the young American composer John Cage delivered a lecture in Seattle on the future of music. In this "Credo" he claimed that he believed that

> the use of noise . . . to make music . . . will continue and increase until we reach a music produced through the aid of electrical instruments . . . which will make available for musical purposes any and all sounds that can be heard.[15] (Cage 1987, 3)

Visionary as they were, Patteson points out that Cage's thoughts should be seen in the context of the many experiments with new musical instruments in the early decades of the twentieth century rather than with the hindsight knowledge of the electronic media that dominated the second half of the century (Patteson 2015, 157). Yet it is remarkable to see with what clarity Cage anticipated what was to come.

An extended overview of electro-acoustic music after 1945 lies beyond the scope of this chapter, but I will highlight broader trends in this tradition that informed the design of the new baroque organ.[16] Cage wrote in his Credo that he expected electrical instruments "to provide complete control of the overtone structure of tones (as opposed to noises) and to make these tones available in any frequency, amplitude, and duration" (Cage 1987, 4). The experiments with electronic sound generation by Werner Meyer-Eppler, Robert Beyer, Fritz Enkel, and Herbert Eimert at the Studio for Electronic Music at the German radio station WDR in Cologne, starting in 1951, materialized this vision. Karlheinz Stockhausen belonged to a new generation of composers who used the new equipment as a means to make electronic music compositions such as *Kontakte* (1958–1960). This new music no longer referred to the sound of traditional instruments and their musical languages; new sonic worlds emerged. These sounds had to be produced by the laborious cutting and splicing of tapes and the electronic equipment that was only available at universities, research centers, and radio stations. This changed when the American engineer Robert Moog developed a voltage-controlled synthesizer in the 1960s. Moog had played the theremin since he was seventeen and used oscillators as pitch generators in his first analog synthesizer that he presented in 1964 at the Audio Engineering Society Convention. Digital synthesizers developed later offered presets or preprogrammed sounds to users who did not share the sense of sonic experimentation of the early years. Technological developments in sound synthesis in the 1970s and '80s included frequency modulation (FM) synthesis developed by John Chowning and real-time granular synthesis that builds up sound textures by combining very short tones, or "grains."

Parallel to electronic sound synthesis, another of Cage's predictions was realized: making available for musical purposes "any and all sounds that can be heard" (Cage 1987, 4). The invention of the tape recorder in the 1930s—AEG presented its Magnetophon in 1935—made it easier to record and edit real-life sounds. This possibility was used by Pierre Schaeffer in his *musique concrète*, compositions like *Etude aux chemins de fers* (1948) in which we hear sound samples from railroad scenes such as the hissing of pistons, the screeching of whistles, and the clacking of wheels on rails. Composers soon started to combine recorded sound samples with electronically generated sounds; in his *Gesang der Jünglinge* (1955/1956), Karlheinz Stockhausen used the singing voice of a young boy and mixed it with electronic sounds. American composer Steve Reich used recorded sound material in his compositions *It's Gonna Rain* (1965) and *Come Out to Show Them* (1966), which explored the principle of gradual musical process through phase shifting, characteristic of his early minimal music (Reich 2002).

Whereas the first decades after the Second World War were characterized by analog sound technologies, from the 1970 onward, computer programs were used to digitally synthesize and sequence sounds. The first experiments with digital sound were done in the context of telecommunications at Bell Labs in New Jersey in 1957. First-generation computer programming languages to synthesize music such as Music-N was developed here. The Musical Instrument Digital Interface, or MIDI, protocol that was introduced in 1983 allowed composers and musicians to exchange remote-control signals between digital musical instruments, computers, and other hardware. It now became possible to write software that can process the MIDI input of electronic instruments not only to generate new sounds but also to play sound sample sets.[17] Desktop and later laptop computers running audio-processing software packages such as Logic Pro, Max/MSP, and Pure Data gave composers the possibility to create complete recording studios at home. Laptops increasingly appeared on stage in the 1990s.

Since the early days of sound synthesis and processing the question of how to interact with new devices such as the Moog synthesizer was on the agenda.[18] With traditional musical instruments, embodied musical gestures and direct interaction with the instrument are needed to produce a sound; this was no longer the case with electronic music. Compositions were made in the studio and performed through loudspeakers. Without a live performance to witness, the theatricality of concerts fundamentally changed, both for performers and audiences (Pinch and Bijsterveld 2003). In the Studio for Electro Instrumental Music (STEIM), founded in 1969, Michel Waisvisz (1949–2008) worked on the question of what it means to perceive a digital object as a musical instrument. Instead of using generic controllers such as MIDI keyboards, he developed a personalized digital music interface that he called The Hands (first exhibited in 1984, last performance in 2008). These were a pair of gloves containing a small keyboard and various types of sensors (Torre and Andersen 2016). Using The Hands, Waisvisz could play the computer as if it were a traditional musical instrument. The Hands of Waisvisz affected two aspects of digital music performances: They gave the performer a tailor-made interface that could respond to movements and gestures, and they provided the audience with something to watch while listening. Following up on the pioneering work by Waisvisz, new interfaces for musical expression (abbreviated as NIME) became an important aspect of digital music cultures. Linking traditional musical instruments to digital instruments was part of this search for new kinds of "instrumentality." The new baroque organ can be seen as an example of this development. Until the moment that it could actually be played, its imagined use took the form of thought experiments and concert performances on the Sauer organ with the digital console.

MOBILIZING THE PERFORMER'S PERSPECTIVE

The performer's perspective on the new baroque organ took center stage at a colloquium on live electronic music in May 2014. The evening before the colloquium, an ad hoc ensemble consisting of a saxophone quartet and theremin performed new compositions by students from the Amsterdam Conservatory. The Sauer organ was played from a laptop controlling the digital console that had been placed out of sight. It was the first time that I heard the

Sauer without seeing an organist. The human performers in the ensemble used a click track[19] to coordinate their timing with the prerecorded organ part. In my field notes of that evening, I reflected on the persona of the computer performer and composer bypassing not only the organist but also the musical conventions of a human interacting live with an organ.

The next day, four musicians who perform regularly at the Orgelpark gave presentations in which they made suggestions for the use of live electronics in the new organ.[20] Jacob Lekkerkerker proposed seeing the new baroque organ from three perspectives: the traditional organist, the contemporary organist, and the composer. Lekkerkerker was the organist in the Oude Kerk (Old Church) in Amsterdam and sees himself as an improviser who seeks to mix the traditional sound of the organ with that of live electronics.[21] He sets up concert situations in which he plays an organ and records its sound to experiment with the mix of acoustic and recorded sound. To do so, he works regularly with other musicians, such as the guitarist Alfredo Genovesi, who uses electronic sound modulation devices such as pitch shifters, delays, and loop stations "the same way an organist uses stops" (Lekkerkerker 2014, §279).

In his presentation, Lekkerkerker drew a parallel to the improvisatory practice of baroque composers like Johann Sebastian Bach, Dietrich Buxtehude, and Jan Pieterszoon Sweelinck, "solitary figures sitting at the console with rows of stop knobs to both left and right" (2014, §293). In line with this image, he envisioned an improviser's digital console for the new baroque organ. He preferred buttons and knobs to a touch screen because "[a] good button has some resistance but a touch screen doesn't and this complicates the production of precise sound-effects. . . . There must be a relationship between motoric memory and equipment" (Lekkerkerker 2014, §282). In addition to the pedal, the digital console should have four feet pistons enabling the performer to create loops—repeating segments of music recorded while playing—and pedals to vary the wind supply, to control microphones that were to be set up permanently in the organ, and manage the overall volume of amplification.

Coming from a background of electro-acoustic music at STEIM, composer, improvisor, and laptop instrumentalist Robert van Heumen sees live sampling as his main tool. He uses the laptop "as a sound-generating device controlled in an instrumental and tactile way, connecting action to sound like an acoustic instrument" (Heumen 2014, §295). Van Heumen presented his vision of the new baroque organ as a hybrid electro-acoustic instrument, merging acoustic and electronic sound. But, Van Heumen underlined in his presentation,

> First of all the most important aspect that should govern all decisions: LIMITATION. In computer-based electronic music, if you do not limit your possibilities, you get nowhere. You play around with programs, with plugins, with effects, with toys, with hardware, but you only scratch the surface, you never go deeper. So: make choices, limit the possibilities in the acoustic design but more importantly in the electronic design. Most composers will not have much time anyway to work with the instrument, so they do not have the time to sculpt the instrument exactly as they like. Provide presets, but also provide the possibility to make one's own presets. Develop an instrument with character. . . . Limitation sparks creativity. (2014, §299)[22]

Van Heumen's vision of an electro-acoustic organ combined traditional acoustics organ pipes with what could be "electronic pipes," sounds from a "sound engine," consisting of virtual analog synthesis, sampling, and physical modeling synthesis, which is to be built inside the organ case. This engine would consist of fixed and flexible microphones inside the organ, recording the sound not only of the acoustic pipes but also of moving mechanical parts such as the key action. These sounds could be played back directly through speakers inside the organ, but also through speakers in the hall.[23] The organ's sounds could be processed through computer software or combined with synthesized sounds using SuperCollider or other sound-programming software, Van Heumen argued. The interface of the new electro-acoustic organ should allow freely assigning registers from both the acoustic and the "electronic" pipes to every manual in every combination. It should have register knobs and dedicated knobs to be used during live performance and also a digital screen interface for more preparatory actions.

Flutist Anne La Berge commented in her presentation on new forms of interaction between the organ and other traditional musical instruments such as the flute and electronic devices such as laptops, keyboards, and various other sound and signal processing equipment.

> The capability to communicate with the organ digitally from various remote locations gives the Orgelpark a leadership role as a modern-day live-electronic instrument. These locations include digital consoles on the main floor and the balcony. The digital communication is not limited to the consoles, it can be via computers or controllers from anywhere in the world. (La Berge 2014, §351)

Although La Berge acknowledges that this may sound like a "network circus," she claims that these interconnections between musical instruments and electronic devices are very common at live electronic concert setups. To make sure that future organists and performers can find their way around in the "circus," she suggests providing a user manual for composers and performers at all levels. In this manual the use of software and hardware would be explained as well as how the organs produces its sounds, both analog and digital. It would describe where fixed and flexible microphones are placed in the organ case and how audio signal paths are routed to processors, computers, and analog audio processors. And it would give basic examples of setting up MIDI or Open Sound Control (OSC)[24] communication among the organ, the console, and computers with sound-processing software such as SuperCollider, Max, and Ableton Live.

The theme of the interconnectivity of the new baroque organ returned in the lecture that the American organist and researcher Randall Harlow gave during the "Electrifying Baroque" symposium in June 2014. Harlow elaborated on the work of Tod Machover at the MIT Media Lab, who experimented with blending acoustic musical instruments with computer hardware, software, and live electronics. The results were what Machover called hyperacoustic instruments such as the hyperviolin or the hyperbow (see chapter 3).[25] According to Harlow, the new baroque organ is a hyperorgan because it offers new performance possibilities through the integration of traditional acoustic sound production and electronic media.

Mixed media not only allow organists and performers to create new sounds, Harlow argued, they enlarge the community of users of the new baroque organ in unexpected ways. The organ can be played not only from the mechanical and digital control but, thanks to the MIDI and OSC communication protocols, it also offers the possibility of nonkeyboard interfacing, including "cross-instrument modulation with other live musicians" who can influence the organ's pitch or stops (Harlow 2018, 160). Dancers or visual artists using video or motion sensors could manipulate the organ through their movements. Live DJs could create a live remix of organ repertoire. The organ could be performed over the internet, enabling international masterclasses or even performances. Audiences could participate in creating sounds through smartphone apps. Mapping devices would allow children to play the organ, solo or together. And finally, "the organ might provide a vehicle for both therapy and artistic expression among individuals for whom acoustic music performance would otherwise be inaccessible" (2018, 160).

For Harlow the hyperorgan, as an ecology of acoustic performance, becomes an exemplar of a hyperconnected society. Making the analogy with augmented reality that provides digital layers over our everyday sensory experiences, thus affording new interactions with the real world, he claims that the new baroque organ, as a hyperorgan, offers what he calls new ecologies for musical expression (2018, 161). He envisions a future where hundreds of pipe organs are connected into a global interactive network of hyperorgans, "engaging and enriching the lives of expert performers and composers, amateur musicians and lay citizens, children and the mobility impaired alike through music, both old and new—becoming immersed in new acoustic music ecologies" (Harlow 2018, 163). Daring as it may sound, Harlow's vision challenges the idea of an organ as an instrument that is, historically and spatially, uniquely situated.

BENDING BAROQUE

To understand musical instruments only as devices that are specifically designed to produce meaningful sounds would be inappropriate for situations in which computers and other electronic equipment are used to make music. When "we see familiar sounds coming out of unfamiliar devices, and unfamiliar sounds coming out of familiar interfaces," the question is how we can define musical instruments (Kvifte 2008, 45). Kim and Seifert (2016) argue that interactivity could be a point of departure for rethinking the traditional concept of a musical instrument. Instrumentality could thus refer to a set of gestures, situations, and repertoires that are not primarily linked to the intrinsic properties of an instrument. Rather it emerges from what could be called musical play, the situated and embodied interactions between musicians and their instruments as "autonomous and artificial social agents" (2016, 79). This critique of traditional ways of analyzing and classifying musical instruments according to their physical and material characteristics is echoed by Théberge (2016), who argues that musical instruments could be better understood if we consider how they are placed in "assemblages," or relational networks with other objects, practices, institutions, and social discourses. Drawing on examples such as the violin, the electric guitar, and phonographic turntable, Théberge shows how these assemblages can be variable and how the meaning of an instrument depends on its place within a particular assemblage.

What these assemblages might be in the case of the new baroque organ with its two consoles, one linking it to historical practices and one connecting it to contemporary musical situations, was the topic of the symposium

"Bending Baroque" in 2015.[26] The title referred to the electronic music practice of circuit benders who render desirable sounds from old, discarded, and broken devices. In his keynote lecture, Trevor Pinch from Cornell University analyzed circuit bending as a model for the kind of musical instrument the new baroque organ could become.

Circuit benders "bend" or "hack" mass-produced electronic devices, often the ones used in toys such as the Furby doll (Pinch 2016). They expose the inner circuits of these devices, using either a contact wire with crocodile clips attached, or their own fingers to make links between different parts of the circuit, until they get an interesting sound. Pinch situated the practice in a broader trend of a "lo-fi" esthetic and of the "technostalgic" revival of vinyl, tape, and vintage synthesizers to capture the elusive desired qualities and idiosyncrasies of these instruments. It is also connected to the practice of experimental electronic music, more specifically of "noise music" (Novak (2013) cited in Pinch (2016, 37)). Circuit bending, Pinch argued, challenges both the conventional meanings of technology and the conventional meanings of musical instruments.

Referring to his own experiences with a modular electronic synthesizer that he built in the 1970s and started using again in 2009, Pinch described how some of its broken circuits could produce sounds that were found desirable. Members of the contemporary rock in which Pinch performed with his instrument would fondly describe one poorly functioning module as "Sector Two." Moving his hands near the patch bay would change the sounds in unpredictable ways, Pinch observed. Both he and his band members preferred this randomness over working circuits that enact more traditional values as predictability and control (2016, 39).

Drawing on his earlier work on the Moog synthesizer, Pinch connected the irony of the desirability of brokenness to the debate on the synthesizer designs of Robert Moog and Don Buchla in the 1960s (Pinch and Trocco 2002). Whereas Moog's use of the volt-per-octave standard defined the synthesizer as a conventional keyboard instrument, Buchla rejected the idea that music was about octaves and favored different types of controllers that allowed for randomness and unpredictability. Precisely their idiosynchrasies gave his synthesizers the quality of an instrument and not a machine, he thought (2016, 39–40). The stochastic sounds of Sector Two brought him closer to the esthetic goals of Buchla, Pinch noted, even though in another musical collaboration he might use a more reliable module.

Pinch's experiences with his home-built synthesizer resonated with the practice of circuit-benders, who favor bricolage and tinkering in their approach of technology:

> They learn new practices and ways of interacting with materials and the innards of machines which favors an immediate and tactile form of knowledge making and a direct encounter with the sonic results. This knowledge making, because it depends upon the interaction of their eyes, hands and ears (sometimes with the aid of a few simple tools) can be described as a form of sonic skill. Some of their skills become standardized such that they can follow a canonical bend for a particular toy—such as the iconic Speak and Spell toy. But even then there is plenty of room for variation and aesthetic judgment in terms of choice of placement and style of knobs and switches and which particular sounds are enacted. (Pinch 2016, 48)

During the symposium, the circuit bender's practice of remaking an ordinary object into a musical instrument was, explicitly or implicitly, related to the new baroque organ. How would the various electronic interfaces it provides musicians open up new possibilities and allow them to tinker with the conventions of organ playing, and of its pipe sounds that had been crafted according to ways of making in the eighteenth century? How to find a balance between control and unpredictability, between convention and experiment in the new electro-acoustic music ecologies it would bring into being? And how to judge the quality of the new sounds that would be produced? Much of the promise of the new baroque organ to render yet unheard sounds depended on the digital console. Its design would be crucial in affording musicians new ways of interacting.

REDESIGNING THE DIGITAL CONSOLE

The ideas and visions that performers like Lekkerkerker, Van Heumen, La Berge, and Harlow presented in 2014 came at a moment when many electronic features of the new organ were still undecided. In the following year, some of these ideas materialized in the design of the digital console and the equipment inside the organ case; others, such as fixed microphones and loudspeakers or a "sound engine" in the organ case, as Van Heumen proposed, did not make it into the final design plan.[27] The core innovation of the project remained the possibility of combining any pipe with any other in the digital mode of the organ. In a mechanical organ, most of the sound combinations that the pipes offer remain out of reach for the organist. In the previous overview of the anatomy of baroque organs and their divisions, it became clear that manuals are linked to fixed ranks of pipes; opening a stop means that every pipe in that rank is linked to its own key.[28] The basic principle behind the system that Sinua had developed for the electropneumatic Romanus Seifert & Sohn organ from 1953 in the St. Peter and Paul Church in Ratingen was that every pipe could be assigned to any key on any of the three manuals (see chapter 3). In addition, the timing of the pipe sound would no longer depend on how long or short the organist presses down a key. Pipe sounds could now be delayed or sustained or repeated.

The existing version of the digital console was developed by the Orgelpark in close consultation with Benedikt Aufterbeck from Sinua in Düsseldorf. Built in 2012 by Hans Elbertse, it was based on the Sinua concept in Ratingen. The digital console also used software developed by Sinua. Yet the redesigned digital console was to have many additional features, even more than were available in the Ratingen organ. Although I knew how mechanical keyboard, pedal, and stop actions work, it took me quite some time to understand the details of the Sinua concept. Two trips of core group members to the console in Ratingen, in March 2015 and February 2016, helped me to grasp its workings and to get an impression of the ways it could control the production of sounds.

LAYERS OF SOUNDS

Like many churches in Germany, the St. Peter and Paul Church in Ratingen, near Düsseldorf, was partly destroyed during the Second World War. After the war, it was rebuilt. Its neogothic interior contains rows of wooden benches and the 1953 Romanus Seifert & Sohn organ on the balcony. The organ was restored in 2006, which was also the year that Benedikt Aufterbeck and Thomas Stöckl asked the church organist Ansgar Wallenhorst if they could experiment with a new system based on *Einzeltonsteuerung*, individual pipe control. As Aufterbeck recalls, he first met Wallenhorst when he studied organ building in 1998: "He was open to our ideas and not afraid that we would break something."[29] Whereas many organs have slider chests, which makes it impossible to have single-pipe control, the organ in Ratingen has electropneumatic cone chests (see chapter 3). This enabled Aufterbeck and Stöckl to link the pedal, which already could control single pipes, the solo division, and the back positive to their system. They wrote most of the

Figure 6.2
Benedikt Aufterbeck from the Sinua company in Düsseldorf demonstrating the console of the Romanus Seifert & Sohn organ in the St. Peter and Paul Church in Ratingen (photo taken by the author).

software for the system themselves, Stöckl says: "We are too small for the big firms to adapt their software for our needs, so we had to 'misuse' things, to make them as we want."[30] The core idea of the Sinua company that Aufterbeck and Stöckl founded is to combine settled sound ideals with new technical possibilities.

Early 2015, shortly after the final plan for the new baroque organ had been finalized, I joined Loek Dijkman and Hans Fidom who traveled to Ratingen to discuss business agreements with Sinua and to visit the St. Peter and Paul Church. Stöckl and Aufterbeck welcomed us at the console of the organ facing the organ at the other end of the church. They introduced us to the console, the heart of which is a computer running Windows (see figure 6.3).

It has four manuals and a pedal, and its stop switches are placed in rounded rows on either side of the manuals. In addition, there are push buttons above the upper manual and also in the cheeks and the wooden blocks at both ends of the keyboard. On the right of the music rack is a touch screen that can also be controlled with a mouse. Sitting behind the console, Aufterbeck explained what it can do. Like any organ, each manual is linked to a given set of stops, but some of the stops on the organ can be linked to any key.

> How do I get a register where it does not belong? With this button I "open" a manual, the switch in the cheek of the manual blinks and I can then add, say, a pedal stop to the manual. If I now press a key on this manual, the pedal stop will sound.[31]

The organist can use the conventional register switches or the touch screen to choose stops. In the register editor mode, the touch screen shows a grid. This enables the organist to build new stops by assigning individual pipes to individual keys. Aufterbeck creates a random pattern on the touch screen. When he plays, every key has its own sound combination. Aufterbeck says that the touch screen is not ideal to change stops combinations while playing. "There is no haptic feedback and it is easy to make mistakes when you touch the screen in the wrong place."[32] Linking pipes to keys gives the organist the freedom to reorder the functionality of the keyboards and combine stops in new ways, whereas the touch screen provides overview and a flexible interface to prepare and store settings.

Another feature of the console is that it allows organists to manipulate the timing of the sounds. On a mechanical organ, the duration of a tone depends on how long a key is pressed down and on the room's acoustics. At the Ratingen console, an organist can use the sostenuto or staccato function.[33] With the first, the tone will sound after the key has been released; with the second, it is the opposite: the sound will be very short, regardless of when the key is released. It is also possible that the pipe will sound only sometime after the key has been pressed. Using these functions, complex sound combinations can be built that go far beyond conventional stop combinations or registrations. To explain how this is done, Aufterbeck refers to a synthesizer.

> I can make a new group on this manual. We know this from the history of synthesizers as a layer, a collection of keys that function as a sound container. For these keys, which could be a whole manual or just one key, I can define several properties. I can decide that only for the left part of the manual the staccato effect is functioning. Because the keys are touch sensitive, I can also decide that when I play with a firm touch, extra sounds will be added.[34]

To give an example, Aufterbeck adds a staccato layer of one register to a normal layer of another. When he presses a key, for a short moment both registers sound, and then only one register remains in the sustained sound. This gives the note an accent. In another example, Aufterbeck uses the "resonanza" function; after pressing a key, five registers sound, one after the other. Creating layers or groups on manuals, or segments of manuals, thus enables an organist to open sound worlds that would be impossible on a conventional console.

After the demonstration, we walk to a nearby café and continue the discussion that started at the console. Loek Dijkman is fascinated by what we heard and he compliments Aufterbeck and Stöckl on their work, but he sees a risk that an organist "would drown in the possibilities, and that we offer something that only a few people will actually use."[35] Although organists are used to improvising and tinkering with sounds, Dijkman says, it might take more to invite them to actually use the tools of the digital console of the new baroque organ. Hans Fidom points out that this is also true for composers: "Organists are used to the fact that an organ has lots of buttons, but the people who write music have little or no idea."[36] Benedikt Aufterbeck argues that many composers do not even write notes for three manuals and pedal anymore. "They compose for computer and use interfaces where it is not the organist but the computer that plays their composition." Loek Dijkman: "But this means, to put it in Dutch, that we use the organ as a *draaiorgel* (barrel organ)? We do not need an organist anymore?" Aufterbeck responds: "It could be, but it doesn't have to be. [Laughs] That's exactly the nice thing. Hans-Ola Ericsson showed us during his Virgil Fox performance that it can be a beautiful moment."[37] Both Hans Fidom and Benedikt Aufterbeck point out that many composers nowadays use laptops to compose and perform their music. Fidom: "Since the organ is a conventional instrument, it is out of sight for these composers. Because of the Sinua software, they may consider again to use the instrument."[38] Loek Dijkman concludes that the new baroque organ can build bridges between the traditional organ world and the instrument and composers and performers who work with electronic devices. "But we will have to invest in that; otherwise it makes no sense, in our research program, but also in our collaboration with conservatories."[39]

HOW LONG LOOPS LIVE

After the visit to Ratingen, I had a better idea of what the new digital console for the baroque organ could be and what type of sounds it could produce. It also became clear to me that the digital console would be a hybrid, mirroring historical stages in the development of the organ interface. It could emulate aspects of the traditional mechanical console, such as coupling manuals and the pedal. Features of the electropneumatic console, such as sequencers, would be added to preprogram series of stop combinations. It would also enable organists to create completely new stop combinations and sound parameters using rotary control knobs, switches, pistons, and pedals. And it would be possible to connect electronic devices such as laptops to the console. How would we offer all these features in a coherent way to musicians coming from different backgrounds? This design brief was on the table during a second visit with Hans Fidom, Johan Luijmes, and Jacob Lekkerkerker to Ratingen in early February 2016. Fidom had prepared a sketch of the layout of the new console. Lekkerkerker wrote a short report on the console and the live electronics connections as he would like to see them.

It is a rainy day when we travel to Ratingen by car. I talk to Lekkerkerker about his work in the Oude Kerk in Amsterdam and his improvisation practice. His ideas about the console derive from this. The existing digital console of the Sauer organ has many possibilities, he says, but for the improvising musician it has its problems. Making new couplings between pipes and keys is not possible while playing, which means that the sound cannot be changed during the performance. In a later report, he writes that this reduces the playing field for an improviser and his suggestions try to solve this. He also proposes to add hand and feet pistons to control the start and finish of a loop, a repeating segment of music. Holding the foot piston to start recording the loop and releasing it to end it would be the most precise way to do this.[40]

In Ratingen, we first go to the St. Peter and Paul Church for a demonstration of the console by Ansgar Wallenhorst. Watching the organist behind the console, swiftly pressing register switches and turning the big rotary knob above the upper console to fine-tune parameters such as sostenuto, delay, and staccato and playing on the keyboards, I realize that Wallenhorst has completely mastered the keyboard and has a clear sense, both mentally and physically, of its possibilities. Listening to his short improvisations, I am surprised how distant they are from conventional organ music, yet how unmistakably present the organ sound is in everything I hear. Fascination mixes with my difficulty to perceive structure and coherence. I realize that the console requires not only new performance skills but also novel ways of listening.

After Wallenhorst has left, we find a table in the nearby café to discuss the design of the console in the Orgelpark. On the table is the sketch by Fidom (see figure 6.3). It shows the three manuals and the stops of the new baroque

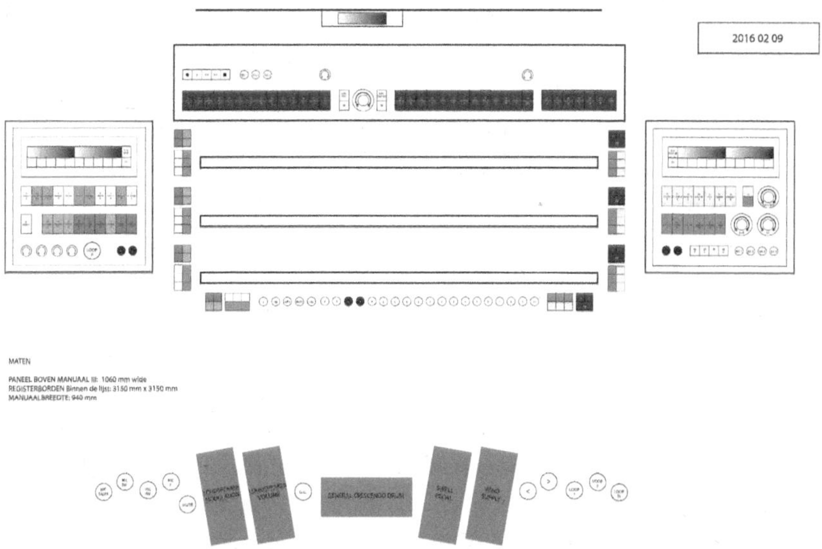

Figure 6.3
Sketch for the digital console by Hans Fidom, February 9, 2016 (reprinted with permission).

organ as black switches above the upper manual. The stops of the Sauer organ are represented by the colored switches on the boards left and right of the manuals. Above these are switches to create new stops (left) and new couplers (right). A rotary knob above the upper manual enables the organist to fine-tune various sound parameters such as sostenuto, staccato, or repeat. Below the lower manual are little round knobs to program the sequencer. Pistons and pedals provide additional control over the loudspeakers, the wind supply, the general crescendo (loudness of the organ sound), and loops.

Johan Luijmes asks how to ensure that the digital console will actually be used. "Would it be possible to add text to explain the principles of creating new stops and couplers? I try to imagine how the average organist will react."[41] In Ratingen, the touch screen shows all available stops for a manual, which gives a complex impression. Luijmes: "We could start from one stop and learn about the possibilities with that stop."[42] Fidom suggests having an instructional video on the touch screen. The console could have a tutorial and a practice mode. In the current design there are several ways to program new stops and couplings, Luijmes remarks, but he wonders if that is too complex. Aufterbeck disagrees. He points out that controls that are left out will be missed by some. Making things simpler implies less functionality, he says: "Ansgar Wallenhorst quoted Olivier Latry, a renowned organ improviser, who said that there should be more than one way to do something. The touch screen is the big friend, easy and flexible, but not for everything. Especially on an instrument where real-time control is essential."[43] Lekkerkerker distinguishes between composers who work a day with the touch screen and program everything, regular organists who make a registration plan and have the time to practice this, and, finally, improvisers like himself who need to find their way on the console organically and in real time.

In his sketch, Hans Fidom used terms like "new stop" and "new coupler" because they sound familiar to organists. In the discussion, Aufterbeck and Stöckl argue that they would rather not use these terms because, strictly speaking, the console does not create new stops and couplers. Its core functionality is that a limitless number of "groups," containers of sounds, can be created and assigned to one or more keys on any manual. New stops and couplers are thus a summary of a series of groups. These groups can be layered; the same keys can combine many groups of sounds. Organist can assemble their own groups and manipulate their characteristics. This can be done before a performance but also in real time. When a group is "open," a light blinks, which means that its characteristics can be changed. Aufterbeck suggests skipping the central rotary knob for controlling sound parameters such as delay, sostenuto, and repeat and instead use a series of smaller push rotary knobs, one for every parameter. Thus, a performer can open a group and change the values of the parameters, such as the delay time of a sostenuto, by turning the small rotary knob. The issue of loops is discussed. By playing multiple loops at the same time, a single performer can create complex musical patterns. Aufterbeck wants to know "how long a loop lives,"—in other words, how long it has to be stored in the console. It is decided that loops will disappear once the console is switched off to prevent too much clutter in its memory. Because the console will have user profiles, however, it can save the groups that a performer has created. These may be reused by other performers, which means that the console can "learn" over time.

On the way back I talk with Jacob Lekkerkerker about the historical temperament of the new organ (see chapter 2). We scrutinize ourselves: Will temperament have meaning only if the pipes are used in a conventional way, as in the diatonic music of baroque harmonies and melodies? And, if these pipe tones are used to make new sounds, what could be the use of the historical temperament? Lekkerkerker says it might be more practical to have the equal temperament of modern instruments, or at least a tuning that would not differ from equal temperament to drastically, also because the Utopa Organ and the Sauer organ will be played together. In my fieldnotes of the day, I reflect on how this tuning issue underlines the hybridity of the organ. In the electroacoustic approach of the digital console, everything revolves around new sounds; temperament does not play an artistic role because it is relevant only in modal and tonal music. Yet one recognizes the acoustic sound of organ pipes, a sound with its idiosyncracies.

That the sonic subtleties of historical temperaments might get lost when the organ was played in the digital mode was a price to be paid for offering new possibilities to make sounds and music that are yet unheard. But if the new console would only offer endless possibilities, it would not work, as Robert van Heumen argued. Even in the seemingly limitless world of electronic and computer-generated music, a musician needs limitations, obstacles, and resistances that trigger creativity. It is precisely because the mechanical console limits the number of pipe combinations that an organist has to develop creative solutions to the problem of what sounds fit a specific musical purpose. For Jacob Lekkerkerker, the practice of improvisation is linked to the tactility of buttons and knobs because they offer the resistance that a touch screen does not. Resistance also plays a role in hacking, as can be seen in the practice of circuit benders. They play with the fixed use inscribed in an object, such as a Furby doll, to open it up to experimental forms of sound production. And finally, during the discussions with Sinua, Loek Dijkman and Johan Luijmes pointed out that the digital console should offer an environment that invites organists to experiment but, at the same time, allows them to use their past experiences. As I hope to make clear, the debate on and design of the digital console can be fruitfully analyzed with the theory of art that John Dewey developed in his book *Art as Experience* (2005 [1934]).[44]

EXPERIENCE AND ITS VARIATIONS

Dewey claims that works of art are usually understood in their external and physical existence—the building, book, painting, or statue. They are detached from the actual life experience from which they emerge and in which they have consequences. This "museum conception of art" (2005 [1934], 4), Dewey argues, is a fallacy. If we want to understand how art works, we should not trace it back to an idiosyncratic experience that is disconnected from everyday experiences. On the contrary, we have to show how it is rooted in these everyday experiences, in the same way that mountain peaks are an integral part of the landscape from which they rise (2). Whereas the "museum conception of art" pairs radical objectivism about artworks with radical subjectivism about experience, Dewey does not see experience as a purely subjective category. Crucial to his reasoning is the ecological idea of the "live creature" that does not live *in* its environment but *in interaction* with it. From these interactions, situations of stability and unrest emerge that alternate in rhythmic patterns resulting in experiences.[45] The live creature not only interacts with the environment, it is also conscious of past, present, and future. In fact, it can use past experiences as a resource in the present to anticipate future possibilities (17). In Dewey's words, "what the live creature retains from the past and what it expects from the future operate as directions in the present. . . . The past absorbed into the present carries on; it presses forward" (18).

Experience resulting from the interaction between live creature and environment is continuous. Sometimes, however, our experiences are composed into *an* experience. This happens when a particular experience is demarcated in the general stream of experience:

> [a] situation, whether that of eating a meal, playing a game of chess, carrying on a conversation, writing a book, or taking part in a political campaign is so rounded out that it is close to a consummation and not a cessation. (37)

Such a consummated experience, Dewey writes, has its own individualizing qualities: It has a plot and becomes a unity that can be told and retold (37). To explain how an experience may have esthetic qualities, he uses the metaphor of a stone rolling down a hill. If it simply rolls, there will be a continuous flow of experiences on its way without any sense of direction and meaning. But when it starts from somewhere, when it is interested in the conditions that hinder or help it to stay on course, is looking forward to reach its final destination, and when, after coming to rest, it relates everything that happened on its way as culminating in its final position—only then, Dewey says, would the stone "have an experience, and one with esthetic qualities." (41)

Interaction with the environment is not just rolling on, in Dewey's metaphor, but takes the form of doing and undergoing. When the live creature acts, say, out of an impulsion such as hunger, it experiences the effects of its actions that result from the conditions in the environment. It is when the live creature becomes aware of the obstacles along the way and the means available to overcome them that this impulsion could become an experience that, as Dewey says, "is clothed with meaning" (62). Energy moving forward is transformed into thoughtful action:

> The junction of the new and old is not a mere composition of forces, but is a re-creation in which the present impulsion gets form and solidity while the old, the "stored," material is literally revived, given new life and soul through having to meet a new situation. (63)

Dewey compares this back-and-forth movement of doing and undergoing, of acting and reflecting, to a painter who brings a painting into existence through the intermittent continuity of control (a brush stroke) and reflection (looking at its effects on the painting). It is the intelligent reflection on this process, taking into account what has already been done and where the work is going, that characterizes the creative work of art (47).

REMAKING PAST EXPERIENCES

Form defined as "something that organizes material into the matter of art" (139) depends on relations, and esthetic form is the "completeness of these relations within a chosen medium" (Dewey 2005 [1934], 139). According to Dewey, art is like nature and life in that these relations "*exist* as actions and reactions in which things are modified." (140, italics in the original) This recalls the brush strokes of the painter as a movement toward consummation. Of all the conditions for form, as the character of every experience that is *an* experience, special attention needs to be paid to the factor of resistance, Dewey argues:

> The existence of resistance defines the place of intelligence in the production of an object of fine art. The difficulties to be overcome in bringing about the proper reciprocal adaptation of parts constitute what in intellectual work are problems. . . . The perceiver as well as the artist has to perceive, meet and overcome problems; otherwise, appreciation is transient and overweighted with sentiment. For, in order to perceive esthetically, he must remake his past experiences so that they can enter integrally into a new pattern. He cannot dismiss his past experiences nor can he dwell among them as they have been in the past. (143–144)

This is a crucial step in Dewey's argument. Precisely because both artist and perceiver have to remake past experiences into new patterns to bring the esthetic experience to a consummation, the outcome of making or perceiving a work of art can never be predetermined because this would only result in mechanical or academic products, Dewey says (144). These would indeed be end products, lacking the element of consummation that allows for presenting something new. Other than with mechanical or academic products, the appreciation of a work of art "carries on" (145).

Artists are experimenters because in every new work they have to express their experiences through the means and materials available in their world. Rather than repeating themselves and becoming "artistically dead," Dewey writes, "only because the artist operates experimentally does he open new fields of experience and disclose new aspects and qualities in familiar scenes and objects" (150). Once created, artworks become part of the environing world, and we may get habituated to them. This may be unfavorable to further esthetic creation, Dewey argues, because

> [t]hings are now too smooth; there is not enough irregularity to create demand for a new manifestation and opportunity for a new rhythm. Art becomes stereotyped, and contented with playing minor variations upon old themes in styles and manners that are agreeable because they are channels of pleasant reminiscence. The environment, in so far, exhausted, worn out, esthetically speaking. (165)

The very familiarity of art that Dewey calls academic can be a source of resistance for some. Great artists absorb the familiar things in a tradition and set up a conflict between it and what is new in themselves (165). If they use something that is, taken in isolation, a repetition, it may gain new meanings if put into a new relationship, one that sums up and carries forward (172). The example that Dewey gives is that of a repeated musical phrase, one that recurs throughout the piece and "gains force because the new contexts in which it is found, color it and give it new value, even if only that of a more insistent, precise and cumulative enunciation of the theme" (172).

The live creature tries to seek a balance between the need for order and the demand for novelty and "craves variety as well as order" (175), Dewey claims. Put more strongly: "The process of organic life *is* variation" (175, italics in the original). Every moment of experience that is brought to consummation—in other words, that completes itself—recurs to its beginning, but the recurrence is "charged with all the differences the journey out and away from the beginning has made" (175). Examples that Dewey gives are returning to the home of one's childhood, meeting an old friend after a long time, or the recurrence of a phrase in music. This, according to Dewey, explains why we can "say of one picture or a story that it is dead, and another that it has life" (183). What endows them with life, not only in a metaphorical sense, is that they, as living beings, are characterized as having a past and a present:

> And I suggest that it is precisely when we get from an art product the feeling of dealing with a *career*, a history, perceived at a particular point of its development, that we have the impression of life. That which is dead does not extend into the past nor arouse any interest in what is to come. (183, italics in the original)

Works of art live actually, not just potentially in some individualized experience. Even though the marble object that is a statue may not change significantly throughout the ages, as a work of art it is recreated when it is esthetically experienced. Great art, such as the Parthenon, is not universal because it remained unchanged through the ages. Rather, it is able to inspire new personal realizations in experience:

> The "works" that fail to become new are not those which are universal, but those which are "dated". The enduring art-product may have been, and probably was, called forth by something occasional, something having its own date and place. But *what* was evoked is a substance so formed that it can enter into experiences of others and enable them to have more intense and more fully rounded experiences of their own. (113, italics in the original)

When old and familiar things are made new in experience, there is imagination. Dewey claims that the concept of imagination is often misused in esthetic theory, for example, as relating to the genius or to specific individuals who bring the new into the world. Yet it is not reserved for art. Every conscious experience has a degree of imagination, for if every experience arises from the interaction between live creature and environment, that experience becomes conscious only when it is imbued with meanings that arise from previous experiences. Imagination is the conscious adjustment of the new and the old (283). According to Dewey, imagination is opposed to habituation. Trains of thought are precisely that, he says. They are too easy (278).

AN EXHAUSTED ENVIRONMENT

My reading of Dewey's esthetic theory brings me back to the exuberant organ console of Virgil Fox and the performance by Hans-Ola Ericsson of Fox's adaptation of a Bach chorale. Both highlight the interaction between an organist and the instrument and the role of a console in this. Following

Dewey, we can now reflect on this interaction by seeing the organist as a "live creature" and the organ and its console as an "environment." Playing an organ now becomes the intermittent movement of doing and undergoing, and as this movement leads to experiences, some of these may be composed into *an* experience. This is how a musical performance may acquire a plot, both for the musicians and the listener. "Clothed with meaning," an experience makes us aware how old material is given new life by integrating it into the present moment.

Why is the digital console of crucial importance for the new baroque organ project and its mission? Following Dewey, we can make the argument that the existing environment, as it was shaped in the Orgelpark, was exhausted or worn out. The very familiarity of what was already there became a resistance. Building on the tradition of new musical instruments and electro-acoustic music in the twentieth century, the digital console offered yet unknown forms of instrumentality and the interactions that these would afford. In a double movement, the digital console also remade the mechanical console and its long history. Through the combination of the two consoles, the new baroque organ would become an "assemblage" (Théberge 2016) or "'electroacoustic ecology" (Harlow 2011).

During the colloquia in 2014, music practitioners and scholars explored the possibilities the new organ would offer, but they also underlined the importance of limitations.[46] This again underlines the importance of resistance. It is precisely because the mechanical console limits the number of pipe combinations that an organist has to develop creative solutions to the problem of what sounds fit a specific musical purpose. Participants at the colloquia underlined that providing unlimited possibilities might lead to artistic paralysis. Van Heumen's request to make an instrument with a "character" can be thought of as an Deweyan environment that offers not only variety but also order. Interacting with the mechanical console renders its own experiences for an organist. Think of how the resistance that an organist feels when pressing a key that opens a valve in the windchest is related to the workings of the action that transfers the movement of the finger through the organ to windchest. Playing on the mechanical console now becomes a choice: Hearing and feeling the mechanical key action offers its own experiences.

The importance of resistance also plays a role in the practice of circuit benders. It takes the shape of a fixed use inscribed in an object such as a doll, which becomes a starting point to literally open it up. The circuit bender's experimental and tinkering approach to sound production also opens new fields of experience and discloses new aspects and qualities in familiar scenes and objects, as Dewey says. Rather than academically predetermining possible outcomes, both the mechanical and the digital console should offer randomness and unexpected forms of interaction. As the synthesizer of Don Buchla became a source of inspiration for Trevor Pinch, the new baroque organ should have the idiosyncrasies that give it the quality of an instrument and not a machine. Machines, or, rather, the mechanical aspects of both instruments and machines, Dewey would say, cannot lead to new experiences that have an element of consummation because they literally repeat the past, lack the relevant variations that are needed to remake the past in the present, and, in doing so, to imagine future directions.

A FOREST OF PIPES

The design of the digital console, as it was discussed in Ratingen in February 2016, continued to evolve. Pistons and knobs were removed, added, or changed to different locations. After going through multiple versions, Hans Elbertse received the final design in the autumn of 2017. Over the summer of that year, he and his employees had dismantled the new organ in the workshop in Soest, transported it to the Orgelpark in Amsterdam, and built it up again. Now they had time to reconstruct the digital console. Some parts would be reused, such as the metal frame and the wooden case, the switchboards on either side of the three manuals, and the music rack. Other parts would have to be changed, such as the cheeks, the large pedals to control the wind pressure and the general crescendo, and the panel above the upper manual that was to be used for the stop switches of the new baroque organ, as well as the knobs and displays for changing the sound parameters.

Elbertse picks me up at the train station in Soest on a gray day in November 2017, and we drive to the workshop. As always, we drink coffee in the office upstairs and discuss the work that has been done in the past months. He explains how the original digital console was constructed and how it will be adapted. First, it had to be taken apart and completely emptied of its inner parts. The new knobs, switches, and displays had to be placed accurately in the panels. The existing wiring had to be rerouted, and new cables had to be connected to the switches and knobs on one end and the Sinua circuit boards on the other end. "Like in the organ as a whole, the cables have to be led neatly through the space we have. We have to understand how this has to be done. That needs concentration, we are organ builders, not computer technicians."[47] Refurbishing the digital console asked for skills that conventional organ builders do not need to develop.

We talk about the music and sounds that all these wires will make possible. When I ask him whether the new baroque organ in its digital mode is still an organ, Elbertse pauses. "Yes, I think so. But it is an organ that we do not know."[48] As a musical instrument it stayed more or less the same since its early beginnings in many ways, he says, but the digital interface of the organ is a big step. It breaks with the age-old convention that pipes are organized in divisions and stop ranks, and that their sound can be accessed only by certain keys on the manual. In the digital mode, every individual pipe can be played by every key, and the traditional sense of ordered sounds disappears. "In fact, you have a forest, a forest of pipes, and you just do something with it."[49] The affordances of the new digital console thus challenge the imagination.

We walk down to the workshop. The frame of the digital console stands in the storage room. I remember how we looked at the stacks of wood that lay around here before the construction of the new organ started in early 2015. Now I see panels with knobs and countless thin colored wires that have paper labels at the end, indicating their destination. I realize that the electronic parts remain almost completely blackboxed to me. In contrast to the mechanical organ, I cannot get a grip on their inner workings other than on a very abstract level. When I visited the organ workshop in the past years, I could see the organ grow and develop an understanding of its parts and their functions by looking at them, touching the surface of the wooden frame or the leather of the bellows, and asking for clarifications. One can see, feel, and hear the workings of the mechanical organ. The printed circuit boards, wires, and cables remain silent in their complexity.

When we enter the main workshop room, it strikes me how empty it is without the organ that stood here two floors tall. Nothing recalls of the hours and hours of work that was done here. Elbertse shows the panels of the digital console and explains how precisely the holes for the knobs and displays have to be milled out. As with the organ itself and the hall in the Orgelpark, space is limited here. It has to be used economically to make things fit. I take pictures of the control panel, which houses the circuit boards that Sinua constructed (see figure 6.4). Realizing that I look at the backstage of the digital console, I marvel at its combination of ordered efficiency and abstract beauty.

Figure 6.4
Backside of the control panel of the digital console of the new baroque organ in the workshop of Hans Elbertse (photo taken by the author).

Before I leave, Elbertse shows me the organ bench that has been made for the new organ. Its blueish-gray color vaguely reminds me of the organ bench that Hildebrandt built in 1723 for his organ in Störmthal. It even has the same trapezium shape. Whereas the Störmthal bench had a flat wooden surface to sit on like most historical and contemporary organ benches, the new bench has a crimson leather upholstery. It also has a turn knob to raise or lower the seat. This will allow the organist to find the perfect position at the mechanical console. I look at the bench and try to imagine what this organist will play, and how the new baroque organ will sound in the hall of the Orgelpark.

7

PLAYING BACH

A SPACE AND ITS ACOUSTICS

In the years that I visited the Orgelpark, I got to know the building and its history. I still have a sense of wonder and expectation when I walk up the stairs to the hall. Writing in my notebook after a day of listening to the final voicing of the new baroque organ, in December 2017, I realized it was not only because of the unusual sight of four large organs in one room, or because of the spaciousness of the dome and its elegantly curved ribs. Or even because of the lively combinations of colors on the paneling and the balconies. When I enter the hall, I anticipate the sounds I will hear there. The ways in which the room shapes these sounds have become as familiar to me as the orange chairs on which the audience is seated, or the windows with their translucent images on the windows, telling the story of the Park Church and its recent restoration. Now that the new baroque organ had been transported from Soest to Amsterdam, its sounds and the room acoustics of the Orgelpark will be attuned during the final voicing of the pipes.

On one of the last days of 2017, I meet Munetaka Yokota and Mariya Kato, who are voicing the Octav 2' stop of the Oberwerk. A metal scaffolding stands in front of the organ, leaving some room to sit at the console. The space around the organ looks like an organ builder's workshop. Tools and paperwork are lying everywhere. Yokota is standing inside the organ, close to the upper windchest. Kato sits on a chair near the console with a row of small pipes on the organ bench before her. She takes up a pipe, measures its windway, blows and overblows the pipe, and puts it back in its row. Following short instructions from Yokota, she moves a piece of lead from one key to the next to keep a pipe sounding. Yokota and Kato use a transceiver to communicate. Its crackling noises mix with steady, sharp, high-pitched tones that fill the room.

I sit down on one of the balcony chairs next to the Van Straten organ to listen. Yokota has descended from his high position inside the organ and joins me to evaluate the Octav 2' at a distance. For him, listening to the newly voiced pipes individually is not enough. He wants to hear them with other stops and assess what they do in a piece of music. He asked Kato to play the first variation from the organ partita *O Gott, du frommer Gott* (BWV 767) by Johann Sebastian Bach. I am excited because it is the first time I hear music on the new organ. After Kato has used the Octav 2' in various combinations of stops, Yokota decides it is too loud. If it would be the only high-pitched stop on the organ, it would be OK, he tells me. But now that it will be combined with other high stops, he will have to "calm it down."[1] He nods when I ask him if this means that he will have to go over every individual pipe in the 2' stop again.

The meticulous work of the final voicing of the new baroque organ had to be done before the organ could be used. Placing the new organ in the hall, however, would have an effect not only on the sound of the instrument itself but also on the Orgelpark as a concert venue. As with any organ innovation, the design and making of the baroque organ can be seen as a mirror of its time. Adapting the sound of the instrument to the hall thus becomes a metaphor for "fitting" the organ into contemporary musical cultures.

In this chapter, I focus on how the new baroque organ was embedded in its environment on several levels. Following the chronology of the new baroque organ project, I will first focus on the design of the case and the ornaments. Discussions about this started early in the project, but the final decisions were only made in the spring of 2016, after the fifth study trip to Saxony and Thuringia (see chapter 5). A recurring theme in the debate was if the organ should present itself as a historical instrument or as an instrument of the twenty-first century. Not only in its appearance but also in its sound and temperament, the organ had to be integrated in the Orgelpark. The debate on the temperament of the new organ was settled with the choice for Hildebrandt as the main historical reference. Yet, a solution had to be found to the problem that emerged when both organs were played at the same time on the digital console: How to combine a historical temperament for early music on the baroque organ with the equal-tempered Sauer organ? After the organ had been shaped, painted, voiced, and tuned, it was ready to present it to its audiences. This was done during a series of concerts that started on March 21, 2018, the 333rd birthday of Johann Sebastian Bach. Combining performances of Bach's music with improvisations on the digital console, these concerts anticipated the future programming that the new organ would make possible.

By describing how the new baroque organ was embedded in its "environment," I also aim to address more fundamental questions on how it relates to its possible futures as a musical instrument. To do so, I will revisit the esthetic and hermeneutic philosophy of Hans-Georg Gadamer that I introduced in chapters 1 and 2.[2] Whereas nineteenth-century Romantic writers and philosophers argued that the esthetic experience should focus on the artwork itself by disregarding the original context of life, according to Gadamer, the work of art cannot be isolated from the contingencies of the conditions under which it presents itself. Rather than being experienced in an alienated esthetic consciousness, art exists it through its acquiring of new uses and adapting to changed circumstances. Being presented under the new conditions of the contemporary, artworks are fundamentally unfinished and their meaning is never exhausted. I will introduce Gadamer's concept of "play" that lies at the heart of his hermeneutic ontology of art. Whereas the new baroque organ project started in 2013 with the ambition to be able to play Bach's music in new ways, I will show how Gadamer's concept of art as play allows us to understand the work that was to be done to achieve this task.

NEGOTIATING THE APPEARANCE OF THE ORGAN

The close relation between a building and an organ not only concerns acoustics and sound. As organ historian Stephen Bicknell (1998) explains, unlike other musical instruments, organs make an architectural contribution to the buildings that house them. Designing an organ case used to be seen as a form of architecture "with its own grammar and traditions, but still allowing free interpretation according to changing fashions, local influences, and the skills of the designer and organ builder" (Bicknell 1998a, 55). Early medieval organs probably did not have a case, but from the fourteenth century onwards, images and remaining historical instruments suggest that cases became common. These had the practical function of protecting the pipes but also of visualizing the symbolic role of the organ. Comparable to altar pieces, the organ doors were painted, and opening them was part of a ceremony on special occasions (1998a, 58). The design of organ cases around 1700 reflected differences between countries and regions as well as changing esthetic preferences. The organs built by Silbermann and Hildebrandt incorporated the taste of the eighteenth century, as can be seen in the use of classical columns, bases and capitals, entablatures, and pediments. With the advent of the organ reform movement in the early twentieth century and especially the postwar neobaroque trend in organ building, the abundant embellishments that characterized baroque organs were traded for stylized shapes. In the late twentieth century, historically informed organ builders started to follow the "pattern languages" of baroque organ builders (see chapter 2). This implied going back to the original esthetic and functional meanings of embellishments such as carved wooden shades. In a parallel development, the late twentieth century saw a wide range of innovative organ designs; organ builders experimented with new case shapes and placements of the pipes, sometimes using new materials such as metal or plastics in the case, and new forms of embellishment, such as colored light effects.

The choice between historical reference and innovative design that characterizes the late twentieth century returned in the discussions about the design of the new baroque organ: Should it present itself as a contemporary organ, or was it better to hint at its historical reference to organs of Hildebrandt? Both approaches can be found in the Orgelpark, as can be seen in the first newly commissioned organ, the French Romantic organ built by the Dutch firm Verschueren Orgelmakers in Heythuysen. Instead of following the style of the nineteenth-century French organ builder Aristide Cavaillé-Coll, it was given a contemporary case designed by the architect Bas van Hille from Oosterbeek, the Netherlands, who was also responsible for the refurbishment of the Orgelpark, including its color schemes. The Verschueren organ, with its freestanding rows of front pipes, a large swell box with vertical slats, and a solid base painted in shades of orange and dark red, blends into the hall. Its colors match those of other places in the building such as the cloakroom, the restrooms, and the foyer downstairs, underlining the coherence between building and instrument. The two other main organs in the hall, the German Romantic Sauer organ and the medieval Van Straten organ, were either brought back to their original appearance and color (Sauer) or followed the original design while omitting historical details such as the pinnacles on the two towers (Van Straten). The existing situation in the hall thus only underlined the design challenge for the new organ.

SHAPE, DETAILS, AND COLORS OF THE CASE

Answering the question of how the new baroque organ should look took several years. The organ's facade is mentioned for the first time in the report that Hans Fidom wrote after the meeting of the core team in Soest in October 2013. It says that Bas van Hille will be responsible for its design. When available, he will get the information about "the conditions that the sound quality of the organ puts to the design."[3] Four months later, in February 2014, Fidom specifies this point: "The structure of the facade of the new baroque organ follows from its specification,"[4] which at this point in time was thought to have forty-five stops. This does not mean, he writes, that the organ will have baroque entablatures and ornaments. On the contrary, the idea is that the organ shows that it is not only baroque but also an instrument for new music. Fidom suggests a cooperation with the VU University in Amsterdam to investigate the developments in organ facade design in recent decades. Two months later, the decision is made to build an organ with a smaller specification because of the weight limitations of the north balcony (see chapter 2). This has consequences for the structure of the organ and its facade. Loek Dijkman says that he plans to consult Bas van Hille because the building is "an architectural unity," but that does not mean that Van Hille will decide on the design.[5] During the meeting, Hans Fidom shows pictures of the Hildebrandt organs in Sangerhausen (1728) and Hettstedt (1749). Of the latter, only the case is original.

In September 2014, the core team visits Hildebrandt organs in Thuringia, such as the organ in Sangerhausen (see chapter 3). In his report of the study trip, Hans Fidom notes that the core team "became enthusiastic about the organ in Sangerhausen and considers to what extent its facade should be an example for that of our own organ."[6] On the one hand, he writes, the idea of an exact copy is attractive; on the other hand, the organ contains new technology and, therefore, is not as thoroughly baroque as a style copy of the facade might suggest. "We can imagine that our organ gets an appearance that follows the example of Sangerhausen and Hettstedt relatively accurately, but that we make our own choices when it comes to ornaments and colors."[7] During the colloquium in October 2014, Fidom shows how the facade of the Hettstedt organ would look in the Orgelpark. At this stage it has not yet been decided to place the bellows inside the organ case, he remarks, and also the spring chests may be larger than the slider chests that Hildebrandt built, which affects the dimensions of the facade.

The idea to model the design of the new baroque organ after the Hildebrandt organs in Sangerhausen and Hettstedt is discussed again in January 2015. Both facades have the same structure and match the dimensions and specification of the new baroque organ. In his minutes of the core team meeting, Hans Fidom summarizes the outcome of the discussion: The facade should show that the organ is meant to play baroque music but also new music. "We decide that the first condition can be met by giving our organ's appearance the structure of the organs in Sangerhausen and Hettstedt; and that the second condition can be met by giving the color and details a 21st century character."[8] In February 2015, Fidom travels to Germany to measure the organ cases in Sangerhausen and Hettstedt. Now that the bellows will be placed inside the organ case, Fidom proposes to take the base of Sangerhausen, the pedal towers from Hettstedt, and the middle part from Sangerhausen, but at the height of that in Hettstedt. The measurements of the structure of the organ case are now fixed in the final plan of the organ. They define the space for the bellows, windchests, pipework, and actions inside the case. The ornaments and colors will be decided upon later.

During a meeting on March 10, 2015, the design of the case is discussed again. Loek Dijkman remarks that the drawings of the facade of the new organ show that "we are still torn between two ideas."[9] He thinks that the structure of the organ is as it should be, including details such as the four columns "that show where the organ comes from."[10] He doubts, however, if the organ should have explicit colors and suggests that it could have a transparent wood finish like the Van Straten organ opposite to it. Johan Luijmes points out that if one refers to historical colors, it would be good to also add other historical elements. "Some people miss the pinnacles on the Van Straten organ."[11] Dijkman repeats that the organ should show its place of origin. "If Bach is sung next to it, the organ sings along. That's nice."[12] Hans Fidom agrees: "One also listens with one's eyes."[13]

The discussion continues in September 2015. Although there is consensus that it refers to Hildebrandt in its overall structure, it should also reveal itself as a new instrument, which raises the question how this should be done. If the organ cannot be a replica, or even follow the pattern language of Hildebrandt, creating a convincing reference that is also esthetically interesting becomes a challenge. Hans Elbertse rejects the idea that the organ would not be painted at all; an unpainted pinewood case would look like a carpenter's box, he says.[14] Hans Fidom summarizes the discussion in his report: the coloring will be reserved and the rather explicit ornaments from the organ in Hettstedt will not be copied. There will be shades to cover the space above the pipes, but these will not be according to historical examples. The rounded entablatures next to the middle tower of the Oberwerk will be deleted, as will be the antique shapes of the capitals above the columns. Hans Elbertse agrees to make a new drawing of the facade, and Hans Fidom will make images that project the organ in the Orgelpark's interior.

Two months later, in November 2015, the core team meets to see a presentation of the results of the research into contemporary organ designs done by Fabienne Chiang, a student from the VU University in Amsterdam supervised by Hans Fidom. She found that the color white is often used as well as light effects. Materials such as glass and plexiglass are common; shades above the pipes are rare. Fidom has projected the drawing by Hans Elbertse in the Orgelpark and gave the organ colors such as ivory white and blue. Loek Dijkman intervenes: "We are now looking at pictures to see who likes what, but I thought we were talking about the shape of the organ. Should there be shades, for instance? But we should not interfere with the color. Because we are not experts in that."[15] Hans Fidom shows pictures of various shade forms, some with red lights behind the pipes instead of shades, following an idea that emerged already in 2013 (see chapter 2).[16] The merits of light effects are discussed as well as the idea of giving the case a stylized appearance where only the rough shapes refer to Hildebrandt. Hans Elbertse points out that he already started working on the four columns in the front, "but we can stop with that of course."[17] After Hans Fidom shows a photomontage of the Hettstedt case in the Orgelpark, Loek Dijkman changes his mind. "We have to determine again what we want. Do we stick to the shape of Hildebrandt's pipe fields which we then dress and undress with details? Or do we say: no we should not do that. And then I make a complete turn and say: it is dangerous to half undress the case. When I see Hettstedt, I think it is a beautiful organ."[18] In his report of the meeting, Hans Fidom summarizes the outcome of the discussion: The organ case will follow the example of Hettstedt in its details. Only in its color scheme will the organ show its twenty-first-century descent. Bas van Hille will be asked to make a proposal for the colors of the case using a new drawing of the facade and photos of the Hettstedt organ.[19]

The issue of the colors of the new baroque organ is on the agenda during the study trip to Germany in March 2016. Hans Fidom, Johan Luijmes, and I talk about the color designs made by Bas van Hille on our way from

Amsterdam to Sotterhausen. The architect prepared three proposals. One leans toward the colors of the organ in Hettstedt. Most horizontal elements in the case are beige, the vertical element such as the four columns are blue-green, and the panels purple-pink. A second proposal combines a reference to Hettstedt following the existing color scheme in the Orgelpark, with green vertical elements. The third proposal fits the colors of the Orgelpark. Here, the base shows a combination of red and crimson red. Both Luijmes and Fidom have strong doubts about the colors, about the sharp contrast between the base and the upper parts of the organ, and about the texture of the paint.

In Dresden, the core group visits the organ in the Hofkirche, built by Silbermann, but finished by Hildebrandt in 1755 after Silbermann's death in 1753. Its case is ivory white with small light laurel green panels and abundant gilded rococo ornaments. In the light, white interior of the church, the organ has a dignified appearance. During lunch, the proposals made by Bas van Hille are discussed. There is consensus that they do not place the new organ as a separate entity in the room and that the colors are not in line with the character of the instrument.[20] At the same time, there is enthusiasm for the combination of ivory white and gilded elements in the facade of the organ in the Hofkirche. Later, the team visits the St. Nicolai Church in Langhennersdorf which houses Hildebrandt's first organ, built in 1722. It is a small instrument in a village church. Its case is painted light gray with grayish-blue panels and gilded entablatures (see figure 7.1). The shades are relatively simple gilded wood carvings with grayish-blue elements. Loek Dijkman is enthusiastic about the colors, and the decision is made to use the Langhennersdorf organ as a point of departure for colors of the new baroque organ.

BUILDING UP AND PAINTING THE ORGAN

Hans Elbertse and his coworkers presented the finished organ during the hundredth anniversary of their firm in June 2017. The capitals of the four columns, ornamented with acanthus leaves, had already been gilded, and some of the entablatures had been painted, but fresh pinewood dominated the look of the organ case. Immediately after the festivities, work on disassembling the organ began. Whereas building the organ took more than a year, taking it apart was done in two weeks. Starting early July, large parts, such as the windchests, the bellows, the caps on the towers, the columns, and the case panels, were transported to the Orgelpark by truck and moved into the building with a telehandler through the old main entrance of the church that is no

Figure 7.1
The Hildebrandt organ in the St. Nicolai Church in Langhennersdorf, built in 1722 (photo taken by the author).

longer in use. The parts were stored in the hall that was almost completely emptied. The smaller parts and the pipes were brought from Soest to Amsterdam in vans. In the end, the floor of the hall was filled with organ parts, as if the components of a construction kit had been neatly laid out.

On the north balcony, preparations had been carried out to accommodate the new organ. The supporting structure beneath the organ had been reinforced and a leveled, flat surface had been created on which the organ case would rest. Around this surface, a wooden scaffolding had been built up, which required some of the chairs and rails to be removed. A metal rolling scaffolding enabled the workers to move around the new structure. In front of the balcony, a hoisting device was attached to the ceiling. With it, heavy parts such as the bellows were lifted to the balcony, rolled to the frame on a dolly, and put into place. "Once the bellows were in place, the heavy work was done," Hans Elbertse says, "but you should not underestimate the windchests. Everything had already been in the right place before, so it should fit."[21] After a month of reassembling, the organ had reached its highest point with the placing of the cap on the middle tower (see figure 7.2).

In August, I see the organ for the first time in its new room and am impressed by its size. It looks much larger than the Van Leeuwen organ. The columns and entablatures as well as most of the panels have been painted. The wooden carved shades in the pedal towers, made by woodcarver Gert van den Dikkenberg following the shades in Langhennersdorf, are gilded and painted as well. Hans Elbertse explains that it was decided to start painting already while the organ was being built, to save time. On the balcony, which looks like a workshop with tool cases everywhere, I walk around the unfinished structure and realize that I am familiar with many of its inner parts after my frequent visits to the workshop in Soest. The space at the console will be painted in the same blue that can be found in Naumburg, and the stop drawknobs already have their name plates. Above the upper manual a rectangular digital display is ready to show the sequencer combinations.

A week later I am back in the Orgelpark. Hans Elbertse called me the day before to inform me that the first organ pipe will be placed by Loek Dijkman and Sylvia de Munck during a small ceremony. On the balcony, I see how the panels in the front are being painted. In the foyer of the Orgelpark, I meet Gerard de Jongh, who is responsible for painting the organ. During its hundred-year history, the firm carried out what De Jongh calls "the better work": his great-grandfather and grandfather worked for the nobility and rich farmers in the province of Utrecht. More recently, the firm specialized in restorative painting of monumental buildings or churches. Being an amateur organist himself, De Jongh started painting organs in the 1970s and became an expert in researching the original layers of paint on an object and reconstructing its historical colors. Sometimes a snippet of paint in a corner is the only reference to the past, he tells me. "You make a small window on the surface and you start to scratch with a scalpel, layer after layer, until you reach the clear wood. You put numbers next to every layer."[22] Going back in time like this, de Jongh lays bare layers of historical paints. He learned how

Figure 7.2
Building up the new baroque organ in the hall of the Orgelpark, August 2017 (photo taken by the author).

to recreate these by using the original components such as pigments, some of which are very toxic, and linseed oil. He still has pigments that his grandfather made, using a rubbing stone and a mortar. Mixing a pigment with linseed oil and adding a siccative, he is able to reconstruct old paints in any color.[23]

How were the colors for the new baroque organ callibrated and made? Hans Fidom recounts what happened after the decision was made in early April 2016 to use the colors of the Langhennersdorf organ as a reference. After initial experiments together with Loek Dijkman to find the right colors by making drawings in Photoshop, he decided to contact organ builder Kristian Wegscheider who had restored the organ in Langhennersdorf between 1990 and 1996. From Wegscheider he learned that the colors on the organ were reconstructed in close cooperation with Hilke Frach-Renner, a color expert from nearby Dresden. Frach-Renner provided the information on the colors of the Langhennersdorf organ; they were taken from the catalogue of Sikkens paints. Using the paint numbers, Fidom was now able to make computer drawings that showed the organ with its exact colors. Given the fact that the organ in Langhennersdorf is much smaller than the new baroque organ and has a very different structure and facade, decisions had to be made about translating the coloring scheme. Hans Fidom: "Loek Dijkman and I agreed that the vertical lines should have an accent color, and also the architraves of the caps on the towers. Loek Dijkman also thought it would be good to have enough gilded objects in the facade."[24] After the color design of the case had been decided, Hans Fidom communicated the colors to De Jongh using the Sikkens colors range. Instead of just using the standard products of the paint company, however, De Jongh mixed his own versions of these colors. Thus, he was able to make paint with the right substance to prevent an industrial, smooth appearance.[25]

Later that day, Loek Dijkman and Sylvia de Munck arrive for the ceremony of placing the first pipe. This has been a tradition in the Elbertse firm since it began this type of work. The pipe, the largest Principal 16' pipe in the left pedal tower, lies on a blanket-covered board on two trestles. An employee of Elbertse is busy engraving the text: "Deze pijp is geplaatst op 22 augustus 2017 / Loek Dijkman and Sylvia de Munck / voorzitter en vice-voorzitter van Stichting Utopa."[26] There are short speeches by Hans Elbertse and Loek Dijkman. Then the pipe is taken from the table, laid on the ground, and attached to special covers to be hoisted to the balcony. Slowly it rises from the floor to a vertical position and hangs in the air. On the balcony, Bert van den Heuvel and Jos Elbertse gently guide the pipe to its place in the left tower. After the blower is turned on, we listen to the sound of the pipe—low and still undefined. When we drink champagne, I cannot take my eyes off the empty organ case with its single pipe. Surrounded by the scaffolding it resembles a mysterious machine. This impression is strengthened by the white, green-blue, and gilded parts that contrast with the other colors in the hall. The arc of the ceiling lies over the structure as a gentle cover. I notice that the organ stands quite far back on the balcony and does not impose itself, as if is completely self-evident that it is there.

VOICING THE ORGAN IN THE HALL

The acoustics in the hall of the Orgelpark are as idiosyncratic as its color scheme. The building has the shape of a Greek cross with one leg flattened and oriented toward its center, in contrast to most churches that have a longitudinal axis. Its dome and rounded ceilings above the balconies act as sound reflectors. Analogous to the challenge of fitting the visual design of the baroque organ in the existing situation, achieving the final sound character of the instrument required a delicate operation. Even though the pipes had been made and prevoiced according to historical Hildebrandt pipes (see chapter 5), their actual sounds would inevitably echo the characteristics of the hall of the former Park Church.

In the early phases of the project, the core team considered if it would be useful to map the acoustics of the Orgelpark, but the decision was made not to do this, because the organ will be voiced "on the hall."[27] This hall has a tendency to overemphasize high-pitched sounds, which was one of the reasons to choose Hildebrandt's organs as a historical reference instead of Schnitger's, which have more high-pitched stops (see chapter 2). In early 2014, the possibility of improving the Orgelpark acoustics with an electronic acoustical control system was explored. This system could emulate the acoustics of the much larger St. Wenzel Church in Naumburg, and it might also inspire performers of new music to experiment. In April 2014, however, the core team decided not to invest in electronically enhanced acoustics.[28]

Another issue was the loudness of the new organ. The core team was keen not to repeat the experience with the newly built Verschueren organ, which was thought to be too loud by many organ experts. Anticipating the moment that he would hear the organ for the first time, Johan Luijmes said in February 2017, "Maybe I think it's terrible and too loud. I have the idea that it is a sort of mild baroque organ. That's what I have in my head."[29] It would take until October 2017, when Munetaka Yokota and his assistant Mariya Kato started to give the organ its voice, before it slowly became clear how the organ would actually sound in the Orgelpark.

FINAL VOICING

In November, I meet Yokota and Kato near the scaffolding in front of the organ. Hans Fidom is also present and invites me to play. I find a book with Bach's *Inventionen und Sinfonias* and start to play the first Invention. Yokota has drawn the newly voiced Prinzipal 8' and Prinzipal 4' and the Viola da Gamba. While playing, I listen to the sound and am keenly aware of the pressure that I have use on the keys. I find the sound of the organ in the acoustics exciting, but for some reason, it proves to be difficult to play the piece fluently. The small mistakes I make annoy me. When I stop, Fidom and Munetaka are standing behind me, listening to the organ sound. Fidom takes over, and I walk with Yokota to the Van Straten organ and sit next to it. We listen long and attentively. The sound of the new organ is beautiful and exciting, and at the same time, I experience how it is still in the making. As I write in my field notes later that day:

> Hans says that the sound is "born" and that is a nice metaphor. I use it in my conversation with Munetaka that afternoon: the idea that you want to get to know a newborn baby by intently looking and listening. He agrees: "It is exactly like that! Trying to find the good character of a pipe and then try to nurture it and let it be, so to speak. That's the voicer's work I guess. And also to teach it some of the coordination with others, and to have a lot of pipes work well together."[30]

In my perception, the organ is neither too loud nor too soft, and its sound seems to blend beautifully in the room.

In December, I learn in more detail how difficult it is to give the organ a coherent sound character. It is not only the hall acoustics that are challenging, but the new organ itself affected its sound in unexpected ways, Yokota found out. In January 2016, he voiced a series of test pipes in the open front of the Van Leeuwen organ. He used these pipes as a reference to do the prevoicing later that year. In the actual situation, however, the pipes of the Hauptwerk division are surrounded by surfaces: The big wooden pipes of the pedal are standing on both sides, the front pipes and the rollerboard for the Oberwerk are before the Hauptwerk pipes, the windchest of the Oberwerk functions as a ceiling for the Hauptwerk pipes, and at the back of the organ is a plywood panel thirteen millimeters thick to prevent sunlight from coming into the organ. Next to this, Hans Elbertse had made wooden frames with thin black cloth to keep the inner parts of the organ out of sight. As a result of all this, the pipes of the Hauptwerk sounded softer than expected, especially the higher-pitched pipes. "It is almost as if the pipes are standing in a swell box," Yokota says.[31]

There are two solutions to the problem of the soft Hauptwerk pipes: opening their toe holes or raising the wind pressure. The second option is impossible, Yokota says, because in contrast to the Hauptwerk, the Oberwerk pipes would turn out to be too loud. Standing above the Hauptwerk, the Oberwerk is completely open. Also, the Oberwerk is positioned below an arched ceiling made out of hard material. Because of that, even though it is softly voiced, the Oberwerk pipes sound "monstrously loud" in the center of the hall, Yokota found out.[32] The sounds of the Oberwerk pipes are reflected in a focused way, which means that they are heard differently in different places in the hall. Raising the wind pressure to amplify the Hauptwerk would make the problem of the loud Oberwerk worse.

Down in the hall, Yokota invites me to experience the peculiar acoustics while Kato plays the organ. Standing in the middle of the hall, I find it difficult to locate where the sound is coming from, as if it has no direction. Walking to the sides, this effect disappears. On the balcony near the Van Straten organ, Yokota points out how the sound of the organ is somehow reflected, possibly by the shutters of the Van Straten organ facing the Utopa Baroque Organ across the hall, which causes an echo.

The final voicing thus becomes a balancing act. Because the Hauptwerk and Oberwerk divisions cannot have different wind pressures, the loudness of the pipes has to be adjusted by opening or closing the toe holes of the pipes. This, however, affects the speech of the pipes, which has to be corrected by slightly opening or narrowing the windways and raising or lowering the languids. These often-minute changes have to be done in conjunction with not only the other pipes in a stop but also the organ as a whole.

> The Rohrflöte 4' and the Gemshorn 4' in the Oberwerk have very different characteristics. Hildebrandt's Gemshorn is distinctive, very different from the Gemshorn built by Schnitger that I know well. The Gemshorn in Naumburg is not built by Hildebrandt himself, so working on this stop was like meeting a new person.[33]

When he adjusts it, Yokota listens to what a stop like the Gemshorn wants to say and how it cooperates with other stops. He compares his work with the casting of a play. A director has to find actors who can give depth to the drama. In the same way, he as a voicer has to give his stops a complex, lively sound so that the organist can perform the score in an artistically interesting way.[34]

DESIGNING THE TEMPERAMENT

The new baroque organ had to be not only voiced but also tuned. The temperament of the organ was an issue at the start of the project when it was still unclear whether it should have a meantone temperament, a well-tempered temperament, or even a combination of both (see chapter 2). The choice for Hildebrandt as the historical reference made it clear that the organ would not have a meantone temperament, as it would if Schnitger's organs had been chosen as models. This still left open a wide range of temperament options. Musicologist, music historian, and expert in historical temperaments Ibo Ortgies, who was also involved in the Swedish GOArt project that built a replica of a Schnitger organ in the Örgryte Church in Gothenburg, was asked to design the temperament.[35] Whereas equal temperament is the tuning standard nowadays, at the time of Hildebrandt and in his region, several temperaments coexisted.[36] Hildebrandt himself used a well-tempered system that "followed Neidhardt" for his organ in Naumburg (Ortgies 2018 73; see also chapter 2).

To design his temperament, Ortgies first established the preconditions and requirements in the Orgelpark situation. The first requirement was that the new organ would have to play with ensembles. In this situation, historical organ builders such as Hildebrandt preferred an unequal temperament, Ortgies argues, which gives a better intonation support to brass instruments such as trumpets and woodwinds such as bassoons, oboes, and flutes. It was also decided to give the organ the "Cammerton" pitch where the note a1 is 415 Hz.[37] The soundscape of the new baroque organ was another point of attention. Equal temperament (see chapter 2) goes well together with fundamentals whereas unequal temperaments allow for stronger higher stops because they tend to be purer. Because the organ would be voiced with both rich fundamentals and partials, an unequal temperament would be better. A third requirement was that the new baroque organ should be combinable with the Sauer organ, which has equal temperament and is tuned in the modern concert pitch of a1 at 442 Hz. When played together from the digital console, the deviations between the two temperaments should be kept to a minimum. The required temperament should thus allow performance in all keys, render all Bach repertoire as artistically convincing as possible, be not too close to equal temperament yet allow the organ to play with the Sauer organ, have thirds that are as harmonic as possible, and have some pure fifths (Ortgies 2018, 79–80). The solution to this puzzle was the "well-tempered tuning Ortgies/Orgelpark 2," which can be seen as a tailor-made answer to a combination of requirements that is totally twenty-first century, yet looks back to the tuning and performance practices at the time of Hildebrandt.[38]

The actual tuning of the pipes was done by Yokota and Kato (see figure 7.3). Although Ortgies claims that it is quite simple to tune his temperament by ear, Yokota also used a tuning device that detects a pipe pitch and measures its deviance from the required pitch. Tuning an organ flue pipe is done by making the air column inside shorter or longer.[39] The first can be done by simply cutting or sawing the metal or wood at the top. The second is more difficult and requires an extension to the pipe.[40] In general, Yokota started to voice the pipes at the relatively low pitch of a1 at 411 Hz, a bit below the required 415.3 Hz of the "Cammerton." The .3 was added because the Sauer organ's pitch is 442 Hz, to meet the pitch most other instruments today have. He explains that the height of the cut-up of a pipe is related to its length: When he shortens the pipe, the height of the cut-up changes. This may result in a sound that is more fundamental or less bright, because that is the effect of a higher cut-up. A pipe that sounds perfect may become dull once it is lowered. Another interdependency is that if a pipe is made to sound louder, its pitch goes up. If Yokota started voicing at the final pitch, he would not be able to make pipes louder anymore. Lowering a pipe by adding an extension to it might be possible, but Yokota wants to avoid that, he says, because it also changes the sound character of the pipe.

TESTING THE DIGITAL CONSOLE

Through their voicing and tuning, Yokota and Kato shaped the sound character of the new baroque organ in its mechanical mode. This mode uses "indirect wind": When a key is pressed, wind flows from the pallet box that is under wind pressure into the tone channel for that key and into the pipes of the stops that have been drawn. This indirect wind has certain qualities that interact with the pipe and, thus, influence the speech and steady sound character. In the digital mode, however, all key valves in the pallet box are opened, and the organist manipulates the smaller pipe valves directly below

Figure 7.3
Mariya Kato inside the new baroque organ tuning a wooden pipe by shortening it with a Japanese saw, Orgelpark, January 2018 (photo taken by the author).

the pipes. As was extensively discussed during the study trip to Germany in September 2014, the two kinds of wind—direct and indirect—might cause differences in speech and tone between the mechanical and the digital modes (see chapter 3). The organ might sound good when played in the mechanical mode, but its refined voicing character might be damaged when played from the digital console. This problem, it was thought, could be resolved once the new Sinua magnets used in the tone drawers of the spring chest could be regulated. By manipulating their force curve, the opening and closing behavior of every individual spring valve could be controlled by a printed circuit board mounted next to it (see chapter 4). Thus, it would be possible to minutely control the wind flow into the pipe and the resulting tone character. In the first months of 2018, Benedikt Aufterbeck and Thomas Stöckl from Sinua worked long days at the Orgelpark to fine-tune their system and its software.

When I arrive in the hall in February 2018, one month before the festive inauguration of the new baroque organ in March, I meet Aufterbeck, Stöckl, Bert van den Heuvel from Elbertse Orgelmakers, and Hans Fidom. Aufterbeck is sitting on a chair behind the console. On the organ bench before him stands a laptop. It displays the software editor that allows him to change the computer code in the console. Fidom shows me how two chairs have been turned into a small test station for the software on the small printed circuit boards next to the spring valves on the tone drawers. He explains that when the magnet drivers were delivered in early 2016, Sinua still had to finish their software. This now means that new software has to be uploaded to every individual circuit board in each of the tone drawers. Without this software, the spring valves do not behave properly: They sometimes open too far and get stuck, causing the tone to continue, or they do not repeat correctly. The new software is transmitted to the circuit boards in an old-fashioned way, Fidom says, using an audio track in the same way modems use sound to transfer digital information. This transfer procedure had to be tested in the improvised test station.

Suddenly, Benedikt Aufterbeck starts pressing all the keys on one of the manuals with a slat while Fidom randomly presses the round knobs of the sequencer. The odd procedure turns out to be a way to overload the organ system with information until it crashes. This will not happen with a human performer, Fidom explains, but when laptops with audio-processing software are connected, the situation is different. Aufterbeck looks at the tablet again to locate the bug.

Seeing Bert van den Heuvel reminds me of the countless hours of work on the tone drawers and their spring valves, connectors, cables, and printed circuit boards in the workshop in Soest. When I ask him what this testing procedure does to the material components inside the spring chest, he says that things move either very quickly or very slowly, which is quite different from what these parts were designed to do in the sixteenth century. Worries about the vulnerability of the mechanical parts inside the spring chest, such as the spring valves and their leather hinges, made Hans Fidom decide to put limits to what the software can do to the mechanical parts through the force of the magnets. Both the attack and the release now have two phases. This results in a smooth opening of the valve and a good speech, and it also reduces the risk of damage.

I accompany Stöckl and Fidom inside the organ where they upload the new software to the circuit boards near the spring valves (see figure 7.4). While Stöckl attaches his smartphone with the software audio track to one of the main switchboards that controls the magnets drivers, Hans Fidom pulls out the tone drawers one after the other to cause the magnet drivers to restart. When the software is installed correctly, every magnet opens and closes in a rhythm that Stöckl made up: two longer and two shorter clicks. Sitting on my knees next to the Hauptwerk windchest, I marvel at the sight of the LEDs on the individual magnet drivers lighting up and hearing Stöckl's rhythmic pattern amplified by the air column on the pipe above the valve.

When we are down in the hall again, I meet Trevor Grahl, who is the Orgelpark's artistic assistant. Grahl is also a composer and a teacher of composition as well as instrumentation at the Royal Conservatory in the Hague. He sits at the digital console and starts to experiment. We listen to the baroque organ doing strange things, making unheard sounds. The speed of the tremulant, normally causing a constant fluctuation in the steady tone to add to its expressivity, can be adjusted. By combining a very slow speed of the tremulant with a short pulse, the sound of the organ seems to stutter in weird patterns. Despite the very long working days, sometimes ending in the middle of the night, the mood is upbeat. While Grahl plays, the hands of Aufterbeck and Fidom are pressing stopknobs or turning rotary knobs; new sound combinations are tried out in which the Sauer organ and the new baroque organ melt into one instrument. Using a very slow tremulant and gradually diminishing the wind pressure with a foot pedal, Grahl plays the Dutch national anthem. The organ sound disintegrates in distorted, eerie noises. "It sounds like it's dying," Grahl says, laughing. When I look up at the new organ, I see a small red light blinking in the circuit board behind an open panel left of the mechanical console. It seems to wink at us.

AN INSTRUMENT TO INVESTIGATE POSSIBILITIES

The Orgelpark's mission is to integrate the organ in contemporary musical life by presenting it in new ways. The work that was done to build up and embed the new baroque organ in the hall of the Orgelpark—negotiating

Figure 7.4
Thomas Stöckl inside the new baroque organ uploading software to the magnet drivers using his smartphone, Orgelpark, February 2018 (photo taken by the author).

its appearance and giving it its voice and temperament—could be seen as a metaphor for this. Once the work of the organ builders was finished, the organ could be played before its audiences. It would become a part of the Orgelpark's concerts, symposia, and other events, but what could be its role?

The magazine *Timbres*, published twice a year by the concert venue since its start in 2007, gives an overview of the history of its programming.[41] From the start, Johan Luijmes developed concert programs and events that depart from the traditional format of organ concerts. Instead of an organist playing a familiar repertoire before an audience of organ aficionados, the focus is on rarely performed organ music, especially from the mid-nineteenth century to the present, and on crossovers and collaborations with other musical genres and art disciplines such as jazz, world music, dance, film, and visual art. In addition, the Orgelpark planned to seek collaboration with higher music education institutes to give young talents a chance to perform during or compose for special concerts.

In the years that followed, the Orgelpark explored existing repertoire in new ways and experimented with new repertoire and performance formats. Repertoire from the early middle ages to the present was performed, new compositions were commissioned, and festivals around minimal music or composers such as Messiaen and Xenakis were organized. Combinations of organ and other instruments such as the piano, saxophone, trumpet, or the theremin were tried out, and repertoire was adapted between the organ and other instruments. Portraits of composers such as Wolfgang Rihm, Hans Kox, Hans Koolmees, and Guus Janssen were organized, and during the colloquia and symposia, recitals and concerts were combined with lectures, presentations, and debates. In addition to the concerts, the Orgelpark issued CDs with performances by Dutch organists playing on its organs. With the completion of the new baroque organ, new concert programs will become possible that might change existing ideas of how an organ could or should sound.

AN ART THING TO CONNECT TO THE DIGITAL

In February 2017, I meet Luijmes in the foyer of the Orgelpark to talk about his work as an organist and concert organizer, his ideas about programming, and his hopes for the new organ and what it might establish. Luijmes has vivid memories of hearing the opening choir of Bach's *St. John Passion* when he was eight. He started playing the piano, but later decided to study organ at the

Conservatory "because I wanted to play Bach's *Trio sonatas* that I recognized as something that is inconceivably smart and well-made."[42] Being able to play the six sonatas was not only a goal in itself but a way to learn more about how to play the organ. His organ teacher Bert Matter introduced him to historically informed performance practice. He continued to play the organ, he says, because he had historical instruments at his disposal that helped him to develop a very good idea of the beauty of organ sound. "At the time when I studied, you did not want to play pneumatic organs. It was a strange notion of purity, the idea that you did not have a direct contact with the keys. The organ became more a machine than an instrument. If you play pneumatically, it feels as if you hand over something."[43]

Luijmes had to get used to the Orgelpark because it is not a historical church with sixteenth- and seventeenth-century instruments, he says. The link between the instrument and the space was broken. "We all know this as organists: that you get the key of the church, that it is dark, and that you walk through a big church, switch the thing on and that you play. You have the feeling that you summon something up, you wake up spirits. That's why you want to be an organist and not a pianist."[44] As an instrument, the organ invites people to inquire, Luijmes thinks, even more so than a piano or a violin. When trying out stop combinations, an organist finds that no organ is the same. "You are always adapting if you play an organ."[45] I remind Luijmes of a conversation we had during the trip to north Germany in October 2013. The idea to combine historical organ-building practices and digital technologies, or even two different temperaments, might lead to what we called a "monster," something that is so hybrid that it does not have a historical character of its own. This is still a challenge, he thinks. Finding people who are able to compose for it will be complicated. "There are not many organists who said: you should build an organ like this. It is the other way around, a bit like the Orgelpark as whole: we started with hundred concerts a year without even knowing if there would be an audience."[46]

Asked how the organ will inspire new music, Luijmes explains why it is essential that the actual sounds are generated by physical pipes instead of electronic devices.

> It gets a kind of humanity. A human proportion. That's why I think the organ is so fantastic, the fact that it really breathes, just like a human voice. . . . At the same time it relates to the superhuman. People value the organ because it is grand and majestic. And because you do not see the organist; it is the organ that plays.[47]

How is the new organ going to change the role of the Orgelpark in musical life? Luijmes notices that the venue has become an independent entity with a unique position both in the organ world and in musical life. That has partly to do with its financial possibilities, he thinks: "We can commission a composer to write for an ensemble or an instrument."[48] Luijmes thinks, however, that many organists find the work of the Orgelpark interesting but will not follow in its trail. He distinguishes between two organ worlds in the Netherlands. On the one hand, there are the young organists who are brought up in parts of the country where the reformed churches are still very much alive. On the other hand, there is the academic organ world that is closely related to the historically informed performance practice. Both worlds are conservative, Luijmes says.

Yet, more than any musical instrument, the new baroque organ opens up to the contemporary musical world and its coevolution with digital technologies, Luijmes argues.

> I can think of no other art thing that is more appropriate to connect to the digital and the world of contemporary music. Think of paintings and photography that eventually became digital photography. These are very different things. Here we still have that thing you have to play, and that keeps extending itself in new directions. There is no more beautiful instrument to investigate all of that, that's fantastic. I mean, a violin and an electric violin and that's it. But this instrument, taking into account that it is still air that comes out of the pipes, manages to include ever new technological extensions—you can use it in so many ways. Yet it retains this sound quality. I hope that what we will have here is similar to why one finds these historical organs so beautiful.[49]

IMPROVISING AS WORKING IN SITU

To have an idea of how the new baroque organ could be used to create new music, I met the organ improviser Jacob Lekkerkerker in the final weeks before the inauguration. In 2016, Lekkerkerker had formulated his proposals for the design of the digital console (see chapter 6), and he was now preparing an improvisation for the opening concert week in March 2018. Every day, Lekkerkerker started working at the digital console in the Orgelpark in the early morning hours to try it out. As Lekkerkerker formulated it: "It is like getting to know someone."[50] After him, others like Wouter Snoei and Robert van Heumen, all regular guests at the Orgelpark for years who had also been invited to compose new music for the organ's inauguration, would have their six-hour time slots to prepare their works. Their trials of the digital console would sometimes cause malfunctions that could be resolved only through debugging the software by Sinua.[51] Artistic and technical learning went hand in hand.

Seated behind the console, Lekkerkerker switches it on using a personalized key that gives him access to the settings he created earlier. These settings provide him with the sound material that he can explore further while playing. He explains that his main influence on the design of the console was to make sure that the improviser could change sounds and settings during the performance. "This means that you do not have to program everything in advance and stick with these settings while performing."[52] He shows me how he makes a "layer" that combines a soft 8' stop from the Sauer organ with a staccato attack from a stop on the baroque organ. Although this sound combination would not be possible on a conventional organ, linking sounds to keys is a first principle in organ building and playing, he says. "But because of the software you can vary a lot more here. You can get intervals that you would not have on an ordinary organ."[53] Lekkerkerker thinks from the tradition of organ playing: "I am still old-fashioned, I look for colors: oh yes, let's start from that Principal, what does it do, and what if I combine it with the Viola di gamba?"[54] When I ask him how he remembers all the settings and sounds, he compares himself with a soccer player who knows his position on the field. "This takes a couple of hours. And you can always check the touch screen. I don't use the screen to make changes. For me it is more like a summary of my settings."[55]

Lekkerkerker has a Calvinist family background, as he calls it. His father and some of his nephews played the organ, so it was part of his daily life. When he moved to Amsterdam, he took up the idea to play the organ himself after he had played the piano already for some time. Because he was born with two fingers on his right hand, the organ did not seem an obvious choice, but thanks to his organ teacher, Jos van der Kooy, he soon learned how to find his way around on the instrument. "The most important thing on the organ is that you know: if I do this, something happens there. That you know how to time and how to use the room acoustics."[56] Because the daily practice studying of organ pieces from the traditional repertoire proved to be too demanding, he finally stopped playing them in 2022. "But the basis of what one has to offer as an improviser can still be broad enough to fill a whole life."[57]

For Lekkerkerker, the Orgelpark proved to be decisive in his artistic career. The first digital console for the Sauer organ provided him with a "playing ground" that the Oude Kerk (Old Church) in Amsterdam, where he was liturgical organist (2007–2016) and organist titulair (2013–2022), could not be for him. Working with the digital console in the Orgelpark led him to the idea of processing the sound of organs in other churches by electronic means and to experiment with placing these sounds in the space. He started collaborating with electronic guitarist Alfredo Genovesi and sound technician Clare Gallagher. "You have the live organ sound that is copied, so it is doubled. This doubled sound is available for the electronic musician."[58] The sound technician can influence the balance between the live and the processed organ sounds. "I always try to make something of which the audience thinks: where does the sound actually come from. Something floats, but what? I find that exciting."[59] Working with others required rethinking his improvising habits because the effect that the electronic musician has on his playing can be quite dramatic, he says.

Improvising is a talent, Lekkerkerker says. It is the ability to develop and evaluate the creative process at the same time without evaluating it too much.

> So you have a kind of hyperconsciousness: it needs something now. Where can I find it and where can I find it in such a way that it is not something completely new, that it fuses. You are thinking on many different levels at once. Harmonically, that you have that in order. That you know which sounds are actually available, that you know the effect of these sounds in the space. It is a combination of all that, but also investing in the site specific data: that you know the organ, that you know the space, something every organist has to do. That you take the time to know where you are.[60]

From the millions of possibilities, Lekkerkerker has to select those that create a balance between space and sound, he says. "It is only now that I can work on it, because I did not know how it sounded. That is how I work: from the sound. I work in situ."[61]

We talk about the importance of having two different interfaces available, the mechanical and the digital consoles. According to Lekkerkerker, they enriched the Orgelpark because it becomes part of a very broad range of musical communities.

> There are organists who make sounds with keys and registers, the early music community for instance. They know how the organ responds when they play it

> in a certain way, and it does what they want because it is top level and all materials are of the highest quality. This community can embrace the organ. People like me can expand the sound at the digital console. Using the organ as a new instrument, but from a background as a traditional organist, perhaps that is actually the smallest group. But they are still there. And then you have a group of people from the world of electronic music. That is also a huge group.[62]

Having two consoles enables everyone to become friends of the instrument, Lekkerkerker says. And this is good, because "music communities nowadays are extremely diverse. People like classical string quartets, jazz, electronic music. Just look at the concert program of the inauguration week. This is also typically Orgelpark."[63]

INAUGURATING THE UTOPA BAROQUE ORGAN

After months of voicing, tuning, testing, fine-tuning, and learning in the Orgelpark, the new baroque organ was presented to the public. The decision was made to call it the Utopa Baroque Organ, after the foundation that funds the Orgelpark.[64] Five days of concerts in March 2018 had been scheduled, starting on March 21, Bach's birthday. The concert programs reflected the two faces of the organ: as an instrument to perform Bach and to make new music. For me, the inauguration marked the end of my fieldwork. This moment was arbitrary; the organ was not completely finished. Further improvements were planned during the summer on the voicing of some pipes and stops and adjusting the software for the magnet drivers to increase the touch sensitivity of the keys of the digital console. Finding out how the organ would be adapted by the diverse communities and audiences it hopes to address would extend my research project indefinitely. Yet observing the preparations and attending the concerts in the inauguration week gave me an impression of how the organ was going to be played and used in the future.

The inauguration week started with a press conference in the afternoon of March 21, 2018. Journalists from Dutch and international newspapers have been invited, as well as editors of various organ periodicals. The white paper bags that they receive contain a copy of the spring issue of *Timbres* with an overview of the concert program, technical information about the Utopa Baroque organ, and a CD with recordings made in the previous weeks. Loek Dijkman welcomes the invitees, and David Franke, who introduced us to the Hildebrandt organ in Naumburg more than four years ago, plays the slow movement from the *Concerto in A* (BWV 593). Jacob Lekkerkerker improvises at the digital console, using variable wind pressure. Hans Fidom summarizes the main ideas behind the project. When Franke plays a baroque improvisation that he called *Komm süsser To(n)*, Fidom shows his registrations (stop combinations) on a projection screen.[65] After a round of questions, mostly focusing on technical details, everyone walks to the balcony to see the new instrument.

When I return to the Orgelpark in the evening, the cloakroom and the foyer are crowded. The sold-out opening concert is broadcast live on the national classical music station, Radio 4. Hans Fidom is interviewed before the concert. When everyone is seated facing the new instrument, Loek Dijkman gives a short speech in which he explains how the organ fits the mission of the Orgelpark "to place the organ in the art world and the music world, and not only in the church."[66] The opening concert by organist Marcel Verheggen and the Studium Chorale choir conducted by Hans Leenders focuses on the world of Lutheran chorales. Organ and choir alternated in performing motets and chorale preludes by Bach and his predecessors. The *Partite diverse sopra 'Sei gegrüsset, Jesu gütig'* (BWV768) exemplifies a tradition that started with Jan Pieterszoon Sweelinck and his pupils to investigate the possibilities of a simple chorale melody by playing variations on it. The idea of musical variation returns in the concert program of the next day. Organist Jan-Willem Jansen and the gamba consort The Spirit of Gambo perform the fugues from Bach's *Kunst der Fuge* (BWV1080), exploring the forms and textures Bach derived from the simple opening theme. The gamba players are seated on the balcony, close to the organ. The sound of their instruments resonates in a wonderful way with the stop combinations that Jansen uses.

Earlier that day, I interview David Franke to hear his comments on the Utopa Baroque Organ. He switched from the piano to the organ when his legs were long enough, he says, and playing on Silbermann organs shaped him as a musician. After he studied organ in Stuttgart, Copenhagen, and Berlin, he applied for the job of organist at the Naumburg organ in the St. Wenzel Church in 2008. He had seen the organ in the 1990s when it was restored, and he was drawn to it because of a book he read as a child about the relation between Silbermann and Hildebrandt. Franke explains that Silbermann built five different types of organs that share the same sound characteristics. "Hildebrandt is different. It is not one model. He always adapted to the situation."[67] Although he followed his teacher Silbermann at first, he gradually developed his own style.

Franke conjectures that Hildebrandt's specifications changed after he met Bach in 1722, in Störmthal. His own experiences on the Naumburg organ support this, he says. When he plays the *Schübler Chorale* or the *Canonic Variations*, which were written between 1745 and 1747, the years in which Hildebrandt built the organ in Naumburg, he can find registrations that perfectly fit the music. "Also in other variations, I found very characteristic registrations. It fits so well, that it cannot only be coincidence."[68] Franke reminds me that Bach was known for his unconventional registrations, and he links this to the new aspects that Hildebrandt introduced into the Naumburg organ. "If you project this idea of trying out new possibilities in the twenty-first century, then you have the electronic console. It gives a new dimension to the organ as an instrument. I like the idea that we look to the past as well as to the future with this instrument."[69]

On the third concert day, David Franke and Ansgar Wallenhorst, organist at the St. Peter and Paul Church in Ratingen (see chapter 6), join forces. Franke plays Bach's *Triosonate in C* (BWV529) and the *Concerto in A* (BWV593) at the mechanical console, offering virtuoso passages that sound the organ's combined stops and slow movements where the musical line progresses from one single pipe to another. Wallenhorst improvises at the digital console. Playing with reduced wind pressure, he creates flageolet-like sounds with a metallic quality, breathing sounds from the lower pipes, and fluttering figures that recall birdsong. He explores the gray zone between speech and tone, and punctures loops of repeating patterns with groups of staccato notes. There is the somewhat alienating experience of hearing sounds that seem to be completely disconnected from the familiar character of an organ tone. Emerging from the soundscape are short fragments from Bach's *Concerto* as a faint memory. After their duo concert, Franke and Wallenhorst improvise at the digital console, sitting down and standing up as a kind of relay, handing over the musical thought to the other to develop it further.

The fourth concert day is Digital Organ Day. Starting at 11 a.m., Jacob Lekkerkerker, Bram Stadhouders, Ansgar Wallenhorst, and Hans-Ola Ericsson play the new organ from the digital console. Accompanied by a small group of musicians, Jacob Lekkerkerker performs an improvisation based on his explorations of the organ in the previous weeks. It strikes me that at some point he places small metal bricks on keys to keep pipes sounding. The simple gesture shows how despite the wealth of software-enabled possibilities, new tools can still be added. After lunch, I listen to Bram Stadhouders.

As a guitarist, he combines improvisation, composition, and electronic music. He plays his electric guitar over the sound system of the Orgelpark. The electric signal of his guitar is led to his laptop where it is processed to control the digital console, which plays not only the Utopa Baroque Organ but also the Sauer organ and the Busy Drone barrel organ. I marvel at the sight of the guitarist surrounded by an ecology of electronic devices, including the idle digital console, and the sound of three organs being played simultaneously with an electric guitar.

The final concert in the afternoon is given by Hans-Ola Ericsson, who wears headphones to monitor the recorded electronic sounds that he combines with his own live playing. When I interview him the day after, Ericsson tells me he expects to learn through the new instrument something he did not know about music. "And that has to do with the touch, with the sound, and how the organ reacts when I melt sounds together, the alloy of sounds. . . . Bach would have been here in the Orgelpark, he would not have been out there rejecting it. No no no. He would have been here." Bach, Ericsson says, was always developing complexity, esthetically and artistically.[70]

The Digital Organ Day ends with an evening concert with works by composers and improvisers who helped to develop the digital console in the years before. At various moments Robert van Heumen, Jacob Lekkerkerker, René Uijlenhoet, and Wouter Snoei give their opinion on the design. Van Heumen, Uijlenhoet, and Snoei all have a background in electro-acoustic music, and they all use laptops in their performances. Snoei also developed his own MIDI input device: a small black box with four black wooden wheels on top acting as controllers (see figure 7.5). He built it twenty years ago during his study years at the Sonology Department of the Royal Conservatory in The Hague. In addition he uses a commercial Keith McMillen QuNeo MIDI drum pad controller that allows additional forms of interfacing. When I interview Snoei the day before the concert, I ask him why he still uses the four wheels. "It is very precise. It detects the smallest movement. I can make 1,024 steps by turning the wheel around a single time. It will never make a huge jump, and I use that in the software to connect the speed of the movement to a sound or a direction."[71]

We discuss the difference between composition and improvisation: "What I do is emphatically not an improvisation, it is a composition."[72] Snoei agrees that what he calls a "composition" is in fact a site-specific combination of a musical conception, a set of documentations, and a technological setup

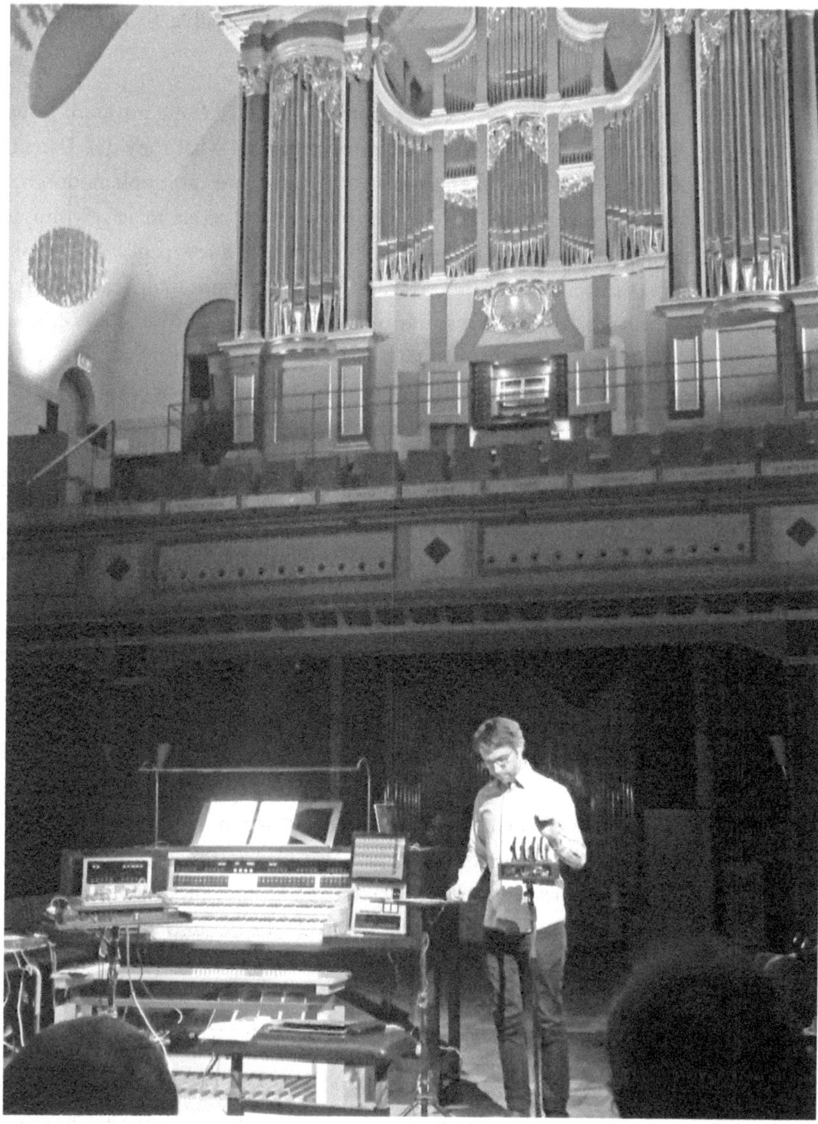

Figure 7.5
Wouter Snoei performing with his MIDI input device during the evening concert of the Digital Organ Day, March 23, 2018 (photo taken by the author).

including specific hardware and software. Will it be possible to perform his compositions in twenty or thirty years?

> Yes, but it will be difficult. We can still perform the electronic music of Luigi Nono because his engineers have documented very well what they did. Even if we do not know what original equipment they used, we can implement what their drawings and schemes tell us. . . . The music of Nono has relatively simple effects that you could summarize in a small diagram of a few pages. If I would make a scheme of my composition, it would be very complicated. I can describe in general terms what I have done, but then it will not be the same piece. Also because there is, of course, twenty years of development and refinement in all the processes that I use. It is a work, but it is also bound to a situation.[73]

As an example, Snoei tells how the piece he made for the inauguration of the Van Straten organ in the Orgelpark in 2012 (see chapter 1) was performed in Poland. In the original situation, he connected contact microphones to various parts of the organ. The bellows were manipulated manually, which was crucial:

> The bellows treader had their own score. It was quite specific. With the Van Straten organ you can influence the air so directly that when the bellows stop moving, the sound suddenly ends. It is about the air circulation in that instrument, that it is a kind of living, breathing instrument. That's why the title of the piece is *Organism*. . . . In Poland it was performed on an organ that I did not consider to be suitable, but I only found that out when I was there. [Laughs] Eventually we had three men lying in the organ and an organist in front of it, and I did the electronics in the hall. We managed to get those bellows effects. People were literally pushing those bellows inside the organ [laughs]. We communicated in half-German, half-Polish, and we had a kind of system with click sounds so that organist knew when he had to play. It was something else, but the thought remained intact. It was unmistakably that piece.[74]

On the program of the final concert in the inaugural week is the Bach cantata *Himmelskönig, sei willkommen* (BWV 182) by the Cappella Amsterdam choir and the Holland Baroque ensemble. It was first performed on exactly the same date 304 years earlier, on Sunday March 25. Bach wrote it for a small ensemble; every instrumental part is performed solo, probably because the chapel in Weimar had only a small balcony.[75] The small choir and soloists stand in a straight line in front of the organ. There is no conductor, and the organ now has the role of keeping the voices and instruments together in the basso continuo playing. Listening to the performance, I think back of the discussions in the core group about having enough space on the balcony for singers and musicians. The proximity between the organ and the musicians works well acoustically, and the sound of the organ modestly blends with the other instruments.

A few weeks later, I look back on the inauguration week with Hans Fidom. In the two months of trials, problems with the digital console were solved. Sometimes the solution was simple, as when the console's computer overwrote previously added Setzer combinations when it was switched on. Sometimes the software itself had to be changed, as in the case where Robert van Heumen's laptop, running the real-time audio synthesis and algorithmic composition software SuperCollider, sent too many information packages per time unit to the console, which led to stuck tones. On the morning of the Sunday concert, Fidom discovered that the organ sounded very out of tune. It turned out that one of the regulators inside the wind channel from the blower to the bellows malfunctioned, causing one of the bellows to open too far. Hans Elbertse and his colleagues solved the problem by adding a safety valve. Fidom: "It was a test. But I am proud of the thing and that it works so well."[76]

Where to go from here? Plans are made to organize a yearly organ academy, Fidom says:

> To bring the best students in the world together around these organs. But also to connect bubbles. I find it super exciting that the organ has this baroque heart. So we can ask people who are fond of playing Böhm or Mendelssohn to be present when Robert van Heumen and his students are working with their laptops and tablets. Because you work with the same sound. Bringing together musicians and researchers around the theme of sound and materiality, trying to understand what it means to make music.[77]

It would be possible to offer a digital copy of the new organ to organists worldwide, Fidom thinks. Using Hauptwerk virtual pipe organ software, players can download sets of organ pipe samples—looped recordings of the pipe sound in its actual acoustic setting—that can be played with MIDI keyboards. The Orgelpark has everything available to make sample sets of the mechanical side of the organ, Fidom says, but sampling the real-time manipulations of the pipe sound would be technically impossible. "And besides that, I think that if you work with this new organ, then you have to be in the acoustics of the space itself. Loudspeakers will never match the real organ."[78]

The instrument reveals an irony. Its combination of the archaic spring chest with twenty-first-century digital technologies allows organists and listeners to use the new instrument anywhere in the world, yet experiencing how the baroque sound material is reworked in new music requires them to travel to the Orgelpark in Amsterdam.

HERMENEUTICS IN ACTION

The inauguration of the Utopa Baroque Organ in March 2018 marked its installation in the Orgelpark. To be played, it had to be fitted into the building in a variety of ways. During earlier stages of the project, the architectural peculiarities of the building limited the weight and, therefore, the specifications of the instrument. The work of Munetaka Yokota and Mariya Kato focused on voicing and tuning the pipes to create a harmonious whole that fit the acoustics of the hall. Doing this meant making modifications to the pipes and to the organ. The debates on the appearance of the organ considered not only the existing color schemes of the Orgelpark but also the message the organ would give about its own historicity. The tuning of the organ anticipated that it had to be played together with the Sauer organ, the building's original instrument. And finally, fitting the organ in the building also implied connecting it to what some envision will become a global network of hyperorgans. All these "fittings" were the subject of extensive discussions in the core group and at the project symposia. They led to practical problems that required the inventiveness and artisanry of the organ builders. And they reflected, and sometimes challenged, assumptions about how the organ would be played.

The Utopa Baroque Organ replaced the Van Leeuwen organ on the north balcony to enable new performances of organ music from the baroque and, more specifically, the compositions by Johann Sebastian Bach. The goal was never to build a "Bach organ," however, but an instrument on which Bach's music would sound convincing, as an exemplar of a world, a tradition. Its aim was not to reconstruct and revisit a sounding past, but rather to have an organ that could take this tradition further. This is how Bach's music was played during the organ's inauguration. It resonates with Snoei's story about *Organism* being a composition and not an improvisation. Snoei traveled to Poland with his piece and performed it under very different circumstances than those in the Orgelpark for which the piece was originally made. To perform it in a new situation, all sorts of work had to be done. Snoei's story is a nutshell of what happened in the Orgelpark with Bach's music. Acknowledging that it cannot be performed under the original conditions, we have to work with it. What kind of work is this? To answer that question, I return to Gadamer's historical and esthetic hermeneutics that I introduced in chapters 1 and 2.[79]

Gadamer's hermeneutical account of how art exists starts from a critique of the "esthetic differentiation," a process of abstraction that disregards the world in which a work is rooted so it becomes visible as a pure work of art (Gadamer 1989 [1960], 74). According to Gadamer, rather than being experienced in an alienated esthetic consciousness, art exists through its acquiring of new uses and being adapted to changed circumstances. How then to overcome the radical subjectivization of the esthetical that started with Kant? In his answer, Gadamer develops an ontology of the artwork that wants to leave room for its hermeneutic significance. At the heart of this ontology lies the concept of play.[80] Art resembles play, he says, in that both do not exist in the subjective experience of the viewer, but in the artwork or the game itself as a "subject." The game has primacy over the consciousness of the player; it offers an order in which the coming and going of the play occurs naturally. A general characteristic of playing, Gadamer writes, is that the fascination of a game consists in the fact that it masters the players: "all playing is being-played" (1989 [1960], 106).

Playing is different from other behavior; it requires a playing field. In its rules and goals, it becomes clear that playing is always playing something, Gadamer writes, and this something presents the player with a task. It is in the performing of the task that the play presents itself.[81] In its presentation the play is brought into existence, not only for the player but also for an audience. The play as a whole comprises players and spectators. In the spectator, the game is raised to its ideality; because the game is played before an audience, it becomes clear that it bears within itself a meaning to be understood and that, therefore, can be detached from the behavior of the player.

At this point, Gadamer makes the crucial comparison between play and art. Like the play, the work of art exists in the ways it is made present before an audience that recognizes in this presentation its meaning. It is important that the work of art cannot be isolated from the contingencies of the conditions under which it presents itself. On the contrary, it can be presented in many ways, yet it has a binding character that again resembles a play like chess: Although every game is different, the game of chess itself retains its identity.

According to Gadamer, the productive variations that result from presenting the artwork under different conditions leads to a tradition (the German uses the more active word *Überlieferung*) to which every new attempt to present it has to relate.

In the case of performative arts, such as theater and music, presenting works of art is not a matter of following earlier acts of creation but of interpretations that keep the future identity and continuity of the artwork open. These interpretations do not give the interpreter unlimited freedom, and on the other hand, Gadamer sees canonizations of particular interpretations, for example, a recording of a musical composition by its composer, as a simple imitation of a model that would deny the real task of interpretation.[82] Gadamer then goes on to argue that every interpretation could be seen as an instantiation of a work, yet it is impossible to distinguish between the work and its mediation (*Vermittlung*). This mediation is total; the work only presents itself through the performance, and it exists in its relations to the life world. This reads as a conundrum: If the work can exist only in the singularity of the situation, how then can we think of a work at all? In other words, if it presents itself differently in changing ages and circumstances, why does it not disintegrate and lose all identity? The answer is that all changing aspects of the identity are contemporaneous (*gleichzeitig*) (1989 [1960], 119).

What kind of contemporaneity is this, Gadamer then asks. What is the temporality of the esthetic? He recapitulates his argument: The work of art is play in that its existence depends on its presentation, in which the unity and identity of a structure emerge. The obligation of every presentation is to relate to the structure and derive a criterion of correctness from it. This might suggest repetition, Gadamer says, but not in the sense of being a repetition of something original. "Rather, every repetition is as original as the work itself" (1989 [1960], 120). This puzzling temporal structure, he claims, reminds of the periodic festival (see also Gadamer [1986]). An upcoming festival is neither a different festival, nor is it a remembrance of the one originally celebrated. That every festival will be different for the people who celebrate it, yet one and the same for the historian, is beside the point, Gadamer argues. The festival is not there in the subjectivity of those who celebrate it; rather, the being of the participant is determined by her "being there present." In other words: "To be present means to participate" (1989 [1960], 121). This is the meaning of contemporaneity: However far the origin of a festival is from us, in its celebration it exists in the participation of those who are present. The same is true for the contemporaneity of the existence of the artwork: It is in its presentation that it achieves full presence, however remote its origin may be (1989 [1960], 123–124).

Gadamer explains the notion of participation through the example of Greek tragedy. Here, the spectator recognizes themself in the actors they see on the stage. Their trials and tribulations have exemplary significance in that they show a metaphysical order that is true for all, including the spectator. This continuity of meaning is what enables the Greek tragedies to exist in the present. Gadamer then shows that this is also true for other art forms, such as visual arts, architecture, and literature. In their various ways they do not exist only in the past, detached from the present. Rather, contrary to the subjectivist esthetic, Gadamer developed the concept of play as the event of art proper (1989 [1960], 138) that requires the active participation of those who present works of art to make them exist at all. The past is continuously renewed in the present. With the example of architecture and historical buildings Gadamer makes this more concrete.

Works of architecture, such as the Park Church, have specific aims to serve and are also taken up in their spatial context. If an architect is successful, a building both fulfills its purpose and adds something new to a town or landscape. Through this dual ordering it can be seen as a work of art in the meaning that Gadamer gives to this term. If its original context of purpose and life—such as being a church—is unrecognizable, the building would become incomprehensible. Yet if it would only be "an object of aesthetic consciousness then it has merely a shadowy reality and lives a distorted life only in the degenerate form of a tourist attraction or a subject for photography" (1989 [1960], 149–150). The building that was once the Park Church and became the Orgelpark exemplifies how "[w]orks of architecture do not stand motionless on the shore of the stream of history, but are borne along by it" (1989 [1960], 150).

What architecture teaches us is that even though every work of art has its own world, this does not mean that when this original world has vanished, it can exist only in an alienated esthetic consciousness. On the contrary, it exists through its acquiring of new uses and adapting to changed circumstances—this is called mediation (*Vermittlung*)—while retaining elements of its original lifeworld. This thoughtful mediation with contemporary life is what is at stake in the playing of artworks as productive variations on what resulted from presenting them under different conditions. In this playing, we refer to a tradition of these earlier presentations that were handed over to us, and we carry on this tradition through our attempts to renew artworks in our own world.

Whereas the project to build a new baroque organ in the Orgelpark started in 2013 from the ambition to play Bach's music in new ways, Gadamer's concept of art as play gives "playing Bach" connotations that the everyday use of the word does not have. Going back to the *Concerto in A* played by David Franke during the inauguration of the new organ, we can now rethink this performance as a presentation (*Darstellung*) of the work. Understood as a thoughtful mediation with contemporary life, his playing rested not only on ideas about the organ or the music that was to sound on it but on all the kinds of work that I have described in this and the previous chapters. There, we saw how this thoughtful mediation materialized: in the travels to historical organs to see and hear historical organs as a reference and inspiration for the new organ; in the reappropriation of knowledge and techniques from the sixteenth and nineteenth century through electronic means; in the understanding and relearning of eighteenth-century craftsmanship through retracing working procedures, circumstances, and tools from that time; and in the ways in which musicians such as Franke and Lekkerkerker rehearsed and added to musical traditions. In all these moments, the goal was never to reconstruct a vanished past but to make Bach's music present in new ways.

In addition to providing us with a hermeneutics-in-action perspective, the concept of playing also allows us to read the story of the Utopa Baroque Organ in a normative register. As Gadamer writes, it is only in the performing of its *task* that a play presents itself or, in other words, that it is able to exist. Translated to art, this means that artworks have to be played to make them work as art. The project that led to the Utopa Baroque Organ sprung from a sense of responsibility toward playing Bach, not only in the narrower everyday meaning of performing his music but also in the Gadamerian sense of handing down Bach's music in new ways and thus contributing to the continuation of its existence. This sense of responsibility was shared by everyone who participated in the project. Although the debate on how the organ should look and sound continued even after its presentation in March 2018, the belief that it had to be built was never challenged.

CODA: NEW VARIATIONS

Seven months after I witnessed the inauguration of the Utopa Baroque Organ, I am back in the Orgelpark. During a Friday afternoon concert, Johan Luijmes will play Bach's chorale variations *Sei gegrüsset, Jesu gütig* (BWV 768), the same work that was performed on the night of the inauguration. Back then, the chorale and the eleven variations were alternated by Lutheran chorales sung by Studium Chorale. Now, the variations that Luijmes plays on the mechanical console will enter into a dialogue with music by Trevor Grahl, who will perform his new work *Of Ancient Days* on the digital console. Sitting in the audience, I have a good view of the console, the score, and Grahl's hands. I decide not to read the program notes that Grahl wrote and to just listen to the music. Luijmes sits on the organ bench on the balcony and looks sideways to the stopknobs. For a short moment, I think that his attentive movements are part of the performance. Down before me, I see the familiar layout of the digital console with its push buttons in many colors, some of them lighted, and the touch screen with its layers of stops. The audience is silent in anticipation.

I hear a sighing sound, and I realize that this must be the blower that provides the organ with wind. More sounds of air currents—it is the piece that has started. In the score in front of Grahl, I see a reproduction of Hokusai's woodblock print *The Great Wave* from c. 1830. The sounds do in fact resemble the rolling of waves, and I cannot tell if they are generated by the organ or the speakers. In the silence that follows, Luijmes plays the chorale and the first variation. I think back of our conversations when he told me how important playing historical organs was for him. It strikes me in his performance how organically he times and articulates the musical lines and motives. The organ sounds beautiful and alive.

I marvel at the back and forth between the mechanical and the digital console, between Bach's music and Grahl's. Whereas the first used a Lutheran chorale to rework it into a series of variations, the latter plays with all the sounds the new organ can produce. I see Grahl's hands on the keys and knobs, handling them with an amazing dexterity. I hear fast tone figures, clusters of sound, the bells of the turning Cymbelster high up in the central tower of the organ, highlighted with a spotlight so that its reflections circle around on the ceiling of the hall as moving stars. Then birdsongs, it seems, including an obstinate cuckoo whose descending interval sounds oddly out of context in the soundscape. I listen mesmerized and realize how difficult it is to describe what I hear. Johan plays the last variation and uses a plenum registration, combining many pipes in the organ to a majestic sound. A beautiful organ, stately, almost stern. It also has entirely its own sound: There is no reference to what I remember from my experience of the sound in Naumburg.

After the piece has ended, I join the audience in a long applause. I read in the program notes that the last part of Grahl's composition is called *Ecce Homo*. It seems an afterimage of music, music that once was there, or is not there yet.

Later, while everyone participates in the after-concert talks in the hall, I walk to the digital console and take a look at the final pages of the score that still stands on the music rack. On the left page, I see lines from the poem "Girls on the Run" by John Ashbery. On the right page, there is a map of global temperature, February 26, 2018. Red for hot, blue for cold.

Back home, I read in Grahl's program notes: "*Of Ancient Days* is my own comment on the story of creation, molded in seven 'days' where the music of Bach represents the nights. One can imagine that the two pieces do not occur simultaneously, but that my piece came about thousands or millions of years before Bach and maybe created the conditions for his work."

When I speak to Grahl the next day and tell him how much his music and his dialogue with Johan had moved me, he says that thinking of his own music as if it was millions of years old enabled him to project Bach's music in a far and distant future, a future therefore hopeful.

CONCLUSION

The story of the Utopa Baroque Organ that I have told in this book began at the console of the organ it was meant to replace, the Van Leeuwen organ. Embodying the esthetics of postwar Dutch organ building, this neobaroque organ was considered to be inadequate to play the music by Bach and other baroque organ composers in a historically inspired way. As we have seen in the previous chapter, replacing one organ by another to play Bach in new ways can be understood as an exemplar of the unfinishedness of art and history that Gadamer has proposed. In a similar vein, Bach's music is never finished—we keep returning to it, performing it, working with it, studying it, using it, and enjoying it. If there are ever new ways to bring Bach's music and the instruments on which it is played into existence, where does this "new" come from? Phrased in this general and even naïve way, the question might be unanswerable. Looking back on the building of the organ, however, I hope to give partial answers by pointing out how ideas came up, experiments were organized, and problems were solved—in short, to look back on the work that was needed to be done to build a new baroque organ.

In these concluding remarks, I return to the question that guided my project: how working with the "pastness" of pipe organs and their music in present situations can lead to the imagination of the new and unfamiliar. In chapter 1, I situated the Orgelpark as a twenty-first-century concert venue, discussed the reasons it initiated the new baroque organ project, and outlined my own research through presenting three interrelated issues. The first issue springs from the fact that the former Park Church near the Vondelpark is now a space where music is performed as an art form. Most of the concerts at the Orgelpark follow a concert ritual that originated in the nineteenth century: A silent and attentive audience contemplates the music that is performed in the hall. This changed the function of both the performance of music and the organs on which it is played, and with this change came esthetic and artistic criteria to assess the value of the music and the qualities of the instruments. The second issue concerns the historicity of organs as musical instruments.

The dual character of the organ, as materialized in its mechanical and digital console, exemplifies how organs evolved and were adapted during the long history of the instrument. Pipe organs embody age-old traditions of music, architecture, artisanry, and science, and yet their builders and performers were always keen to adopt the latest novelties, be they musical or technical. This made pipe organs and the buildings that house them sites of conservation and restoration as well as experiment and innovation. The third issue I introduced in chapter 1 focuses on how pipe organs, and the new baroque organ in the Orgelpark in particular, can be researched, and what we can learn from this research, not only about musical cultures of today but about how the new comes into the music world. Put more generally, studying questions of knowledge, artisanry, and esthetics; of historical significance and cultural transmission; of tradition and innovation; and of renewing crafts and musical performance can contribute to our understanding of the practical, collaborative, and imaginative work of handing down the past into the present to make new things.

In this conclusion, I will analyze the three issues through the philosophical and theoretical apparatus that I developed throughout the book. In chapter 2, I focused on bringing the historical and esthetical hermeneutics of Hans-Georg Gadamer to the study of musical innovation. Whereas Gadamer's views on the ontology of the work of art and its presentations in a tradition provide a way to conceptualize the "pastness" of music and its instruments, I referred to the pragmatist work of Richard Sennett on the problem-solving of artisans and John Dewey on esthetic experience to understand how reiterations of the past in the present depend on the workings of the imagination and its ability to produce open futures. In addition to these hermeneutic and pragmatist literatures, I have used science and technology studies and music sociological approaches to musical performance and practice, notably the work of Bruno Latour and Antoine Hennion and their reading of the work of Etienne Souriau and William James, using concepts such as "instauration" and "work to be done." Retelling the story of the Utopa Baroque Organ through the lens of these three theoretical perspectives is necessary, I believe, to make it relevant in other contexts.[1]

THE SAME BUT DIFFERENTLY

Knowing the situation in the Orgelpark around 2012 helps to explain how the idea to build a new baroque organ emerged. Since the start of the concert venue, in 2007, the focus had been on organ music after 1850. Once the Van Straten organ was unveiled next to the two Romantic organs in the hall, a gap was felt because organ music from the baroque could not be performed on an organ well suited for this repertoire. The fact that Amsterdam and its surroundings already have great historical baroque organs led to the decision to add an extra relevance to the new instrument: the possibility to play it with state-of-the-art digital interfaces. Combining the two faces of the organ into one musical instrument became the dominant design challenge.

Initially, the core team focused on the historical face; through visiting organs in Germany, the members assembled the knowledge, criteria, and experiences that together shaped a discursive space in which visions, preferences, and problems could be debated and developed. Using Gadamer's vocabulary, I showed how the excursions to the organ worlds of Bach shaped the hermeneutical situation that helped to orient the project and its goals. The insight emerged that it would be impossible to make a copy of a historical organ, yet this did not mean total freedom. Negotiating a middle ground between being true to past as well as present-day practices characterized the project in the early days. Participating in a tradition of organ building that spanned centuries implied that dilemmas had to be faced, such as the choice for Schnitger or Hildebrandt as the project's reference builder, which had consequences for the seemingly mundane issue of how the new organ should be tuned.

What does it actually take to design something new while being aware of the hermeneutical situation? How can we think of design as working with history, or even histories? To combine the baroque sound of historical organs with a digital interface, the sixteenth-century technology of the spring chest was used in combination with electromagnets that operated small spring valves below the pipes. As became clear in chapter 3, determining the right dimensions of the tone channels and tone sliders of the windchest, and developing the electromagnetic devices that could operate the pipe valves, depended on organ-building knowledge from different historical periods: the sixteenth, the nineteenth, late twentieth, and the twenty-first centuries. The two prototypes of the windchest that were built materialized various historical horizons and, at the same time, enabled the organ builders to develop new knowledge and understanding. Thus, the hybrid histories that came together in the prototypes of the windchest were put to the test, a test that regarded not only the mechanical and electronic functioning of the device but also the characteristics and qualities of the pipe sounds.

As my reading of Latour's work suggests, drawing things together from different times in a design or a prototype relied on assembling heterogeneous artifacts, knowledge, skills, and technologies of different periods. The test organ, as a "quasi-object," exemplified not simply a movement back in time to the archaic spring chest mechanism, bypassing the innovations of modern organs, but rather a reworking of this mechanism through the principles of electropneumatic key action as well as digital software and interfacing. Design, Latour argues, is always redesign: Everything builds on what was there already, not just in the abstract sense of the historical texts that are to be interpreted in Gadamer's hermeneutical tradition but rather as *things* or entities that exist in the complex trajectories through which they are continuously assembled and reassembled. History in Latour's view is not a timeline to conveniently order the events in the lives of humans but the variable ontology of artifacts such as organ pipes as well. What a pipe is, or even can be, depends on everything that affected its existence.

Once the windchest design had been tested and the size and specification of the organ had been determined, the final plan of March 2015 reduced the room for trying out solutions. Whereas the two prototypes of the windchest allowed the organ builders to experiment with and test its design, and thus keep their options open to a certain extent, the start of the actual building of the new baroque organ meant that choices had to be made. If we think of history as the ontological plurality of things that are drawn together in a design, then actually making a new thing implies the work of "fitting" these ontologies. In chapter 4, we saw how various elements of the new instrument had to be made to fit together. This lead to problematic situations, and sometimes the solutions resulted in new problems, as in the case of the wind system. Whereas in historical situations the bellows for an organ were often placed in the church tower, it was decided that the bellows of the new baroque organ would be housed inside the organ case. This, in combination with the choice to give the organ wedge bellows to improve the quality of the wind, meant that there was no room in the organ case for the wooden mechanisms of a traditional stop action. This had to be solved by electromagnetic transmission between stopknobs and the windchests. Another example of fitting were the electronic components in the windchest. Once the wood for the tone sliders had been sawed by Hans Elbertse, their dimensions determined the size of the printed circuits that Sinua had to design.

Making things fit in the actual organ thus led to a whole series of smaller and bigger problems that had to be solved in a collaborative learning process.

Generally speaking, these problems followed from the dual design of the new baroque organ and its mechanisms of transmission. Understood in the pragmatist vocabulary of Dewey and Sennett, solving problems went hand in hand with finding new ones. Time and again, the organ builders tried to create space, which was sometimes limited, as in the case of the windchest and the tone sliders where every millimeter of wood and microcircuits counted. In the actual making of the organ the builders learned how to win degrees of freedom to solve the problems set by the limits within which they had to work. Their learning in these problematic or "doubtful" situations through inquiry, reflection, curiosity, and imagination enabled them to make something truly new.

In the pragmatist philosophy of both Dewey and Sennett, there is a clear relation between problem-solving in a doubtful situation and the evaluation of its outcomes and consequences. For Sennett, craftsmanship involves the ethics of a job well done; for Dewey, any doubtful or problematic situation that requires judgment and choice before taking an action can be qualified as a moral situation as well. This element of evaluation and judgment also became evident in the making of the pipes for the new baroque organ. Ensuring that these would sound as good as the historical pipes built by Hildebrandt required skilled hands, eyes, and ears. In the design, making and voicing of the pipes, technical, sonic, historical, artisanal and esthetic knowledge, skills, and judgment went hand in hand. Chapter 5 showed how materiality mattered in this: The qualities of materials such as alloys, wood, and cloth, as well as the machines and tools to shape and work on these materials, were paramount in Munetaka Yokota's attempts to understand Hildebrandt's ways of making, as well as the pipe sounds that resulted from it. The example of casting the sheets of metal from which the pipes would be made revealed the intricate work to be done in order for it to have the "right" qualities. Bringing the texture of the linen on the stone casting bench, the speed with which the melted metal flowed from the single casting box onto the linen, the walking speed of the men pushing the box forward, and the time it took for the metal to cool down all affected the outcome. Every casting attempt might result in failure or success. Following the ecological approach of the role of materiality and skill in making that Gibson and Ingold developed, it was in the correspondence with tools and materials that making pipes assumed making judgments.

Once the pipes had been made, they would sound well only after they had been voiced. Here we encounter the same conundrum that characterizes the whole project: Even though we have remaining historical pipes that can teach us about the sound that an organ builder such as Hildebrandt achieved, these can never simply be copied. Rather, they provide a reference for the kinds of qualities that the present-day pipe voicer works toward. Observing the voicing work of Yokotaka, it became clear that the sound of Hildebrandt's pipes was not created or constructed but instaured. As Latour and Hennion argue, the work of Étienne Souriau, and his concept of instauration in particular, is helpful to understand how the agency between the maker and the made oscillates; voicing the pipes, Yokota not only shaped the pipes, their sounds shaped his actions as well. To paraphrase Souriau, here it is the expectation of the sound to come that hangs on the shoulders of the voicer, who in the end has no model but can only take the next step, without knowing where to go next. The sound of the pipes could not be judged by simply comparing them to historical models, such as Hildebrandt's pipes. Rather, it was through his experimentations that Yokota attempted to bring the pipes and their sounds to "more" existence. This work was risky—it could go wrong, and it implied trials and tests and a constant judging of the outcomes.

Even though Hildebrandt's pipes could not be copied, measuring, analyzing, and studying them was important because they provided an exemplar of the work that was done in the eighteenth century to achieve the qualities that are now highly valued. But how to offer musicians an organ that has sounds we do not yet know? Following the argument in the previous chapter, these new sounds cannot be the continuation of something that already exists. We could argue that making new sounds requires imagination, but where does this originate? Comparing the two consoles of the new baroque organ that are the topic of chapter 6 provides an answer to that question. Once the newly made, voiced and tuned pipes were placed on the organ's windchests, their sounds could be accessed and controlled both from the mechanical and the digital consoles. Representing the two faces of the organ, the consoles offer the musician different ways of interacting with the organ's sounds. To understand how these interactions can lead to imagination, John Dewey's ideas about experience and esthetics are helpful.

For Dewey, art and our experience of it can never be isolated from everyday life. Key to his understanding of art as experience is the continuous interaction between the life creature and the environment. Even though our experiencing is continuous, some experiences become consummated and have their own individual quality. They can be told and retold as stories with a plot. An experience is "clothed with meaning" if it contemplates how what happened in the past is revived to act in a new situation. In art, as in an experience, the past is made into new patterns. Whereas mechanical or academic products lack the element of consummation that allows for presenting something new, the work of art "carries on." Variation as recurrence with a difference is what gives an experience or an art product a sense of life, because it extends into the past and arouses an interest in what is to come. When old and familiar things are made new in experience, Dewey argues, there is imagination. The new baroque organ and its two consoles would help to create new musical experiences because they allow interactions with the pipe sounds that are unexpected and idiosyncratic. As a musical instrument rather than a machine, it would not repeat but vary, the same but differently.[2]

Dewey's philosophy of art, and the crucial importance it gives to a reworking of the past in the present, resonates with the work of Gadamer on historical and esthetic hermeneutics. Like Dewey, Gadamer sees canonizations of particular interpretations of art, such as a recording of a musical composition by its composer, as a simple imitation of a model that would deny the real task of interpretation. Artworks do not exist in an alienated esthetic consciousness but through their acquiring of new uses and being adapting to changed circumstances, while retaining elements of its original lifeworld. This thoughtful mediation with contemporary life is what is at stake in the playing of artworks as productive variations on what resulted from presenting them under different conditions. When we "play" an artwork, we are also played by its earlier renditions to which we add yet another presentation, or *Darstellung*, that in turn adds to its tradition. In doing so, we keep the future identity and continuity of the artwork open.

By transporting the new baroque organ from the organ builders workshop in Soest to the Orgelpark, it was embedded in contemporary musical life. As the Utopa Baroque Organ, its mission is to make new music and musical performance possible. But we saw in chapter 7, also in a more practical sense, the appearance and sounds of the instrument had to be made fit to the hall in which it would reside. The debates in the core team on how the organ had to look and what it would thus tell about itself took years. Voicing the pipes meant that their sounds had to be brought to life in the peculiar acoustic of the Orgelpark's hall. During the inauguration week in March 2018, the instrument was used in very different ways to demonstrate its possibilities as a musical instrument. Placing it in its environment, the people at the

Orgelpark hope to attract new communities of organ musicians, composers, performers, improvisers, and other artists, as well as new audiences. It is through their playing that the organ will meet its mission of contributing to the future of the instrument and its music.

WORKING WITH ART AND HISTORY

The Utopa Baroque Organ is an exception to the rule that organs are usually named after their builders. Its name refers to the foundation that funded the project, and it is on the plaque right above the mechanical console that the names of the builders are mentioned: Hans Elbertse, Hermann Eule Orgelbau, Munetaka Yokota, Sinua. In the years that I was a member of the core team, I have witnessed the collaborative nature of the project. During meetings, the members of the team discussed ideas and proposals. Their excursions helped them to learn more about what the organ could or should be. Inviting experts to the regular symposia offered context to the project and allowed the Orgelpark to update relevant communities about their work. Being part of the project, I have tried to pay close attention to what happened in the situations in which I participated. My goal was to understand how the Utopa Baroque Organ was built, and how it became an exemplar of the work to be done to bring music into existence.

As Johan Luijmes pointed out in the previous chapter, the organ is more appropriate to connect to the digital and the world of contemporary music than any other "art thing" because it can be used in so many ways. To see how the new baroque organ became an art thing, we need a concept of art that moves away from essentializing it in a product, be it a musical score, an instrument, or a performance. Despite the composer name plates on the balconies in the Orgelpark—as we saw in chapter 1, these are an ironic reference to the nearby Concertgebouw—an art concept originating in the nineteenth century does not help to understand how and why the new baroque organ became an art thing. This requires a keen eye for the hermeneutic, practical, and imaginative work to be done, rather than a notion of artworks as having an unchanging essence. What the three theoretical approaches that I worked with in the preceding chapter have in common is that they depart from what Dewey called the "museum conception of art," where it has its own compartments, both in experience and in society. For Gadamer, artworks do not travel through time unchanged, as works of genius, only to be contemplated upon through the subjective judgments that we give of them. Rather, we present works of art under the conditions of the present and thus contribute to what they became in previous renditions. When we "play" a work of art, it plays us through the tradition of its earlier renditions. And finally, reasoning in the conceptual register of Latour and Hennion we can see how artworks call upon us, and how the responsibility of bringing them into more existence can be felt in what they describe as the *faire faire* of instauration.

To summarize, concerning the first issue that I introduced—the disruptive moment in which the organs in the Orgelpark came to be seen as art things after they lost their function in a religious context—the work of art is characterized by the reciprocity of Dewey's doing and undergoing and Gadamer's playing and being played. Even though there were dreams of a new organ at the start, its actual existence could not be planned or predicted, and its making was not a matter of constructing something. Making a truly new organ that would work as an art thing did not mean executing a plan; rather, its art qualities emerged in the very making process through which they could be recognized and put to the test.

The second issue, concerning the long history of organs, their technologies, and their music, was an integral element in the making of what became the Utopa Baroque Organ. The work that was to be done in the project always implied working with history. How can we conceptualize this work to address the second issue, the historicity of pipe organs and how they embody age-old traditions not only of music but also artisanry, science, and architecture? To understand how the new comes into the world, we have seen on many occasions how the past was reiterated in the present and, in the process, created the conditions for the new.

When the musician Spinvis introduced the concert programmed around Sweelinck's variations on *Mein junges Leben hat ein End* in chapter 2, he used the felicitous metaphor of the Orgelpark being founded on basements, and basements below these, filled with organs and their organists. Speaking about Sweelinck's time he said: "1586. What a strange time this must have been," and in doing so, he implicitly referred to Gadamer's metaphor of the historical horizons that determine our understanding of what was. The past is both strange and familiar to us, and it is from this tension that comes the task of historical hermeneutics. Arguing against the historicist project of understanding the past by reconstructing it through the eyes of contemporaries such as Sweelinck, Gadamer shows how it can only live on by mediating it in the present, that is, through embedding it in our own world. Acknowledging the impossibility of building a copy of the Hildebrandt organ in Naumburg in the Orgelpark, or any historical organ for that matter, the members of the core team were keenly aware of the hermeneutical situation that they worked in.

Latour's critical analysis of the concept of the Modern that I explored in chapter 3 is helpful to understand that histories are hybrid, and that working with these histories always implies redesigning what already exists. Or to put it differently, to be aware of how this existence is shaped by what happened to a *thing* such as an organ pipe, and to extend its trajectory through reassembling it with other things. This assemblage is fundamentally fragile; to exist is to be tested. This idea resonates with what Souriau called instauration: The attempt to bring an artwork to more existence is risky, because there is no model to follow, as we saw in chapter 5 about the making and voicing of the pipes for the new baroque organ.

Dewey's "life creature" adopts its past and uses past experience to inform the present and to know what to expect from the future. When routine and mechanical recurrence occur, Dewey says, this only means that past and present fit exactly together. Yet the coming together of past and present is not simply a matter of mobilizing resources, but a re-creation in which the "stored material" of past experience is revived by having to meet a new situation where it kindles the imagination. Even though this may remind of Gadamer's "thoughtful mediation of the past in the present," Dewey's focus is more on the practical problems that emerge in problematic, doubtful, troubled, or confused situations; when there is resistance and a situation is out of balance, solving the problems asks for inquiry that draws on previous knowledge and experience, yet also requires experiments the outcome of which is uncertain.

Building the Utopa Baroque Organ meant working with history in the three ways I have summarized here: by being aware of the strangeness and familiarity of the organ worlds that inspired the project, by reassembling technologies or *things* from various historical periods and in doing so adding to their varied ontologies, and, finally, by finding and solving the many problems that emerged in the shaping of these new assemblages. If music and its instruments exist through the work that has to be done to bring them into existence, and if this work always means reiterating the past in its manifold occurrences, and if this work cannot be planned but emerges in situations, the question can be raised how to study these situations. Or, coming back to the third issue, how we can learn from research in and through these situations, not only about musical cultures of today but about how the new comes into the music world.

UNDERSTANDING ART THINGS

When invited by Loek Dijkman to write about our dreams of the project in 2013, Hans Fidom, Johan Luijmes, and I reflected on the idea that building a new organ would also mean doing research. For Hans, one of the goals of this research was to position the new organ and its project in the history of technology and music. For Johan, the research would have not only a technical instrument as its object but also an artistic instrument. Closely following how the new organ was built would show that "the specific ways of thinking and working of artists can be greatly beneficial to society" (see chapter 1). For me, the research aim back then was to participate in the project from the first day to see if and how it might lead to a musical instrument that could be used in innovative ways.

Looking back on the project, I realize that the term "research" that we used at the start has limited value to understand what actually happened during the project from an epistemological and methodological angle. This is because the concept of research often has connotations with the positivist trope of methods as planned and validated procedures that guarantee the neutral observation of clearly defined phenomena with the goal to systematically intervene. How, then, can we think of the work that was carried out to bring the Utopa Baroque Organ into existence as a knowledge practice?

To start answering that question, I go back to the early morning in 2015 when, in the organ workshop of Hermann Eule Orgelbau, the metal for the casting of the pipe sheets was slowly heated to the right temperature (see chapter 5). When Yokota had suggested using a single-chamber casting box, he drew on his knowledge and understanding of sources on pipe casting from the eighteenth century as well as his own experiments with historical casting procedures. His epistemic authority, however, was implicitly challenged by the employees at Hermann Eule Orgelbau, who drew on their own experiential knowledge and doubted if the proposed procedure would make a meaningful difference. The members of the core team were watching the experiment and understood its relevance from their own knowledge interests: making sure that the pipes would sound as good as Hildebrandt's pipes, or even making sure that the organ would actually be built. The situation in the organ-building workshop in Bautzen, Germany, shows how historical, practical, experiential, and sensory knowledge were mobilized in a situation that was epistemologically hybrid. As such, the casting of the pipe metal, as a hybrid knowledge practice, is an exemplar of many situations that I have presented in this book.

My own work as an organ anthropologist consisted of participating in these situations. I did so by attending project meetings, going on study trips, talking to people, presenting my field work at symposia and conferences, reading relevant literature, and basically trying to understand how the work of the organ builders was done. I collected materials, traces of the experiences that I had, such as field notes, recordings of meetings, and interviews, images, and short videos. Sketching narrative and argumentative orderings of what had happened in a first draft of this book made me acutely aware of what it takes to do justice to a very complex making process that evolved over five years. Referring to Tim Ingold's "art of inquiry" that I introduced in chapter 1, anthropology's aim is not documentary and learning *about* a practice but transformational because it intends to learn *from* a practice by corresponding to it. If, for Ingold, anthropology seeks to work with its materials the way an organ builder would work with pipe metal—noticing what is going on as an effect of one's actions and responding to how the material behaves—what does it mean to study and understand art things such as the Utopa Baroque Organ? To answer that question, I will return to the theoretical and philosophical apparatus that I summarized earlier.

In *Truth and Method*, Gadamer argues that our experience of society and history cannot be raised to a science by following the methodological procedures of the natural or social sciences. A historical perspective means that we cannot see a concrete phenomenon as an instance of a universal rule. Understanding a tradition is not a matter of constructing principles. We do not place ourselves outside this history but hand it down in our situated understandings and applications of it. This handing-down requires hermeneutic work that Gadamer compares to an ongoing conversation driven by the polarity of familiarity and strangeness of a historical text or object. Reasoning in a hermeneutic register, understanding the Utopa Baroque Organ as an art thing, means that the organ anthropologist cannot present it as an instance of a universal rule. Rather, it requires *positioning* oneself by being conscious of the hermeneutical situation in which one participates.

As I have shown in the previous chapters, thoughtfully mediating the past in the present led to situations in which historical truth or falsehood, authenticity or fake were not at stake as a goal in itself. The question was not whether a new pipe sounded as good as one made by Hildebrandt but rather how its qualities resulted from a complex of occurring and related events, such as the casting of the pipe metal. To understand the character of such events, we can paraphrase pragmatist philosopher William James and say that the sound quality of a pipe is not a stagnant property inherent in it; it happens to it. It *becomes* good; it is *made* good by events.[3] Understanding organ pipes as art things requires paying careful attention and finding ways of *noticing* what happened.[4] As Antoine Hennion has argued, this is not about following a social science method "in general," but rather being interested in what is going on here and now. Reasoning with Hennion's vocabulary, understanding art things is never neutral in that it resembles instauration: it aims at sustaining emergent beings, and if their existence remains fragile, as researchers we can only make their emergence graspable through our own engagement (Hennion 2017a, 79).

In many of the situations that I described in this book, my noticing of what went on was often not enough to understand the issue that was at stake. This happened in situations that were problematic or "doubtful," as John Dewey would say. In these situations, there was collaborative thinking that took the form of inquiry. For Dewey, inquiry is a process of looking into things, of investigating. Acquiring knowledge is instrumental to the act of inquiring that he defines as seeking for something that is not at hand (Dewey 2008 [1916], 155). Reconstructing this seeking for what was not at hand often happened after the fact. Even though Yokota had explained to me why he wanted to use a single-chamber casting box for the casting of the pipe metal, it was through editing a video from the fragments that I made in the workshop that the steps in his argument made sense to me, as well as the doubts that the employees had about the procedure. Working with the traces that I had collected, I was able to reconstruct the inquiry that happened, and this work resulted not only in knowledge about the project but in *learning* from it.

Positioning myself in the hermeneutic situation of the project, noticing how events unfolded by engaging with them, and learning from reconstructing the many moments of inquiry in problematic situations, I tried to understand the Utopa Baroque Organ as an art thing that exists in the complex trajectories through which it was and will be continuously assembled and reassembled.

If the term *researcher* does not adequately cover my role in the new baroque organ project, what word would work better? If anything, writing this book has been an attempt to bring the journey that started in June 2013 to consummation in the way John Dewey has described it: "Every moment of experience in completing itself recurs to its beginning" (Dewey 2005 [1934], 175).[5] This reminds us of the Romantic trope of a foot traveler who looks back to see the path that emerged while walking. As their travel companion, I was able to learn along with the organ builders and to experience parts of what happened during the five years it took to build a new musical instrument. It will be played, and it will generate ever new experiences and stories. It is by narrating this story, my story of the Utopa Baroque Organ, that it has a plot that might extend its existence to other situations, where it will do its work as an art thing in yet unexpected ways.

NOTES

INTRODUCTION

1. These examples of canonical "object lessons" are taken from Bijker (1995), Latour (1992), Winner (1999), Shapin and Schaffer (1985), De Laet and Mol (2000), and Pinch and Trocco (2002). The list could easily be expanded.

2. As Goehr writes: "I shall claim specifically of the concept of a musical work (i) that is is an *open* concept with *original* and *derivative* employment; (ii) that is is correlated to the *ideals* of a practice; (iii) that it is a *regulative* concept; (iv) that it is *projective*; and (v) that it is an *emergent* concept" (Goehr 1992, 89–90).

3. See also Pinch and Bijsterveld (2004). Examples of research on innovation of musical instruments are Bijsterveld and Schulp (2004), Bijsterveld and Peters (2010), and Pinch and Bijsterveld (2015). In their book on the Moog synthesizer, Trevor Pinch and Frank Trocco explain why contemporary synthesizers have interfaces that use conventional keyboards. One of the synthesizer pioneers, Robert Moog, decided that this familiar interface would make it easier for musicians to actually use the new device (Pinch and Trocco 2002). See also Bates (2012) on the social life of musical instruments. For an overview of work in the field of sound studies, see Pinch and Bijsterveld (2012).

4. The dialogue between science and technology studies and artistic research exemplifies a broader interest in the humanities for STS-informed approaches to the study of art and literature (Felski 2020; Felski and Muecke 2020).

5. I refer here to how Bruno Latour's use of the term "putting to the test" in his explanation of what he calls "matters of concern": "A matter of concern is what happens to a matter of fact when you add to it its whole scenography, much like you would do by shifting your attention from the stage to the whole machinery of a theatre. . . . Instead of simply being there, matters of fact begin to look different, to render a different sound, they start to move in all directions, they overflow their boundaries, they include a complete set of new actors, they reveal the fragile envelopes in which they are housed. Instead of "being there whether you like it or not" they still have to be, yes (this is one of the huge differences), they have to be liked, appreciated, tasted, experimented upon, prepared, put to the test" (Latour 2008b, 39).

6. Artistic research, or practice-based research in art and design, has gained currency since the 1990s in higher arts education. Artist-scholars in this field focus on

the knowledge, understanding, and experiences enacted in creative processes and embodied in artistic products such as art works, compositions, and performances.

7. Artistic research is contested, both in the art world and in academia, because it undermines the modernist dichotomy of autonomy and instrumentalism. The debate on art as research deals with fundamental philosophical questions of epistemology and methodology and issues of artistic agency and autonomy, as well as institutional and educational strategies. Publications that map the debate on artistic research or take positions in it are Biggs and Karlsson (2010), Borgdorff (2010), Schwab (2014), and Benschop (2020).

CHAPTER 1

1. In this book, I do not translate the names of the stops, as a Dutch "Nachthoorn," for example, sounds quite different from an English "Night horn," which is virtually nonexistent as a stop.

2. On the concert program were compositions by baroque composers such as Bach, Böhm, and Muffat, and also the twentieth-century composers Kropfreiter and Heiller.

3. The Verschueren organ was built according to the principles of the French builder Aristide Cavaillé-Coll (1811–1899), but its case was designed by the architect, Bas van Hille, to fit the overall design of the hall.

4. Recording at the Orgelpark, Amsterdam, June 2, 2016.

5. Orgelpark, "About the Building," https://www.orgelpark.nl/en/about-the-building.

6. Orgelpark, "Documentary Baroque Organ," https://www.orgelpark.nl/en/documentary-baroque-organ.

7. Recording interview with Loek Dijkman and Sylvia de Munck, April 27, 2016, in Leiden.

8. Finding the Park Church resulted from a happy coincidence, Dijkman recalls. When paying a visit to Stadsherstel, the company for city renovation in Amsterdam, he talked about his plan to do "something for organists." His interlocutor told him about the former church near the Vondelpark, which had been acquired by Stadsherstel in 1994. Dijkman said that he was not looking for a church, but he agreed to see the building. Being raised as a Catholic and, thus, not used to Protestant churches, Dijkman immediately experienced the church, with its square layout and balconies, as a theater. As his wife, Sylvia de Munck, recalls, the fact that the building already had an important organ contributed to the fact that the decisions to pay for the renovation of the building and start a concert venue there was taken overnight. One of the staff members of the Foundation had the idea for the name "Orgelpark."

9. Architect Bas van Hille, who supervised the renovation of the Park Church, took his inspiration for the color scheme from the facade of the Sauer organ that had been given its original colors.

10. Dijkman recalls how attempts to organize a more structural collaboration between the two concert halls stranded in the early years of the Orgelpark. Recording interview with Loek Dijkman and Sylvia de Munck, April 27, 2016, in Leiden.

11. In 2023, Elbertse Orgelmakers BV merged with Gebroeders Van Vulpen to form the new company, Elbertse en Van Vulpen Orgelmakers, in Utrecht.

12. Recording interview with Loek Dijkman and Sylvia de Munck, April 27, 2016, in Leiden.

13. Paraphrasing of the "word of welcome" signed by Loek Dijkman, Johan Luijmes, Hans Fidom, and Sonja Duimel in the first issue of *Timbres* that was published by the Orgelpark between January 2007 and March 2018. The magazine featured articles on a broad range of topics related to the concert programs, the research program, as well as the organ world in general.

14. Johann Sebastian Bach (1685–1750) knew the musical qualities of his organ music, yet he considered it in service of his religious beliefs, artful to the best of his abilities, but not primarily reflecting artistic aims or esthetic priorities (Wolff 2020).

15. This section is based on research published in Peters and Cressman (2016).

16. "In een zondagochtendbeurt in de maand Mei 1885 kon het nog de voorzang begeleiden, maar het was den organist niet meer mogelijk om voort gaan; zijn gewoon krijschend en kreunend geluid kon het oude instrument zelfs niet meer voortbrengen" (translated by Peter Frank Peters).

17. For an overview of the history of the Gerritsz organ and the dating of the remaining parts, see Diepenhorst 2009. Van Dijk and Diepenhorst built on the work of the Dutch musicologist Jan van Biezen (1995) and Onno Wiersma (1946–2004), who was organ advisor at the Cultural Heritage Agency (CHA) of the Dutch Ministry of Education, Culture, and Science.

18. Three decades later, an initiative was undertaken to return the Gerritsz organ to its original home in the Nicolaï Church in Utrecht. These plans never materialized, however.

19. For the CHA, the parts of the Gerritsz organ are important artifacts of Dutch cultural heritage because, taken as a whole, they are the only example of a fifteenth-century organ in the Netherlands.

20. In his book *Musik als Klangrede* published in 1983, the conductor Harnoncourt countered this criticism. In an age of airplanes, televisions, and computers, he argued, early music has lost its meaning. People are no longer able to understand it as a language. Music from previous epochs had become an ornament that was meant to be "beautiful." The goal of musician-researchers such as Harnoncourt was not to banish early music to the museum but instead to create modern interpretations of masterpieces in which the performance and listening conventions shaped in the past decades were questioned (Harnoncourt 1983, 9–10).

21. The organ makers of Reil Organ Builders (Heerde, the Netherland), who were commissioned to replicate the Gerritsz in the Van Straten organ, had to rely on

their own knowledge and expertise but had to learn to think as fifteenth-century builders as well.

22. A recording of this piece is available on SoundCloud: https://soundcloud.com/woutersnoei/organism.

23. Conversation recorded during the car trip to Hermann Eule Orgelbau in Bautzen, Germany, November 11, 2015.

24. The original Gerritsz organ is thought to have had two *blokwerks*, one for each manual.

25. Constructing this slider mechanism required very precise woodwork; any irregularity might cause it to block or air to leak out.

26. On September 23, 2013, the core team supervising the design and building of the new baroque organ visited the 1655 organ in the Nieuwe Kerk in Amsterdam, built by Schonat, and expanded by Hagerbeer. It is one of the biggest remaining organs in the world that has original spring chests.

27. In his book *Reassembling the Social: An Introduction to Actor-Network Theory* (2005), Latour presents actor-network theory as a heuristic to trace the heterogeneous assemblages that make up "the social." When it was developed at the Centre de la Sociologie de l'Innovation (CSI) in Paris in the context of the sociology of innovation in the 1980s, actor-network theory was a critique of traditional and critical sociology. Rather than using social relations as explanations, actor-network theorists argued for a "sociology of translations" that allows to take the agency of human as well as nonhuman actors into account.

28. In their analysis of the introduction of new technologies into music, Pinch and Bijsterveld draw on the work of Harold Garfinkel who introduced "breaching experiments" in sociology as a way to demonstrate the unspoken rules and conventions that are enacted in everyday life (Garfinkel 1967).

29. Originally published in French in 1993, Hennion's book reflects the context of the early actor-network theory as it was developed at the CSI in Paris around 1990. Hennion translated the material-semiotic approach of actor-network theory, which brought back objects and things in sociology, to the study of music.

30. For an account of how the concept of "attachment" emerged in the context of the CSI in the 1990s, see Hennion (2017b).

31. See Cameron (2008) for an example of a philosopical analysis of the ontology of musical works.

32. I want to thank Henk Borgdorff for pointing out to me the Dutch translation of *Wahrheit und Methode* and encouraging me to read it. My reading of Gadamer has been informed by Davey (2013) and Warnke (2013).

33. "What we call a work of art and experience (*erleben*) esthetices depends on a process of abstraction. By disregarding everything in which a work is rooted (its

original context of life, and the religious or secular function that gave it significance), it becomes visible as the 'pure work of art.' In performing this abstraction, esthetic consciousness performs a task that is positive in itself. It shows what a pure work of art is, and allows it to exist in its own right. I call this 'esthetic differentiation'" (Gadamer 1989 [1960], 74).

34. My own learning process evolved over the years of observing and participating in the project, but also in the analysis of the materials I collected. Paul Atkinson found a similarity between the craft worker and the ethnographer, who both engage "directly with her or his materials, physically and imaginatively. Embodied skills and educated eyes—the gaze of the craft—interact. Esthetics and intellectual imaginations, traditions, and innovations interact in the craft of ethnography, just as they do in the ethnography of craft" (Atkinson 2013b, 63).

35. All recordings have been transcribed by myself for the same three reasons that Edwin Hutchins mentions in his book *Cognition in the Wild*, an ethnography of ship navigation as distributed and embodied cognition, and in many ways an inspiration for my own writing (Hutchins 1995). Like the practice of ship navigation, organ building is a specialist's area, and other transcribers lack the knowledge and context that I developed through the years to understand what is said. My familiarity with the situations in which I made the recordings as well as the people that were present made it easier to distinguish who was talking at any moment. And finally, as Hutchins rightly observes, closely listening to the same material several times is an important way of learning.

36. For a pianist, many of Bach's compositions have the problem that sustained notes eventually die out; on an organ, however, a note can sound forever.

37. Ingold elaborates on the concept of "correspondence" in Ingold (2021) and Ingold (2022).

38. Text written for the preparatory meeting on June 22, 2013, at the Utopa Foundation in Leiden.

39. Text written for the preparatory meeting on June 22, 2013, at the Utopa Foundation in Leiden.

CHAPTER 2

1. Concert recording at the Orgelpark on February 21, 2016. Translation of the Dutch text by Peter Frank Peters.

2. Recording of the core group meeting at the Utopa Foundation, Leiden, September 9, 2013.

3. In addition, a weblog was developed including the logos of the main institutional partners in the project at the time: the Orgelpark, the VU University Amsterdam, and Maastricht University.

4. Draft press release on the New Baroque Organ project, September 2013.

5. Draft press release on the New Baroque Organ project, September 2013.

6. Töpfer sought to rationalize the scaling of pipes by grounding it in a mathematical formula. See McNeil (2012).

7. An example is the large organ in the Martini Church in Groningen, the Netherlands, dating from the mid-fifteenth century, expanded by Rodolphus Agricola in 1479 and Arp Schnitger in 1691–1692. In the late 1930s, organ builder J. de Koff & Zoon made extensive changes to the historical instrument, such as an electrical, detached console, and new stops, and revoiced their pipe work in a Romantic style (see Fidom [2019]).

8. Next to this organ, a French Romantic instrument was to be built as well as an eighteenth-century Swedish organ. To teach new generations of organists, it was thought to be better to have three organs in different historical styles than one universal organ (Speerstra 2003b, 15).

9. The Hamburg organ had a facade with long pedal pipes that would not have fit in the church in Örgryte. The Lübeck Dom organ case was chosen because it had the right size for the church in Sweden.

10. For a detailed documentation of the project, see Speerstra (2003a).

11. Eastman School of Music, University of Rochester, "The Craighead-Saunders Organ," https://www.esm.rochester.edu/organ/instruments/craighead-saunders/.

12. Cornell Center for Historical Keyboards, "Baroque Organ," https://historicalkeyboards.as.cornell.edu/baroque-organ.

13. Cornell Arts and Sciences, "Design of Cornell University's New Baroque Organ," YouTube, https://www.youtube.com/watch?v=EudpSCYL9NM.

14. NOMINE e.V., "Arp Schnitger Orgel Norden—Thiemo Janssen im Gespräch," YouTube, https://www.youtube.com/watch?v=QWPb65wRXQE.

15. On September 23, 2013, the core group visited the 1655 organ in the Nieuwe Kerk in Amsterdam, built by Schonat and expanded by Hagerbeer. It is one of the biggest remaining organs in the world that has original spring chests. This specific type of spring chest is not very practical: In the case of malfunctions, many pipes have to be taken off the chest to reach the small valves directly underneath the pipe feet.

16. Field notes made during study trip to north Germany, October 20–21, 2013.

17. During one of the meetings of the core group, Hans Elbertse also highlighted technical issues related to a dual temperament. It would have consequences for the design of the windchests, he argues, because next to the twelve pipes per octave, six to eight additional pipes would be necessary. As a result, the windchest would be broader. If one places a C next to a well-tempered C, this may lead to unwanted

inferences between the two pipes. A solution could be to place them on different sides of the windchest, but this may lead to a decreased homogeneity in the sound.

18. This overshoot is almost a quarter of a semitone, or of 23.46 cents in the modern measurement system where an equally tempered semitone is 100 cents (Kent 1998). The physical properties of resonating strings result in the fact that we cannot have both pure fifths and pure or major thirds.

19. In the modern system of equal temperament, this is done by "tempering," or flattening, each fifth with one-twelfth of the Pythagorean comma and adapting other intervals such as the major third and sixth as well. This means that all intervals except the octave are slightly out of tune, but we accept this disadvantage because all keys can be used.

20. Variations on the quarter-comma meantone system were developed throughout the sixteenth century, such as fifth-comma meantone and sixth-comma meantone. Cf. Kent (1998).

21. For an emphatic defence of this position, see Duffin (2008).

22. Orgelmuseum Borgentreich, "Herzlich willkommen in der Orgelstadt Borgentreich," http://www.barockorgel-borgentreich.de/

23. Recording of the visit of the core group to the Hildebrandt organ in the St. Wenzel Church on December 5, 2013.

24. Recording of a meeting of the core group after the visit to the Hildebrandt organ in the St. Wenzel Church on December 5, 2013.

25. Recording of the meeting of the core group at the Utopa office in Leiden, December 13, 2013.

26. Recording of the meeting of the core group at the Utopa office in Leiden, December 13, 2013.

27. Recording of the meeting of the core group at the Utopa office in Leiden, December 13, 2013.

28. The origin of this quote is not clear.

29. Hermann Eule Orgelbau from Bautzen built a new organ in the reconstructed Dreikönigskirche in 1992 that followed aspects of Hildebrandts building style in the mechanical parts, the scaling of the pipes, and the design of the organ case designed by Burkhart Goethe.

30. Recording of the meeting of the core group of the New Baroque Organ project at the Utopa Foundation, Leiden, March 12, 2014.

31. Recording of the meeting of the core group of the New Baroque Organ project at the Utopa Foundation, Leiden, March 12, 2014.

32. Recording of the meeting of the core group of the New Baroque Organ project at the Utopa Foundation, Leiden, March 12, 2014.

33. In the 1970s and 1980s, records of historically informed interpretations of well-known compositions by Bach and others were often labeled by record companies as "authentic performances." The moral and ethical overtones in the concept were thus linked to commercial interest. C.f. Haynes (2007), 10: "Authenticity became a hexed word and served for a while as a kind of lightning rod for anybody who was dissatisfied with some aspects of the Movement [early music movement—PP]. Despite this, the idea that the word represents refuses to go away. The reason is clear: Authenticity is simple, it's logical, and (as we have seen) it's central and essential to the concept called HIP."

34. Kivy is not alone in his critical analysis of the concept of authenticity in musical practice. For Stephen Davies, authenticity represents an *ideal* to be aimed at, rather than an attempt to reproduce the *actual* past performance (Davies 1987). James O. Young argues that the concept of authentic performance is not a useful one. There cannot be an "authentic" and ideal performance of a composition but, rather, many. "[An early music performance] is valuable not because it bears some relation to past performances, but because present listeners find it artistically appealing" (Young 2002, 77).

35. Thomas (2018) has applied the historical hermeneutics of Gadamer to early music to argue that it allows a middle ground between historical reconstruction and leaving open different possibilities for applying historical knowledge in performance.

36. "However much experiential universals are involved, the aim is not to confirm and extend these universalized experiences in order to attain knowledge of a law—e.g., how men, peoples, and states evolve—but to understand how this man, this people, or this state is what it has become or, more generally, how it happened that it is so" (Gadamer 1989 [1960], 4).

37. The German word used here is *Überlieferung*. The English language does not have a verb or adjective that does justice to the active connotation of this word. Whereas the word "tradition" suggests something that lies behind us, or that is repeated without reflection, the German word suggests a much more active process that requires questioning and self-reflection.

CHAPTER 3

1. Johannes Josephus Elbertse became the master organ builder in the workshop of Maarschalkerweerd and started his own firm in 1917, which was later taken over by his son Jan Elbertse. Under the direction of Hans Elbertse, the company built new mechanical organs and chest organs. Since the 1980s, Elbertse has focused on restoration and maintenance of mechanical organs mainly from the nineteenth century and, since 1990, also on pneumatic organs. In 2009, his son Jos Elbertse entered the firm (source: Elbertse Orgelmakers, http://www.orgelmakers.nl/). In 2023, Elbertse Orgelmakers BV merged with Gebroeders Van Vulpen to form the new company Elbertse en Van Vulpen Orgelmakers in Utrecht.

2. Recording of the visit to Elbertse Orgelmakers BV, Soest, October 7, 2013.

3. Recording of the visit to Elbertse Orgelmakers BV, Soest, October 7, 2013.

4. Recording of the visit to Elbertse Orgelmakers BV, Soest, October 7, 2013.

5. Recording of the visit to Elbertse Orgelmakers BV, Soest, October 7, 2013.

6. Töpfer's "Normalmensur" is a rank of pipes based on an internal diameter of 155.5 mm (6.12 in) at 8' C (the lowest note of the modern organ compass) and a mouth width which is one-quarter of the circumference of such a pipe. The scale provides for a reduction in diameter of the pipes by half at every succeeding seventeenth pipe (McNeil 2012).

7. For the Austrian composer Arnold Schönberg (1874–1951), the lack of fluent dynamics, the gradations between loud and soft, were a problem he needed to solve to make the organ a relevant instrument for him. Instead of sixty or seventy different tone colors, his organ of the future would have only two to six stops, but the new instrument "would have to include the entire range (7–8 octaves) and a range of expression from the softest pianissimo to the greatest fortissimo" (Arnold Schönberg cited in Patteson [2015]). The Hungarian composer György Ligeti (1923–2006) called the organ a "giant prosthesis." In an article in the musical journal *Melos* in 1966, he wrote that he was fascinated by learning to walk again with this tool (Fidom 2014a; Giesen 2005; Watkins 1965). For Ligeti the shortcomings of the organ—its "Unbeholfenheit, Steifheit und Eckigkeit" [awkwardness, stiffness and angularity—transl. PP] (Ligeti 1966, 313)—were attractive at a time when he explored the possibilities of micropolyphony through combining complex moving lines that outline close clusters. His composition *Volumina* (1962) counts as an important example of this style; here Ligeti created stationary sound spaces that contrast the use in melody and rhythm, which are more time based (Fidom 2014a).

8. Latour wrote his book around 1990, briefly after the fall of the Berlin Wall led to debates on a new global world order. Social theorists such as Anthony Giddens and Ulrich Beck linked these debates to modernity theories that went back to Max Weber, who had analyzed forms of rationality and the rise of "'value spheres" in Western societies. Modernity theory, Latour argues, rests on sharp distinctions between society and culture on the one hand and nature on the other. The first was the domain of politics, values, and interpretation, the second of the natural sciences and their hard facts (Vries, de 2016).

9. My use of the terms "Baroques" and "Moderns" draws on how Hennion (2015) labels the proponents of the historically informed performance practice and the modern performance styles that emerged after the Second World War. Hennion describes this debate as a rare opportunity to conduct an experiment in vivo: making visible the mediators that connect the music from the baroque with the historical period in which it was conceived. As a controversy, it could be explained as a difference in taste—the Moderns preferring modern, powerful instruments to the feeble sound of period instruments; the Baroques being interested in restoring the "authentic" sound

of baroque music. Hennion, however, presents the debate as competing forms of mediation, bridging a gap of more than two centuries (Hennion (2015).

10. The German organ builder Eugenio Casparini (1623–1706) built windchests that resembled the nineteenth-century stop channels. Walcker's cone-valve chests "refreshed" a technology that was already known and used.

11. In fact, the idea to use cone-shaped valves was not new. Johann Carl Sigmund Hausdörffer (1714–1767) had successfully used cones in a stop-channel windchest (Owen 2002).

12. According to Shannon (2009), the disadvantages of the electric action went beyond the speech of the pipes. Because it was now possible to place pipework anywhere in the organ, from where it could be connected to the console with electric wires, organ builders, particularly in the United States, destroyed "the architectural integrity of the instrument, an integrity that was over five hundred years old" (Shannon 2009, 45).

13. Fokker, who had studied with Albert Einstein, Max Planck, and Hendrik Lorentz, was a physics professor in Leiden and also worked as a curator in the Teyler's Museum in Haarlem.

14. Although many composers from the Netherlands and abroad have written compositions for the instrument, its design was never repeated elsewhere, and hence, this music remains largely unknown. During its restoration in 2009, after which it resides in the Muziekgebouw in Amsterdam, the organ has been given a MIDI interface, so that it can also be played from a computer.

15. As a church organist, Bares was interested in innovating the practice of liturgical music. When this lead to problems with his superiors in 1985, he moved to the Kunst-Station.

16. As Harlow remarks, "there is nothing about these instruments that could not have been built three hundred years ago" (Harlow 2011, 24)

17. The primary reason that the French composer Pierre Boulez "has not written for the organ, and continues to refuse to do so, is because of the instrument's lack of touch-controlled timbre and dynamic flexibility" (Harlow 2011).

18. Open Sound Control (OSC) is a relatively new protocol for communication among computers, sound, and other multimedia devices, optimized for modern network technology. It is specifically used in situations where real time control of sound and other media is needed (OpenSoundControl.org, http://opensoundcontrol.org).

19. A sample set of an organ consists of recordings of every pipe in the original acoustic environment. Using virtual organ software such as Hauptwerk, the sample sets can be played on MIDI keyboards.

20. Latour uses the term "matters of concern" in opposition to "matters of fact." Whereas in the modernist view facts are seen as objective and independent from the

theories and experiments that disclose them, Latour argues that we should instead think of objects as "the warm, interested, controversial building sites of matter of concern" (Latour 2005, 125): "A matter of concern is what happens to a matter of fact when you add to it its whole scenography, much like you would do by shifting your attention from the stage to the whole machinery of a theatre. . . . Instead of simply being there, matters of fact begin to look different, to render a different sound, they start to move in all directions, they overflow their boundaries, they include a complete set of new actors, they reveal the fragile envelopes in which they are housed. Instead of "being there whether you like it or not" they still have to be, yes (this is one of the huge differences), they have to be liked, appreciated, tasted, experimented upon, prepared, put to the test" (Latour 2008b, 39). See also Suchman, Trigg, and Blomberg (2002) on protypes as exploratory technologies "designed to effect alignment between the multiple interests and working practices of technology research and development, and sites of technologies-in-use" (2002, 163).

21. During the first half of 2014, the Orgelpark organized three colloquia and one international symposium on topics related to the new baroque organ. The first colloquium in March focused on the question of how the building of the new baroque organ could contribute to artistic research in musical practices. During two colloquia in May, issues concerning the new organ's temperament and the use of live electronics were discussed. In June, a group of Dutch and international organists and organ experts was invited to reflect on various aspects of what Hans Fidom had titled "Electrifying Baroque." In their lectures, debates, and concerts, they explored the history of historicizing organ building, the history of the spring chest, the organs of Gottfried Silbermann and Zacharias Hildebrandt, and also the use of live electronics in organ improvisations and the artistic and even social possibilities of the "hyperorgan." From the group that initiated the GOArt organ research center at Gothenburg University, Sweden, Hans Davidsson, Carl-Johan Bergsten, Ibo Ortgies, and Munetaka Yokota were present. From Sinua, Benedikt Aufterbeck and Thomas Stöckl were present, as well as Ansgar Walllenhorst, the organist in the St. Peter and Paul Church in Ratingen. Randall Harlow presented on hyperorgans, and Hans-Ola Ericsson gave a concert and talked about the organ in Piteå, Sweden. From the Dutch organ world, Wim Diepenhorst, adviser for the Van Straten organ, Jacob Lekkerkerker, and Piet Kee had been invited.

22. Three sounds were used, a 4' Principal, a 4' Gedackt, and a combination of both.

23. "Orgelpark Listening Experiment 6 and 7 June 2014," compiled by Hans Fidom, which provides an overview of all written responses from symposium participants for all comparative situations.

24. Short report on the symposium "Electrifying Baroque,"' June 2–4, 2014, at the Orgelpark in Amsterdam, written by Hans Fidom for the core group. Fidom later expanded on his conclusions in a text for the Orgelpark Research Report 5/2, dedicated to the Utopa Baroque Organ (Fidom 2018, 28–31).

25. Recording made at a diner during the visit of the core group to Bautzen, Germany, on September 3, 2014.

26. Recording made at a diner during the visit of the core group to Bautzen, Germany, on September 3, 2014.

27. Recording made at a diner during the visit of the core group to Bautzen, Germany, on September 3, 2014.

28. Recording made at a diner during the visit of the core group to Bautzen, Germany, on September 3, 2014.

29. Recording made at a diner during the visit of the core group to Bautzen, Germany, on September 3, 2014.

30. On this occasion, three comparisons are made: (1) mechanical ⇔ electronic with Laukhuff magnet, (2) electronic with Laukhuff magnet ⇔ electronic with Sinua magnet, and (3) electronic with Sinua magnet ⇔ mechanical.

31. Recording during the listening test on January 15, 2015, at Elbertse Orgelmakers in Soest.

32. This approach can be situated in a longer tradition of thought on how models are related to the world, see, for instance, Nelson Goodman's canonical *Languages of Art* (1976).

33. Spring chests are normally build wider than slider chests, which in the case of the new baroque organ would mean that the pipes would not stand in the same way on the chests as in Hildebrandt organs. Bert van den Heuvel solved this problem by grouping the tone sliders, that is, by skipping the dividers that isolate the sliders in the Stade organ.

34. A tremulant is a device inside the main wind channels of an organ that creates small fluctuations in the wind pressure. This results in a slight vibrato in the organ sound. The tremulant is often used in baroque music in slow pieces with a melody voice.

35. This arrangement means that the organ will be built by the Orgelpark itself, with the help of four companies, which is quite different from other organ-building projects, where one company takes responsibility, and the others become subcontractors.

CHAPTER 4

1. Van den Heuvel used Hans Fidom's drawing of the front of the organ. All details—profiles, curves, measurements—are based on the case of the Hildebrandt organ in Hettstedt, Germany.

2. Interview Bert van den Heuvel, May 21, 2015, Soest.

3. I thank Tullio Viola for pointing out to me the importance of problem "articulation"' in Dewey's philosophy.

4. Originally quoted in Robbins (1994), 126.

5. Van den Heuvel based his drawings on computer-generated drawings made by Hans Fidom, who determined the appearance of the new organ in consultation with the members of the core team.

6. Mitering is a technique to change the shape of a wooden organ pipe if they cannot stand upright due to space limitations (Shannon 2009, 135).

7. Recording at Elbertse Orgelmakers, Soest, May 27, 2017. At an other moment, Hans Elbertse explained that because his firm does not make many new organs, using a computer would not make sense, because the software would not be used very often.

8. Recording at Elbertse Orgelmakers, Soest, May 27, 2017.

9. In Dutch, these drawings are called "*uitslag*," a word that has no English translation.

10. Recording at Elbertse Orgelmakers, Soest, May 27, 2017.

11. Recording at Elbertse Orgelmakers, Soest, May 27, 2017

12. In the new baroque organ, the inner sides of the upper and lower boards of the bellows are covered with white sheepskin leather to make them airtight. The two sides that form the hinge of the bellows are connected by ropes. A board is three meters long and one and a half meter wide. It weighs 130 kg, which means that the four bellows, each consisting of two boards, will weigh more than 1,000 kg.

13. These holes usually have flaps to maintain the direction of the wind and prevent it from flowing in and out of the bellows through the wrong holes.

14. Manipulating the bellows by hand is done with the Van Straten organ in the Orgelpark (see chapter 1).

15. Shannon (2009) writes that other means of power the centrifugal blower were attempted, such as steam, gasoline, and water (Shannon 2009, 18).

16. According to Forkel, one of J. S. Bach's first biographers, the composer had the habit of testing the "lungs" of a new instrument by drawing as many stops as possible and then playing many keys at the same time. This story is quoted by another Bach biographer, Albert Schweitzer; see Schweitzer (1908).

17. An example is the restoration of the A. A. Hinsz organ from 1780 in Roden, the Netherlands, in 2006. Three wedge bellows were built that can be operated both manually and by a blower that uses the bellows as a reservoir (see Peters [2009]). As Stef Tuinstra, the organ advisor who oversaw the project, explained to me during an interview in 2007, this was done to give the organ "breathing wind" (Peters 2009).

18. In his report on the core group meeting on March 7, 2014, Fidom referred to the research that has been done by Carl-Johan Bergsten in the GOArt project on wind behavior. "His research shows that the quality of the baroque sound is strongly determined by the way the pressure waves in the wind system emerge and how they are controlled: this is one of the factors that determine the speech of the organ pipes, and

therefore the timbre and dynamics of organs as a whole" (Report core group meeting March 7, 2014. Translation PP).

19. Interview Bert van den Heuvel, May 21, 2015, Soest.

20. Final plan New Baroque Organ, March 2015.

21. Interview Bert van den Heuvel, February 2, 2017, Soest.

22. To prevent this, Hans Elbertse advised to build in a safety margin in the software to prevent risky pipe-to-key combinations.

23. It has been calculated that these treaders can do their job adequately if they weigh around 70 to 80 kg themselves.

24. Interview Hans Fidom, May 5, 2016, Amsterdam.

25. Recording core group meeting March 10, 2015, Leiden.

26. Field notes January 7, 2016.

27. The issue was discussed again during the meeting of the core group on March 31, 2016. In his report of the meeting, Hans Fidom wrote that Elbertse would investigate the possibility of placing the blower in the Molzer organ, which is situated below the balcony on the ground floor and in the basement. Anticipating this decision, Hans Elbertse bought a blower that had enough capacity to transport the wind vertically over a longer distance (Report from meeting of the core group on March 31 2016, written by Hans Fidom).

28. Hans Elbertse calculated that these holes would be 30 square centimeter each.

29. Interestingly, Noortje Marres argues that the pragmatist notion of the "problematic situation" can be linked to "the ANT idea that there can be "no re-composition without de-composition" (Marres 2020, 112). See also Marchand (2017) on craftwork as problem-solving.

30. Recording meeting of the core group on March 12, 2014, in Leiden.

31. Recording meeting of the core group on March 12, 2014, in Leiden.

32. This became evident during the visit that the core team made in September 2013 to the 1655 organ in Amsterdam's Nieuwe Kerk. This is one of the biggest remaining organs in the world that has original spring chests (see chapter 2, note 15).

33. Interview Hans Elbertse, September 29, 2016, Soest.

34. As Shannon (2009, 26) points out, one of the reasons for this is that mechanical instruments are almost indestructible in comparison to electropneumatic organs, which are more vulnerable to environmental damage.

35. Stickers normally push vertically and are relatively short wooden dowels. Trackers pull and are made of thin strips of wood, approximately one centimeter wide and four millimeters thick. In the case of the new baroque organ, each key has a hinge at its end. When pressed, a tracker right behind the visible part of the key is pulled

down. The direction of a movement can be reversed from upward to downward by using a backfall (a lever mechanism), and changed by ninety degrees through squares (wooden components with a right angle) (Shannon 2009, 36–37).

36. Interview Hans Elbertse, October 26, 2016, Soest.

37. This number corresponds to the number of stops on the windchest.

38. In his minutes of the first meeting of the project on June 20, 2013, Hans Fidom wrote: "Ook moeten er tien musici zichtbaar bij de klaviatuur kunnen musiceren. Tevens moet er ruimte zijn voor een koor." [Ten musicians visible from the mechanical console have to be able to perform. In addition there has to be space for a choir. (transl. PP)] The issue of room for musicians on the balcony remained on the agenda until January 2016, when Loek Dijkman decided that the balcony structure should not be changed.

39. Minutes of the meeting of the core group on February 5, 2014, written by Hans Fidom.

40. Interview with Benedikt Aufterbeck and Thomas Stöckl, June 4, 2016, Amsterdam.

41. Interview with Benedikt Aufterbeck and Thomas Stöckl, June 4, 2016, Amsterdam.

42. Interview with Benedikt Aufterbeck and Thomas Stöckl, June 4, 2016, Amsterdam.

43. As the relative humidity in the Orgelpark is controlled and kept stable at about 50 percent at all times, and as the temperature is kept stable at about 20 degrees Celsius—all to keep the organs healthy—corrosion risks are minimized.

44. Interview Hans Elbertse, September 9, 2015, Soest.

45. For some of the pedal pipes, Laukhuff magnets are used.

46. Sennett gives the example of raised front stoops of tenements of New York's Lower East Side that were designed as passages in and out of the building but were used to dry clothes, sell goods, or just sit on (Sennett, 2008, 236).

47. The mass-produced circuits that were built into the tone sliders were numbered v1.0.

48. Recording interview with Sinua, June 4, 2016, Amsterdam.

49. The effect of this is comparable to traditional, or the physical, manipulation of a pipe to give it its final sound characteristics (see chapter 5).

50. Recording interview Sinua, June 4, 2016, Amsterdam.

51. Recording interview Sinua, June 4, 2016, Amsterdam.

52. The magnet has a metal sticker in its center that moves up and down over a certain range. In the mechanical mode, the magnet is off and the sticker can move freely. It then simply extends the sticker that is inside the slat.

53. As Hans Elbertse pointed out, the two strategies to enable control also come at their own micro-economic price. Adding springs and adjusting them has to be done 1,700 times, once for every pipe in the organ. This has to be done with great precision, because if the valve springs are too powerful, the magnet will not function correctly (Recording interview Hans Elbertse, December 9, 2016, Soest).

54. Recording interview Sinua, June 4, 2016, Amsterdam.

55. Recording interview Sinua, June 4, 2016, Amsterdam.

56. "While all thinking results in knowledge, ultimately the value of knowledge is subordinate to its use in thinking. For we live not in a settled and finished world, but in one which is going on, and where our main task is prospective, and where retrospect—and all knowledge as distinct from thought is retrospect—is of value in the solidity, security, and fertility it affords our dealings with the future" (Dewey 2008 [1916], 158).

CHAPTER 5

1. The Dutch historian Johan Huizinga defined a historical sensation as "an *Ahnung* (hunch) just as much of roads and houses and fields, of sounds and colors, as of simulated and simulating people. This contact with the past which is accompanied by an utter conviction of genuineness and truth, can be evoked by a line from a document or a chronicle, by a print, by a few notes of an old song. . . . One does not realize the historical sensation as a re-experiencing, but as an understanding that is closely related to the understanding of music, or rather of the world by means of music" (Huizinga 1959, 54).

2. Gothic, "Stormthal, Germany: Hildebrandt Organ (1723)," http://www.gothic-catalog.com/St_rmthal_Germany_Hildebrandt_organ_1723_s/884.htm.

3. A common term for the bellows treaders at the time of Hildebrandt.

4. My overview of the general information on organ pipes in this section draws on Shannon (2009) and McNeil (2012).

5. This principle enables organ builders to build 8' stops even when there is no room in the case for pipes that are 8' long.

6. Some organ stops such as the harmonic flute use this principle to sound an octave higher than their length would suggest.

7. The pipes that can be seen in the organ front are usually principal pipes, hence their German name *Praestant* (standing in front) or French name *Montre* (showing).

8. The principle of the mixture as a combination of pipes at multiple pitches goes back to the medieval *blokwerk* organs such as the Peter Gerritsz organ from 1479 (see chapter 1).

9. The ratio being 1:1,68, which means that if the diameter of the lowest pipe of a register is 168 mm, then the pipe that sounds one octave higher has the same timbre

with a diameter of 100 mm, and the pipe that is another octave higher with a diameter of 100:1, 68 = 68 mm (Fidom 2006a, 15).

10. In the standard scale, the width of the pipe mouth is a quarter of the circumference of the pipe.

11. Interview Munetaka Yokota, May 16, 2014, Orgelpark in Amsterdam. It is striking that local materials were used for the organ in Chico as well. To produce the alloy for the front pipes, lead from spent bullets from the Los Angeles Police Department shooting range (Yearsley 2010).

12. Interview Munetaka Yokota, May 16, 2014, Orgelpark in Amsterdam.

13. The Hermann Eule Orgelbau firm was founded in 1872 and focused initially on building mechanical cone valve chest organs and, from 1900 onward, on pneumatic "Taschenladen" organs. When Hermann Eule's daughter Johanna took over the firm in 1929, 173 new organs had been built. In 1933, the firm restored its first organ, built by Gottfried Silbermann in Crostau. The early 1950s saw a general transition to slider chest organs with mechanical actions. Studies of historical construction principles and scalings provided the knowledge for historically informed organ restorations in the Saxony region. Hans Eule who took over the firm in 1957 built organs with a neobaroque character and started export activities to Sweden, West Germany, and the Soviet Union. After his death in 1971, his wife, Ingeborg Eule, took over the firm. A year later, the firm was nationalized as a "Volkeigene Betrieb" (people's workshop). In 1990, the firm returned to family ownership again. A flagship project in the first decade after the end of the German Democratic Republic was the restoration of the Hildebrandt organ in Naumburg between 1993 and 2000. In 2005, Anne-Christin Eule, granddaughter of Ingeborg Eule, took over the firm. On May 1, 2013, master organ builder Dirk Eule was appointed as managing director (Hermann Eule Orgelbau, https://www.euleorgelbau.de/de/werkstatt/tradition/).

14. Recording of the meeting at Hermann Eule Orgelbau, Bautzen, Germany, on June 11, 2015. German translated by Peter Frank Peters.

15. This practice of making nicks to "smoothen" the pipe sound was not common before the eighteenth century but became a more or less standard procedure in late-eighteenth- and nineteenth-century organ building.

16. Recording of the meeting at Hermann Eule Orgelbau, Bautzen, Germany, on June 11, 2015 (translation Peter Frank Peters).

17. In this approach, scaling follows temperament. Temperament is dependent on the lengths of the pipes, and these in turn are related to their diameter.

18. This monumental book was published between 1766 and 1778, and is still used as a reference by modern organ builders (for details on the book and its author, see Kursell and Peters [2020]).

19. Making metal resonators for reed pipes resembles the way flue pipes are made. The general description of the pipe-making process in this paragraph is based on Blanton (2003) and Shannon (2009).

20. The proportions of tin and lead can vary. Tin strengthens the alloy and is considered more beautiful than lead. Tin is also far more expensive than lead which is why the pipes standing in the facade usually have a slightly higher percentage of tin than pipes standing inside the organ case. Because of their weight, pipes made of lead have to be carefully supported to prevent them from collapsing under their own weight (Shannon 2009, 130).

21. Some organ pipes are made of lead only. The sheets for these pipes are usually hammered after they have been cast to make the metal structurally stable.

22. The pipe toe and its body are soldered with two straight seams. A protective fluid is applied to prevent the soldering metal from damaging the surrounding pipe surface or flowing onto it. Then the hole at the pipe's toe is given a cone shape using a toe cone, which partly closes the toe hole. The upper and lower lips of the pipe are made by burnishing the metal on a lip iron, where soapy water is used to reduce the friction between burnishing iron and organ metal. After the languid has been cast and given its final shape and bevel angle, it is soldered to the foot. Then the foot with its languid is soldered to the body.

23. Examples of historical organ building treatises that were used in the GOArt project are J. P. Bendeler, *Organopoeia* (1609), Salomon de Caus, *Les raisons des forces movvantes* (1615), and Dom Bedos, *L'art du facteur d'orgues* (1766–1778).

24. In the Netherlands, similar casting experiments were done by the organ builder Reil in Heerde and Van der Putten in Winschoten.

25. Recording visit to Hermann Eule Orgelbau, Bautzen, November 12, 2015.

26. "Aber es geht so auch." Recording visit to Hermann Eule Orgelbau, Bautzen, November 12, 2015.

27. My description of the pipe casting at the Hermann Eule Orgelbau organ workshop draws on Peters (2019).

28. Video recording visit to Hermann Eule Orgelbau, Bautzen, November 12, 2015.

29. The Lausitz region is world famous for its centuries-old linen industry.

30. Video recording visit to Hermann Eule Orgelbau, Bautzen, November 13, 2015.

31. Video recording visit to Hermann Eule Orgelbau, Bautzen, November 13, 2015.

32. Video recording visit to Hermann Eule Orgelbau, Bautzen, November 13, 2015.

33. Video recording visit to Hermann Eule Orgelbau, Bautzen, November 12, 2015.

34. Published between 1766 and 1778, its four volumes were part of the *Déscriptions des Arts et Métiers* that were commissioned by *the Académie Royale des Sciences*. Bédos's treatise is in line with the overarching project of a representation of craftsmanship. As is characteristic for the volumes of the *Descriptions des Arts et Metiers*, Bédos's treatise is accompanied by rich illustrations of excellent quality. The plates that are added

to each of its volumes show detailed illustrations of organ parts, tools, and working procedures. In this combination of text and images, the publication shared the same stylistic characteristics that were adopted in the *Encyclopédie* (Kursell and Peters 2020).

35. See O'Connor (2005) on learning the craft of glass blowing as embodied knowledge.

36. Paul Atkinson has published extensively on the sensory ethnography of crafts such as glass blowing glassblowing, woodworking, silversmithing, photography, life drawing, and perfume blending (2013b, 2022).

37. This argument resonates with Sennett's analysis of the role of crafts in modern societies (see chapter 4).

38. Reed pipes are voiced by manipulating the size of the toe hole, the position of the shallot, the shape of the shallot opening, as well as the size, shape, thickness, and curvature of the tongues (Shannon 2009, 136).

39. In this chapter, I focus on the prevoicing of the pipes of the new baroque organ that was done by Munetaka Yokota in 2016. The final voicing will be discussed in chapter 7.

40. Dom Bedos only wrote in his *L'art du facteur d'orgues* (see note 18) that "since organs are made only to be heard, they must be given a suitable, pleasing tone" (cited in Segurado 2015, 12).

41. Segurado's own research focused on the relation between voicing an organ and making music on it. Himself being a musician, he followed the voicing work of Gerald Woehl on the *Studio Acusticum* organ in Pitea, Sweden (see chapter 3). Because he not only gives descriptions of the voicing but also reflected on the topic by making musical compositions, Segurdao considers his studies as a form of artistic research. Other recent studies of voicing include (Rioux 2001) who studied what he called "the art of voicing" from an applied acoustics perspective.

42. Field notes, January 4, 2016.

43. Tuning pipes by cutting metal away has to be done careful and in steps. If too much metal is removed, the pitch becomes too high and this cannot be repaired easily.

44. According to Yokota, in present-day historicizing of organ-building practices it is often the case that information about the wind pressure in a historical organ is followed to the letter. In contrast to the tacit and, hence, unwritten knowledge of historical pipe voicing, the numbers that indicate the wind pressure in a contract or other historical document are tangible. Strictly following these numbers in a smaller space might result in an organ that is too loud, or a voicing approach that has to change the characteristics of a pipe to such an extent that it is no longer in the style of a historical organ builder (Interviews Munetaka Yokota, Soest, September 29, 2016; Soest, October 25, 2016).

45. Interview Hans Elbertse, Soest, September 29, 2016.

46. Interview Munetaka Yokota, Soest, October 25, 2016.

47. Recording of the listening session in the Orgelpark, Amsterdam, on January 7, 2016.

48. Recording of the listening session in the Orgelpark, Amsterdam, on January 7, 2016.

49. Recording of the listening session in the Orgelpark, Amsterdam, on January 7, 2016.

50. Recording of the listening session in the Orgelpark, Amsterdam, on January 7, 2016.

51. Field notes made on January 7, 2016.

52. Recording at Elbertse Orgelmakers, Soes, September 29, 2016.

53. Recording at Elbertse Orgelmakers, Soes, September 29, 2016.

54. "Overbite" is a characteristic of Hildebrandt pipes. The upper lip is not aligned with the lower lip, but slightly pronounced.

55. Recording interview Munetaka Yokota, Orgelpark Amsterdam, December 22, 2017.

56. Recording interview Munetaka Yokota, Orgelpark Amsterdam, December 22, 2017.

57. Recording interview Munetaka Yokota, Orgelpark Amsterdam, December 22, 2017.

58. In the world of historicizing organ building the use of nicks to change the speech of a pipe is contested. Pipes from he eighteenth century often have been nicked during later restorations, thus adapting their sound character to the current esthetics (Recording interview Munetaka Yokota, Orgelpark Amsterdam, December 22, 2017).

59. Recording at Elbertse Orgelmakers, Soes, September 29, 2016.

60. Recording at Elbertse Orgelmakers, Soest, September 29, 2016.

61. See Kursell and Peters (2020)for an elaboration of this point.

CHAPTER 6

1. The information in this paragraph is mostly based on the website of The Virgil Fox Society: http://www.virgilfox.com/.

2. The organ that he played there was built for him by G. Donald Harrison, who also built the organ in the Mormon Tabernacle in Salt Lake City. Fox continued to expand his instrument until it became one of the largest organs in the United States.

3. An example is the *Revelation Lights* concert in 1974. An impression of the concert can be found on the Virgil Fox webste at http://www.virgilfoxlegacy.com/welcome.html.

Under the heading "Heavy Organ," *Time* magazine (January 7, 1974) described Fox's concerts as follows: "The organ is lit up like the stage at Radio City Music Hall. Overblown poppies bloom in Oriental splendor in the organist's iridescent paisley jacket. At the keyboard, he rocks vigorously in gigue time, his rhinestone-decorated black suede shoes dancing over the pedals. Cascading waves of sound shake Manhattan's Carnegie Hall. Then, with a puff of smoke, the organist disappears. Overhead, a glowing portrait of a rotund face with crimped curls and dimpled chin flashes on a screen. The overflow audience explodes in cheers for Virgil Fox and Johann Sebastian Bach."

4. See his statements on the topic in the short documentary on the *Revelation Lights* concert in 1974, on the Virgil Fox website at http://www.virgilfoxlegacy.com/welcome.html.

5. The 1479 Peter Gerritsz organ is an example of this development. In addition to the *blokwerk* it had a second rank of flute pipes that can be used independently, called "Doof" (from the Dutch word "*dof*" or dull). The Van Straten replica in the Orgelpark has two manuals, one for the *blokwerk* and one for the Doof. These can be coupled to combine the two sounds.

6. The reason that organs have two or more divisions is that tone channel windchests can "carry"' only up to thirteen stops. If more sounds were needed, the only way was to build more chests. The easiest way to operate these chests was to give them a keyboard each; otherwise, playing the organ would require too much power.

7. This anatomy was retrospectively described in the twentieth century as the *Werkprinzip* (Bicknell 1998a, 64–66).

8. The Hildebrandt organ in Naumburg is an exception because Hildebrandt used the existing case of the organ by Zacharias Thayssner (built between 1696–1705).

9. The church and the organ were destroyed in 1945.

10. An Unda Maris stop consists of two principal 8' pipes on one key that are slightly out of tune, which causes a vibrating tone, hence the metaphor of the "ocean wave." The decision to add this stop was made relatively late in the process. During a study trip in March 2016, the core team visited the organ in the Hofkirche in Dresden that was designed and started by Silbermann and finished by Hildebrandt. Hans Elbertse had anticipated the possibility that an extra stop might be added and kept space on the windchest of the Hauptwerk. He was commissioned to build a wooden Unda Maris according to the style of Hildebrandt's contemporary Tobias Trost (Fidom 2018, 39).

11. The Cymbelstern and the Nachtigall were not newly designed but ordered from the catalog of the Laukhuff company. The Nachtigall was used unchanged, but the Cymbelstern was tuned slightly differently according to the specifications of Ibo Orgies, who was the project's tuning adviser. The eight little bells first play the notes B(flat)-A-C-H and then G-sharp, C-sharp, G, and D.

12. The practice of making many stop changes while playing became more common only in the nineteenth century, when composers wrote down more detailed

instructions for the registration. Baroque composers such as Buxtehude, Bruhns, and Bach often used "dialogical" registration, where two or three manuals with their own stop combinations would be used alternatingly in one piece.

13. Based on a recording made during the car trip to Hermann Eule Orgelbau, November 11, 2015.

14. Synthmuseum.com, "Thaddeus Cahill's Teleharmonium," http://synthmuseum.com/magazine/0102jw.html.

15. In the original typography of the "Credo" these words are printed as capitals and alternated with elaborations on, respectively, the notion of noise, music as the "organization of sound," and electrical musical instruments as attempts not to imitate the past but to construct the future.

16. This section draws on various sources, including Chadabe (1997), Manning (2004), Pinch and Trocco (2002), Oskamp (2011), Holmes (2016), and Cox and Warner (2017).

17. An example is the software package Hauptwerk by Milan Audio. This can be used to play sample sets of pipe organs, where the sound of every individual pipe is recorded in situ and can be played back using a digital keyboard.

18. In the context of this chapter, it is interesting to note that, as Pinch and Trocco (2002) show, an important reason for the success of the Moog synthesizer was the use of a traditional keyboard to control it.

19. A metronome sound via earphones.

20. The texts of these presentations were published in the Orgelpark Research Report 5/1 (Fidom 2014b).

21. As an improviser, Lekkerkerker followed in the footsteps of Jan Pieterszoon Sweelinck (1562–1621), who was organist at the Oude Kerk in Amsterdam between 1577 and 1621 (see also chapter 2).

22. Van Heumen made this point also during a round table discussion at the colloquium in October 2014. Recording October 11, 2014, in the Orgelpark, Amsterdam.

23. "Imagine a perfect blend of great acoustic pipes and versatile electronic sound, being able to live-sample an acoustic register and play that back instantly with just a little change in pitch or timbre, merging that with other acoustic registers and strangely familiar organ-like sounds generated by the physical modelling module, then adding gesturally controlled extremely pitched-up and -down sampled material from the mechanism of the organ, grounded by an extremely low-pitched throbbing sine tone. Wow, that would be definitely amazing" (Heumen 2014, §303).

24. Open Sound Control (OSC) is a relatively new protocol for communication among computers, sound, and other multimedia devices, optimized for modern network technology. It is specifically used in situations where real time control of sound and other media is needed. OpenSoundControl.org, http://opensoundcontrol.org.

25. An example that has been commercially successful is Yamaha's Disklavier, which uses MIDI to record and play piano performances, not unlike mechanical player pianos and their piano rolls.

26. This three-day event was set up as a meeting between scholars and practitioners working in the fields of artistic research, science and technology studies, sound studies, and musicology.

27. Instead of building electronic devices into the organ, the decision was made to offer an abundance of connectors (patch bays), allowing musicians to hook up their favorite equipment to the organ.

28. Coupling manuals can extend the number of pipes that can be played from one manual, but this does not change the principle that keys are linked to pipes with a certain pitch.

29. Recording, Skype interview, February 23, 2016, with Benedikt Aufterbeck and Thomas Stöckl.

30. Recording, Skype interview, February 23, 2016, with Benedikt Aufterbeck and Thomas Stöckl.

31. Recording of the visit to the St. Peter and Paul Church, Ratingen, March 31, 2015.

32. Recording of the visit to the St. Peter and Paul Church, Ratingen, March 31, 2015.

33. The Italian term "sostenuto" refers to playing sustained and prolonged notes. In "staccato" playing, notes are articulated shortly and separated from each other.

34. Recording of the visit to the St. Peter and Paul Church, Ratingen, March 31, 2015.

35. Recording of the visit to the St. Peter and Paul Church, Ratingen, March 31, 2015.

36. Recording of the visit to the St. Peter and Paul Church, Ratingen, March 31, 2015.

37. Recording of the visit to the St. Peter and Paul Church, Ratingen, March 31, 2015.

38. Recording of the visit to the St. Peter and Paul Church, Ratingen, March 31, 2015.

39. Recording of the visit to the St. Peter and Paul Church, Ratingen, March 31, 2015.

40. Internal document, "Externe speeltafel Nieuw Barok orgel—Orgelpark Amsterdam Schets samenvoeging orgel en electronica," by Jacob Lekkerkerker, December 2015.

41. Recording of the meeting in Ratingen, February 9, 2016.

42. Recording of the meeting in Ratingen, February 9, 2016.

43. Recording of the meeting in Ratingen, February 9, 2016.

44. My reading of *Art as Experience* is informed by Eldridge (2010).

45. As Perricone (2006) has argued, Dewey's esthetical theory is strongly influenced by Darwinism and its evolutionary and biological notions of organic life that grows in and through interactions with its environment.

46. In a short essay on the recording studio as a musical instrument, Brian Eno voiced his frustration about new technologies that offer so many possibilities that they hinder the act of music making that they aim to support. "I recently spent three days working with what is possibly the most advanced recording console in the world, and I have to report that it was a horribly unmusical experience. The console, which has more than 10,000 controls on its surface and a computer inside, was designed in such a way that music-making tasks once requiring a single physical switch now require a several-step mental negotiation. My engineer kept saying 'Wait a minute' and then had to duck out of the musical conversation we were having so he could go into secretarial mode to execute complex computer-like operations. It's as though a new layer of bureaucracy has interposed itself between me and the music we want to make. After days of tooth-gnashing frustration, I had to admit that something has gone wrong with the design of technology—and I was paying $2,000 a day in studio fees to discover it" (Eno 1999).

47. Recording at Hans Elbertse Orgelmakers, Soest, November 28, 2017.

48. Recording at Hans Elbertse Orgelmakers, Soest, November 28, 2017.

49. Recording at Hans Elbertse Orgelmakers, Soest, November 28, 2017.

CHAPTER 7

1. Recording interview with Munetaka Yokota, December 22, 2017, at the Orgelpark in Amsterdam.

2. As I explained earlier, being conscious of the fact that our understanding is affected by history means that we are aware of the hermeneutical situation. Within the historical consciousness of the hermeneutical situation a tension between the historical object and the present is experienced. Only by projecting a historical horizon that is different from the present can we mediate between historical ideas and our own thinking, which is what determines the historical life of a tradition (see chapter 2).

3. Report by Hans Fidom of the meeting of the core team, October 7, 2013.

4. Report by Hans Fidom of the meeting of the core team, February 5, 2014.

5. Recording of the meeting of the core team on April 8, 2014, in Leiden. During a meeting of the core team on April 22, the role of Van Hille was also defined by Loek Dijkman in terms of guarding the architectural unity of the building.

6. Report by Hans Fidom of the study trip by the core team to Germany, September 2–5, 2014.

7. Report by Hans Fidom of the study trip by the core team to Germany, September 2–5, 2014.

8. Report by Hans Fidom of the listening sessions, January 12–15, 2015.

9. Recording, meeting of the core team, March 10, 2015, in Leiden.

10. Recording, meeting of the core team, March 10, 2015, in Leiden.

11. Recording, meeting of the core team, March 10, 2015, in Leiden.

12. Recording, meeting of the core team, March 10, 2015, in Leiden.

13. Recording, meeting of the core team, March 10, 2015, in Leiden.

14. Recording, meeting of the core team, September 30, 2015, in Leiden.

15. Recording, meeting of the core team, November 4, 2015, in Leiden.

16. Fidom shows four types of shades: slats (both in wood color and gilded), a copy of the shades in Hettstedt, platforms and finally tubes as elongations of the pipes. He also shows the option without shades in two variants: without any intervention and with colored light behind the upper end of the front pipes (Report by Hans Fidom meeting of the core team, November 4, 2015, Leiden).

17. Recording of the meeting of the core team, November 4, 2015, Leiden.

18. Recording of the meeting of the core team, November 4, 2015, Leiden.

19. "We see how the curves in the front, the columns and the carving together form a meaningful and harmonious whole that manifests itself as 'aus einem Guß' [in one flow—PP]. We decide that we will build our organ in the details in accordance with Hettstedt's example: including caps, curls at the end of the lists, capitals and wood carved shades" (Report by Hans Fidom meeting of the core team, November 4, 2015, Leiden).

20. Report of the study trip to Germany from March 31 to April 2, 2016, by Hans Fidom.

21. Interview with Hans Elbertse on August 22, 2017, in the Orgelpark, Amsterdam.

22. Interview with Gerard de Jongh, August 22, 2017, in the Orgelpark in Amsterdam.

23. Interview with Gerard de Jongh, August 22, 2017, in the Orgelpark in Amsterdam.

24. Interview with Hans Fidom, August 22, 2017, in the Orgelpark in Amsterdam.

25. Interview with Hans Fidom, August 22, 2017, in the Orgelpark in Amsterdam.

26. 'This pipe has been placed on August 22, 2017, Loek Dijkman and Sylvia de Munck, president and vice president of the Utopa Foundation'.

27. Report by Hans Fidom of the meeting of the core team, October 7, 2013, Soest.

28. Reports by Hans Fidom of the meetings of the core team, February 5, March 12, and April 8, 2014.

29. Interview with Johan Luijmes on February 8, 2017, at the Orgelpark in Amsterdam.

30. Recording interview with Munetaka Yokota, November 28, 2017, at the Orgelpark in Amsterdam.

31. Recording interview with Munetaka Yokota, December 22, 2017, at the Orgelpark in Amsterdam.

32. Recording interview with Munetaka Yokota, December 22, 2017, at the Orgelpark in Amsterdam.

33. Recording interview with Munetaka Yokota, December 22, 2017, at the Orgelpark in Amsterdam.

34. Recording interview with Munetaka Yokota, December 22, 2017, at the Orgelpark in Amsterdam.

35. Ortgies was also responsible for the tonal design of the Cymbelstern stop (Ortgies 2018).

36. Equal temperament was not unknown, however, as Ortgies points out. On the contrary, the preference for unequal temperaments is to be associated with twentieth-century historically informed performance practice, he argues: "ET is certainly the best choice if one considers the task of transposing, since equal steps will not change in quality. The historical argument that leans somewhat toward favoring ET is however contradicted by the wish for a variation in the harmonic quality of the chords. The more unequal, the more character in the device. While writers in the eighteenth century that favored unequal circulating temperaments did not form a majority, there is especially *today* a strong conviction about the quality of non-ET temperaments among musicians that play according to historically informed performance" (Ortgies 2018, 75).

37. The note indiciation "a1" refers to the nomenclature that has been developed in Europe over the centuries and for which the organ was the source of inspiration. In Anglo-Saxon organology, this a1 would be named a4, because—especially in the United States—the piano is used as a reference, and the note is the fourth A above the lowest C on the piano keyboard.

38. Technically, the well-tempered tuning Ortgies/Orgelpark 2 has four fifths smaller by one-fifth Pythagorean comma: C-G-D-A-E. Two fifths are smaller by one-tenth Pythagorean comma: B flat-F and B-F sharp. The remaining six fifths are pure. "The structure of the circle of 5ths is symmetrical with a symmetry axis D-A flat (G sharp: major keys with the same amount of accidentals each have the same major 3rd. . . . The quality of major 3rds increases stepwise from the only Pythagorean major 3rd F sharp-A sharp to the close-to-pure 3rd C-E; the major keys on the flat side of the circle of 5ths, however enjoy purer 5ths" (Ortgies 2018, 83).

39. The pitch of reed pipes can also be adjusted by raising or lowering its tuning wire.

40. Smaller adjustments can also be done by using a tuning cone to bend the upper edge of the pipe inward to make the sound lower or outward to make it higher. Using a tuning slide, a pipe can be shortened or extended. A third method is slotting. Here, part of the upper end of the pipe is cut to create a slot that can be rolled down to shorten the air column in the pipe and raise its pitch, or up to lower it. Wooden pipes can have sliders in one side of the pipe that can be raised or lowered (Shannon 2009, 133–134).

41. The periodical has been edited by Hans Fidom from the start in 2007. Contributions were written by him or by invited music journalists. It was sent to subscribers or "Guest Friends" of the Orgelpark and distributed for free during the concerts. The 23rd Spring 2018 issue, featuring the inauguration of the new baroque organ, was to be the last. The Utopa Foundation decided to end the publication on paper and focus on extended web presence instead. All issues of *Timbres* are available on the Orgelpark's website at www.orgelpark.nl.

42. Recording interview with Johan Luijmes, February 8, 2017, in the Orgelpark in Amsterdam.

43. Recording interview with Johan Luijmes, February 8, 2017, in the Orgelpark in Amsterdam.

44. Recording interview with Johan Luijmes, February 8, 2017, in the Orgelpark in Amsterdam.

45. Recording interview with Johan Luijmes, February 8, 2017, in the Orgelpark in Amsterdam.

46. Recording interview with Johan Luijmes, February 8, 2017, in the Orgelpark in Amsterdam.

47. Recording interview with Johan Luijmes, February 8, 2017, in the Orgelpark in Amsterdam.

48. Recording interview with Johan Luijmes, February 8, 2017, in the Orgelpark in Amsterdam.

49. Recording interview with Johan Luijmes, February 8, 2017, at the Orgelpark in Amsterdam.

50. Recording interview with Jacob Lekkerkerker, February 25, 2018, at the Orgelpark in Amsterdam.

51. The preparations for the concerts thus became a way to free the system of bugs. Sometimes the interaction between composer and console derailed because of misunderstanding. Both Lekkerkerker and Snoei read the text "Waiting" on the touch screen as meaning that the computer was busy and they should wait until it was ready with its processing. It turned out that it was actually waiting for their input at the console.

52. Recording interview with Jacob Lekkerkerker, February 25, 2018, at the Orgelpark in Amsterdam.

53. Recording interview with Jacob Lekkerkerker, February 25, 2018, at the Orgelpark in Amsterdam.

54. Recording interview with Jacob Lekkerkerker, February 25, 2018, at the Orgelpark in Amsterdam.

55. Recording interview with Jacob Lekkerkerker, February 25, 2018, at the Orgelpark in Amsterdam.

56. Recording interview with Jacob Lekkerkerker, February 25, 2018, at the Orgelpark in Amsterdam.

57. Recording interview with Jacob Lekkerkerker, February 25, 2018, at the Orgelpark in Amsterdam.

58. Recording interview with Jacob Lekkerkerker, February 25, 2018, at the Orgelpark in Amsterdam.

59. Recording interview with Jacob Lekkerkerker, February 25, 2018, at the Orgelpark in Amsterdam.

60. Recording interview with Jacob Lekkerkerker, February 25, 2018, at the Orgelpark in Amsterdam.

61. Recording interview with Jacob Lekkerkerker, February 25, 2018, at the Orgelpark in Amsterdam.

62. Recording interview with Jacob Lekkerkerker, February 25, 2018, at the Orgelpark in Amsterdam.

63. Recording interview with Jacob Lekkerkerker, February 25, 2018, at the Orgelpark in Amsterdam.

64. Organs are usually named after their builders. Given the fact that four firms collaborated in the building of the new baroque organ, this was not possible in this case. The suggestion to call it Utopa Baroque Organ was discussed during the study trip to Germany in 2016, but the final decision was made only in the months before the inauguration.

65. The title of Franke's improvisation is a pun on the almost homonymous Bach song '*Komm, süsser Tod*' (BWV478).

66. Radio recording of the opening concert, March 21, 2018. NPO Klassiek, "Gemiste uitzendingen," https://www.nporadio4.nl/avondconcert/gemist?start=85 [Accessed on September 22, 2018].

67. Recording interview with David Franke, March 22, 2018, at the Orgelpark in Amsterdam.

68. Recording interview with David Franke, March 22, 2018, at the Orgelpark in Amsterdam.

69. Recording interview with David Franke, March 22, 2018, at the Orgelpark in Amsterdam.

70. Recording interview with Hans-Ola Ericsson, March 23, 2018, at the Orgelpark in Amsterdam.

71. Recording interview with Wouter Snoei, March 22, 2018, at the Orgelpark in Amsterdam.

72. Recording interview with Wouter Snoei, March 22, 2018 ,at the Orgelpark in Amsterdam.

73. Recording interview with Wouter Snoei on March 22, 2018, at the Orgelpark in Amsterdam.

74. Recording interview with Wouter Snoei on March 22, 2018, at the Orgelpark in Amsterdam.

75. Program notes by Gerard van der Leeuw in *Timbres* 23 (Spring 2018): 30.

76. Recording interview with Hans Fidom, April 13, 2018, at the Orgelpark in Amsterdam.

77. Recording interview with Hans Fidom, April 13, 2018, at the Orgelpark in Amsterdam.

78. Recording interview with Hans Fidom, April 13, 2018, at the Orgelpark in Amsterdam.

79. McAuliffe (2024) explores Gadamer's hermeneutics in a musical context. Christensen (2018) expands Gadamer's theory on the esthetics of play to an ontology of the process of musicking (Small 2011) rather than of fixed musical objects.

80. This concept also has a role in the esthetics of both Kant and Schiller, but Gadamer gives it a different meaning. Tracing the genealogy of the concept of "play" in philosophical literature, Reichling (1997) argues that "[p]lay as a way of knowing is an exercise of the imagination, central to education and, I emphasize, to music education" (1997, 42).

81. The German word here is *Darstellung*, which has many connotations, such as representation, presentation, depiction, description, account, or performance.

82. Interestingly, Gadamer seems to criticize what later became known as the historically informed performance practice: "Thus, for example, historicizing presentations—e.g., of music played on old instruments—are not as faithful as they seem. Rather, they are an imitation of an imitation and are thus in danger 'of standing at a third remove from the truth' (Plato)" (Gadamer 1989 [1960], 118).

CONCLUSION

1. I could not have developed the arguments in this conclusion without the work of and discussions with my colleagues in the Maastricht Centre for the Innovation of Classical Music (MCICM): Ruth Benschop, Imogen Eve, Karoly Molina, Denise Petzold, Neil T. Smith, Veerle Spronck, and Ties van de Werff. Their research projects

have shaped my own thinking in many ways, notably Benschop's work on the dialogues between artistic research and science and technology studies (Benschop 2009; Benschop, Peters, and Lemmens 2014; Benschop 2020); Eve's ideas on concert design (Eve 2021); Petzold's use of the concept of the "archive" to analyze the relationship between innovation and tradition in classical music and the broader question of how to conserve its artistic heritage for the future (Petzold 2023); Spronck's ethnographic research on in the study of audience participation in symphonic music in the practice of four Dutch symphony orchestras (Spronck 2022); Smith's work on conventional and unconventional concert spaces (Smith 2024); and Van de Werff's work on observing and intervening during concert experiments aimed at artistically relevant forms of audience participation (Werff, Eve, and Spronck 2023). I also learned from Linnea Semmerling's research on exhibiting sounding artworks in Germany and the US since the 1960s (Semmerling 2020).

2. *The Same but Differently* is the title of a book written by Imogen Eve on the innovation of classical music practices and how conservatories could educate their students for an innovative profession as musicians (Eve 2021). The book was written in the context of the four-year "Artful Participation" project caried out by the Maastricht Centre for the Innovation of Classical Music (MCICM; https://www.artfulparticipation.nl).

3. Italics are as in the original text. In his book *The Meaning of Truth* (1909), James wrote: "The truth of an idea is not the stagnant property inherent in it. Truth happens to an idea. It *becomes* true, is *made* true by events. Its verity *is* in fact an event, a process, the process namely of verifying itself, its veri-*fication*. Its validity is the process of its valid-*ation*" (James 1975, 201).

4. In her book *The Mushroom at the End of the World* (2015), in which she describes how the valuable mushroom matsutake is entangled with global capitalism in ecological assemblages, Anna Lowenhaupt Tsing introduces what she calls "arts of noticing" with a reference to polyphonic music: "When I first learned polyphony, it was a revelation in listening; I was forced to pick out separate, simultaneous melodies *and* listen for the moments of harmony and dissonance they created together. This kind of noticing is just what is needed to appreciate the multiple temporal rhythms and trajectories of the assemblage" (Lowenhaupt Tsing 2015, 24). I am grateful to Anna Harris for pointing out this passage to me.

5. "Every moment of experience in completing itself recurs to its beginning, since it is a satisfaction of the prompting initial need. But the recurrence is with a difference; it is charged with all the differences the journey out and away from the beginning has made. For random samples, take the return after many years to childhood's home; the proposition that is proved through a course of reasoning and the proposition as first enunciated; the meeting with an old friend after separation; the recurrence of a phrase in music; of a refrain in poetry" (Dewey 2005 [1934], 175).

REFERENCES

Acord, Sophia Krzys, and Tia DeNora. 2008. "Culture and the Arts: From Art Worlds to Arts-in-Action." *The Annals of the American Academy of Political and Social Science* 619 (1): 223–237.

Aguilar, Ananay, Ross Cole, Matthew Pritchard, and Eric Clarke, eds. 2021. *Remixing Music Studies: Essays in Honour of Nicholas Cook*. London and New York: Routledge.

Alexander, Christopher. 1979. *The Timeless Way of Building*. Oxford: Oxford University Press.

Alexander, Christopher, Sara Ishikawa, and Murray Silverstein. 1977. *A Pattern Language: Towns, Buildings, Construction, Center for Environmental Structure Series*. Oxford: Open University Press.

Atkinson, Paul. 2013a. "Blowing Hot: The Ethnography of Craft and the Craft of Ethnography." *Qualitative Inquiry* 19 (5): 397–404.

Atkinson, Paul. 2013b. "Ethnography and Craft Knowledge." *Qualitative Sociology Review* 9 (2): 56–63.

Atkinson, Paul. 2022. *Crafting Ethnography*. London: Sage.

Bates, Eliot. 2012. "The Social Life of Musical Instruments." *Ethnomusicology* 56 (363–395).

Benschop, Ruth. 2009. "STS on Art and the Art of STS: An Introduction." *Krisis. Journal for Contemporary Philosophy* 2009 (1): 1–5.

Benschop, Ruth. 2020. "A Thought Experiment on Artistic Research as High-Risk Ethnography." In *Dialogues Between Artistic Research and Science and Technology Studies*, edited by Henk Borgdorff, Peter Frank Peters, and Trevor Pinch, 46–60.

Benschop, Ruth, Peter Frank Peters, and Brita Lemmens. 2014. "Artistic Researching: Expositions as Matters of Concern" In *The Exposition of Artistic Research: Publishing Art in Academia*, edited by Michael Schwab and Henk Borgdorff, 34–51. Leiden: Leiden University Press.

Bicknell, Stephen. 1998a. "The Organ Case." In *The Cambridge Companion to the Organ*, edited by Nicolas Thislewaith and Geoffrey Webber, 55–81. Cambridge: Cambridge University Press.

Bicknell, Stephen. 1998b. "Organ Construction." In *The Cambridge Companion to the Organ*, edited by Nicolas Thislewaith and Geoffrey Webber, 18–30. Cambridge: Cambridge University Press.

Bie, Jos de. 2014. "The New Baroque Organ at the Orgelpark: Temperament." In *The New Baroque Organ at the Orgelpark*, edited by Hans Fidom, §140–§182. Amsterdam: The Orgelpark.

Biggs, Michael, and Henrik Karlsson, eds. 2010. *The Routledge Companion to Research in the Arts*. London: Routledge.

Bijker, Wiebe E. 1995. *Of Bicycles, Bakelites, and Bulbs: Toward a Theory of Sociotechnical Change*. Edited by Wiebe E. Bijker, W. Bernard Carlson, and Trevor Pinch, *Inside Technology*. Cambridge, MA: The MIT Press.

Bijsterveld, Karin. 2008. *Mechanical Sound: Technology, Culture, and Public Problems of Noise in the Twentieth Century*. Cambridge, MA: The MIT press.

Bijsterveld, Karin, and Marten Schulp. 2004. "Breaking into a World of Perfection Innovation in Today's Classical Musical Instruments." *Social Studies of Science* 34 (5): 649–674.

Bijsterveld, Karin, and Peter Frank Peters. 2010. "Composing Claims on Musical Instrument Development: A Science and Technology Studies' Contribution." *Interdisciplinary Science Reviews* 35 (2): 106–121.

Blanton, Robin. 2003. "Workshop Documentation: I: Casting, II: Metal Preparation, III: Pipe Construction." In *The North German Organ Research Project at Göteborg University*, edited by Joel Speerstra, 203–262. Göteborg: The Göteborg Organ Art Center.

Blok, Anders, Ignacio Farías, and Celia Roberts, eds. 2020. *The Routledge Companion to Actor-Network Theory*. London and New York: Routledge.

Borgdorff, Henk. 2010. "The Production of Knowledge in Artistic Research." In *The Routledge Companion to Research in the Arts*, edited by Michael Biggs and Henrik Karlsson, 44–63. London: Routledge.

Borgdorff, Henk, Peter Frank Peters, and Trevor Pinch, eds. 2020. *Dialogues Between Artistic Research and Science and Technology Studies*. New York and London: Routledge.

Bormann, Karl. 1966. *Die gotische Orgel zu Halberstadt: Eine Studie über mittelalterlichen Orgelbau*. Berlin: Merseburger.

Born, Georgina. 2005. "On Musical Mediation: Ontology, Technology and Creativity." *Twentieth-Century Music* 2 (01): 7–36.

Born, Georgina. 2010. "For a Relational Musicology: Music and Interdisciplinarity, Beyond the Practice Turn: The 2007 Dent Medal Address." *Journal of the Royal Musical Association* 135 (2): 205–243.

Burstyn, Shai. 1995. "Authenticity in Interpretion." *Early Music* 23 (4): 721–723.

Butt, John. 2004. *Playing with History: The Historical Approach to Musical Performance*. Cambridge: Cambridge University Press.

Cage, John. 1987. *Silence: Lectures and Writings*. London: Marion Boyars.

Cameron, Ross P. 2008. "There Are No Things That Are Musical Works." *The British Journal of Aesthetics* 48 (3): 295–314.

Chadabe, Joel. 1997. *Electric Sound: The Past and Promise of Electronic Music*. Upper Saddle River, NJ: Prentice Hall.

Christensen, Justin. 2018. *Sound and the Aesthetics of Play: A Musical Ontology of Constructed Emotions*. Cham: Palgrave Macmillan.

Clarke, Eric, and Nicholas Cook, eds. 2004. *Empirical Musicology: Aims, Methods, Prospects*. Oxford: Oxford University Press.

Cook, Nicholas. 2014. *Beyond the Score: Music as Performance*: Oxford: Oxford University Press.

Cox, Christoph, and Daniel Warner. 2017. *Audio Culture: Readings in Modern Music*. Revised Edition. London: Bloomsbury Academic.

Dähnert, Ulrich. 1962. *Der Orgel-und Instrumentenbauer Zacharias Hildebrandt; sein Verhaltnis zu Gottfried Silbermann und Johann Sebastian Bach*. Leipzig: VEB Breitkopf und Hartel Musikverlag.

Davey, Nicholas. 2013. *Unfinished Worlds: Hermeneutics, Aesthetics and Gadamer*. Edinburgh: Edinburgh University Press.

Davidsson, Hans. 1993. "The North German Organ Research Project at the School of Music and Musicology, University of Göteborg." *I Svensk tidskrift för musikforskning* 1: 7–27.

Davidsson, Hans. 2003. "Organ Building in Northern Europe since 1969: Historical Revival and Renewal." In *The North German Organ Research Project at Göteborg University*, edited by Joel Speerstra, 329–340. Göteborg: Göteborg Organ Art Center.

Davies, Stephen. 1987. "Authenticity in Musical Performances." *British Journal of Aesthetics* 27 (1): 39–50.

Davies, Stephen. 1991. "The Ontology of Musical Works and the Authenticity of their Performances." *Noûs* 25 (1): 21–41.

De Laet, Marianne, and Annemarie Mol. 2000. "The Zimbabwe Bush Pump: Mechanics of a Fluid Technology." *Social studies of science* 30 (2): 225–263.

DeNora, Tia. 2003. *After Adorno: Rethinking Music Sociology*. Cambridge: Cambridge University Press.

DeNora, Tia. 2016. *Music-in-Action: Selected Essays in Sonic Ecology, Ashgate Contemporary Thinkers on Critical Musicology Series*. London: Routledge.

Devine, Kyle, and Alexandrine Boudreault-Fournier, eds. 2021. *Audible Infrastructures: Music, Sound Media*. Oxford: Oxford University Press.

Dewey, John. 2005 [1934]. *Art as Experience*. New York: Penguin Group.

Dewey, John. 2008 [1916]. *Democracy and Education*: Simon & Schuster.

Diepenhorst, Wim. 2009. "Bechrijving van het Orgel." In *Het oude orgel van de Nicolaikerk te Utrecht. Kroongetuige van de Nederlandse muziekgeschiedenis*, edited by Henk Verhoef, 204–247. Zutphen: Walburg Pers, Stichting Nederlandse Orgelmonografieën, Stichting Peter Gerritsz orgel, Rijksdienst voor het Cultureel Erfgoed.

Dijk, Rogier van. 2009. "Het Peter Gerritz-orgel van de Nicolaikerk te Utrecht." In *Het oude orgel van de Nicolaikerk te Utrecht. Kroongetuige van de Nederlandse muziekgeschiedenis*, edited by Henk Verhoef, 11–153. Zutphen: Walburg Pers, Stichting Nederlandse Orgelmonografieën, Stichting Peter Gerritsz orgel, Rijksdienst voor het Cultureel Erfgoed.

Dijkman, Loek. 2019 [2013]. *50 Jaar Anders Ondernemen*. Second revised ed. Leiden: Stichting Utopa.

Duffin, Ross W. 2008. *How Equal Temperament Ruined Harmony (and Why You Should Care)*. New York and London: WW Norton & Company.

Eldridge, Richard. 2010. "Dewey's Aesthetics." In *The Cambridge Companion to Dewey*, edited by Molly Cochran, 242–263. Cambridge: Cambridge University Press.

Eno, Brian. 1999. "The Revenge of the Intuitive: Turn off the Options, and Turn up the Intimacy." *Wired (January)* 176.

Eve, Imogen. 2021. *The Same but Differently*. Maastricht: Lectoraat What Art Knows/Zuyd University for Applied Sciences.

Fabian, Dorottya. 2001. "The Meaning of Authenticity and The Early Music Movement: A Historical Review." *International Review of the Aesthetics and Sociology of Music* 32 (2): 153–167.

Felski, Rita. 2020. *Hooked: Art and attachment*. Chicago: University of Chicago Press.

Felski, Rita, and Stephen Muecke, eds. 2020. *Latour and the Humanities*. Baltimore: Johns Hopkins University Press.

Fidom, Hans, ed. 2000. *Het orgel: een historisch instrument. Vier lezingen over historiserende orgelbouw*. Groningen: Stichting Groninger Orgelland.

Fidom, Hans. 2002a. "Diversity in Unity: Discussions in Organ Building in Germany between 1880 and 1918." PhD diss., Vrije Universiteit Amsterdam.

Fidom, Hans. 2002b. "Herhaalt de geschiedenis zich? De betekenis van vroeg 20-eeuwse orgels." *Het Orgel* 98 (4): 24–39.

Fidom, Hans. 2006a. *Miskend, verguisd & afgedankt: Nederlandse orgels uit de vroege 20ste eeuw*. Zaltbommel: Uitgeverij Aprilis.

Fidom, Hans, ed. 2006b. *Orgels van de Wederopbouw: Het orgel van de Nicolaïkerk in Utrecht en andere orgels van na 1945, Nederlandse orgelmonografieën*. Zutphen: Uitgeversmaatschappij Walburg Pers.

Fidom, Hans. 2014a. "Digital Historicism: The New Baroque Organ at the Orgelpark." In *The New Baroque Organ at the Orgelpark*, edited by Hans Fidom, §21–§54. Amsterdam: The Orgelpark.

Fidom, Hans, ed. 2014b. *The New Baroque Organ at the Orgelpark*. Edited by Hans Fidom. *Vol. 5/1, Orgelpark Research Report*. Amsterdam: The Orgelpark.

Fidom, Hans. 2018. "The Utopa Baroque Organ at the Orgelpark." In *The Utopa Baroque Organ*, edited by Hans Fidom, 15–71. Amsterdam: VU University Press / Orgelpark.

Fidom, Hans. 2019. *Het Maakzel van Agricola. De orgels van de Martinikerk te Groningen*. Zutphen: Walburg Pers.

Fidom, Hans, ed. 2023. *Helmut Werner Orgelbauer*. Edited by Hans Fidom, *Orgelpark Research Reports*. Amsterdam: VU University Press.

Gadamer, Hans-Georg. 1989. *Truth and Method*. Translated by J. Weinsheimer and D. G. Marshall. Second revised ed. New York: Continuum. Original edition, 1960.

Gadamer, Hans-Georg. 1986. *The Relevance of the Beautiful and Other Essays*. Cambridge: Cambridge University Press.

Garfinkel, Harold. 1967. *Studies in Ethnomethodology*. Upper Sadle River, NJ: Prentice Hall.

Gibson, James J. 1979. *The Ecological Approach to Visual Perception*. Boston: Houghton Mifflin.

Giesen, Matthias. 2005. "Arnold Schönberg und die Orgel." *Österreichisches Orgelforum* 2005 (7): 12–35.

Goehr, Lydia. 1992. *The Imaginary Museum of Musical Works: An Essay in the Philosophy of Music: An Essay in the Philosophy of Music*. Oxford: Oxford University Press.

Goodman, Nelson. 1976. *Languages of Art; an Approach to a Theory of Symbols*. 2nd ed. Indianapolis/Cambridge: Hackett Publishing Company, Inc.

Harlow, Randall. 2011. "Recent Organ Design Innovations and the 21st-Century 'Hyperorgan.'" Last Modified May 2011. https://www.huygens-fokker.org/docs/Harlow%20-%20Recent%20Organ%20Design%20Innovations%20and%20the%2021st%20Century%20Hyperorgan.pdf.

Harlow, Randall. 2018. "Hyperorgan Mediation Technology for New Acoustic Music Ecologies." In *The Utopa Baroque Organ at the Orgelpark*, edited by Hans Fidom, 153–165. Amsterdam: VU University Press / Orgelpark.

Harnoncourt, Nikolaus. 1983. *Musik als Klangrede: Wege zu einem neuen Musikverständnis*. Salzburg/Wien: Residenz Verlag.

Haynes, Bruce. 2007. *The End of Early Music: A Period Performer's History of Music for the Twenty-First Century: A Period Performer's History of Music for the Twenty-First Century*. Oxford: Oxford University Press, USA.

Hennion, Antoine. 2015. *The Passion for Music: A Sociology of Mediation*. Translated by Margaret Rigaud and Peter Collier. Surrey, UK: Ashgate.

Hennion, Antoine. 2017a. "From Valuation to Instauration: On the Double Pluralism of Values." *Valuation Studies* 5 (1): 69–81.

Hennion, Antoine. 2017b. "Attachments, You Say? . . . How a Concept Collectively Emerges in One Research Group." *Journal of Cultural Economy* 10 (1): 112–121.

Hennion, Antoine. 2019. "Objects, Belief, and the Sociologist: The Sociology of Art as a Work-To-Be-Done." In *Roads to Music Sociology*, edited by A. Smudits, 41–60. Springer.

Hennion, Antoine, and Christophe Levaux, eds. 2021. *Rethinking Music through Science and Technology Studies*. London and New York: Routledge.

Heumen, Robert van. 2014. "The New Baroque Organ as a Hybrid Electro-Acoustic Instrument." In *The New Baroque Organ at the Orgelpark*, edited by Hans Fidom, §294–§320. Amsterdam: The Orgelpark.

Holmes, Thom. 2016. *Electronic and Experimental Music: Technology, Music, and Culture*. 5th ed. London and New York: Routledge.

Huizinga, Johan. 1959. "The Task of Cultural History." In *Men and Ideas*, Johan Huizinga, 51–76. New York: Meridian Books.

Hutchins, Edwin. 1995. *Cognition in the Wild*. Cambridge, MA: The MIT Press.

Ingold, Tim. 2000. *The Perception of the Environment: Essays on Livelihood, Dwelling and Skill*. London and New York: Routledge.

Ingold, Tim. 2013. *Making: Anthropology, Archaeology, Art and Architecture*. London and New York: Routledge.

Ingold, Tim. 2021. *Correspondences*. Cambridge: Polity.

Ingold, Tim. 2022. *Imagining for Real: Essays on Creation, Attention and Correspondence*. New York: Routledge.

James, William. 1975. *The Meaning of Truth: A Sequel to Pragmatism*. Cambridge, MA: Harvard University Press. Original edition, 1909.

Keller, Hans 1984. "Whose Authenticity?" *Early Music* 12 (4): 517–519.

Kent, Christopher. 1998. "Temperament and Pitch." In *The Cambridge Companion to the Organ*, edited by Nicolas Thistlethwaite and Geoffrey Webber, 42–54. Cambridge: Cambridge University Press.

Kenyon, Nicholas. 1988. *Authenticity and Early Music*. Oxford: Oxford University Press.

Kim, Jin Hyun, and Uwe Seifert. 2016. "Interactivity of Digital Musical Instruments: Implications of Classifying Musical Instruments on Basic Music Research." In *Musical Instruments in the 21st Century. Identities, Configurations, Practices*, edited by Till Bovermann, Alberto de Campo, Hauke Egermann, Sarah-Indriyati Hardjowirogo, and Stefan Weinzierl. Singapore: Springer.

Kivy, Peter. 1998. *Authenticities: Philosophical Reflections on Musical Performance*. Ithaca and London: Cornell University Press.

Kraul, Peter. 2006. "Three Prototypes—Pushing the Technical Boundaries in Organ Building." In *Fresh Wind*, edited by Michael Eidenbenz, Daniel Glauss, and Peter Kraut, 70–78. Saarbrücken: PFAU-Verlag.

Kronengold, Charles. 2018. "Harpsichords (and People) at the Limits of Mediation Theory." *Contemporary Music Review* 37 (5–6): 575–605.

Kursell, Julia, and Peter Frank Peters. 2020. "Making Sound Present: Re-enactment and Reconstruction in Historical Organ Building Practices." In *Reconstruction, Replication and Re-enactment in the Humanities and Social Sciences*, edited by Sven Dupré, Anna Harris, Julia Kursell, Patricia Lulof, and Maartje Stols-Witlox, 115–139. Amsterdam: Amsterdam University Press.

Kvifte, Tellef. 2008. "What is a Musical Instrument." *Svensk tidskrift för musikforskning* 1 (2008): 45–56.

La Berge, Anne. 2014. "Proximity and Communication with the New Baroque Organ." In *The New Baroque Organ at the Orgelpark*, edited by Hans Fidom, §321–§367. Amsterdam: The Orgelpark.

Landy, Leigh. 2007. *Understanding the Art of Sound Organization*. Cambridge, MA: The MIT Press.

Latour, Bruno. 1992. *Aramis, ou l'amour des techniques, Textes a l'appui, Anthropologie des sciences et des techniques*. Paris: Editions la Decouverte.

Latour, Bruno. 1993. *We Have Never Been Modern*. Translated by Catherine Porter. Harvard: Harvard University Press.

Latour, Bruno. 2005. *Reassembling the Social: An Introduction to Actor-Network-Theory*. Oxford: Oxford University Press.

Latour, Bruno. 2008a. "A Cautious Prometheus? A Few Steps toward a Philosophy of Design (with special attention to Peter Sloterdijk)." *Proceedings of the 2008 Annual International Conference of the Design History Society*, Falmouth, Cornwall, 3 September 2008.

Latour, Bruno. 2008b. *What Is the Style of Matters of Concern? Two Lectures in Empirical Philosophy*. Assen: Van Gorcum.

Latour, Bruno. 2013. *An Inquiry into Modes of Existence*. Cambridge, MA: Harvard University Press.

Lawson, Colin, and Robin Stowell. 1999. *The Historical Performance of Music: An Introduction*. Cambridge: Cambridge University Press.

Lekkerkerker, Jacob. 2014. "Am I an Organist or do I play the Organ?" In *The New Baroque Organ at the Orgelpark*, edited by Hans Fidom, §271–§293. Amsterdam: The Orgelpark.

Lelie, Christoph. 2012. "Van Straten-Orgel is Grensverleggend, Subliem en Inspirerend." *Trouw*. https://www.trouw.nl/home/van-straten-orgel-is-grensverleggend-subliem-en-inspirerend~a5d9f2b0/.

Leppard, Raymond, and Reinhard G. Pauly. 1988. *Authenticity in Music*. Cleckheaton: The Amadeus Press.

Ligeti, György. 1966. "Die Orgel sprengt die Tradition." *Melos* 33 (10): 311–313.

Lipski, Thomas. 1997. *Hans Henny Jahnns Einfluß auf den Orgelbau*. Hildesheim: Olms.

Lowenhaupt Tsing, Anna. 2015. *The Mushroom at the End of the World: On the Possibility of Life in Capitalist Ruins*. Princeton and Oxford: Princeton University Press.

Luttmer, Biëlla. 2016. "Spinvis en consorten zetten vier eeuwen oud werk van organist voort met nieuwe noten, teksten en instrumenten." *Volkskrant*, 23 February 2016.

Manning, Peter. 2004. *Electronic and Computer Music*. Oxford and New York: Oxford University Press.

Marchand, Trevor H. J., ed. 2017. *Craftwork as Problem Solving: Ethnographic Studies of Design and Making*. London and New York: Routledge.

Marres, Noortje. 2020. "As ANT is getting undone, can Pragmatism help us re-do it?" In *The Routledge Companion to Actor-Network Theory*, edited by Anders Blok, Ignacio Farías, and Celia Roberts, 112–120. London and New York: Routledge.

Marres, Noortje, and David Stark. 2020. "Put to the Test: For a New Sociology of Testing." *British Journal of Sociology* 71 (3): 423–443.

Martin, Peter J. 2006. *Music and the Sociological Gaze: Art Worlds and Cultural Production*. Manchester: Manchester University Press.

Mattozzi, Alvise, and Laura Lucia Parolin. 2021. "Bodies Translating Bodies: Methodological Reflections on How to Tackle Aesthetic Practices from an ANT Point of View." *Science & Technology Studies* 34 (4): 2–29.

McAuliffe, Sam, ed. 2024. *Gadamer, Music, and Philosophical Hermeneutics*. Cham: Springer.

McNeil, Michael. 2012. *The Sound of Pipe Organs*. Mead, CO: CC&A.

Novak, David. 2013. *Japanoise: Music at the Edge of Circulation*. Durham: Duke University Press.

O'Connor, Erin. 2005. "Embodied Knowledge in Glassblowing: The Experience of Meaning and the Struggle Towards Proficiency." *Ethnography* 6 (2): 183–204.

Ortgies, Ibo. 2014. "Johann Sebastian Bach and Temperament." In *Orgelpark Research Report 5/1*, edited by Hans Fidom, §224–§269. Amsterdam: Orgelpark.

Ortgies, Ibo. 2018. "The Temperament of the Utopa Baroque Organ." In *The Utopa Baroque Organ*, edited by Hans Fidom, 73–90. Amsterdam: VU University Press / Orgelpark.

Oskamp, Jacqueline. 2011. *Onder stroom. Geschiedenis van de elektronische muziek in Nederland*. Amsterdam: Ambo.

Owen, Barbara. 1997. *The Registration of Baroque Organ Music*. Bloomington and Indianapolis: Indiana University Press.

Owen, Barbara. 2002. "Technology and the Organ in the Nineteenth Century." In *The Organ as a Mirror of Its Time: North European Reflections, 1610–2000*, edited by Kerala J. Snyder, 213–229. Oxford: Oxford University Press.

Patteson, Thomas. 2015. *Instruments for New Music: Sound, Technology, and Modernism*. Oakland, CA: University of California Press.

Peeters, Flor, and Maarten Albert Vente. 1984. *De orgelkunst in de Nederlanden van de 16de tot de 18de eeuw*. Amerongen: Gaade.

Perricone, Christopher. 2006. "The Influence of Darwinism on John Dewey's Philosophy of Art." *The Journal of Speculative Philosophy* 20 (1): 20–41.

Peters, Peter Frank. 2009. "Retracing Old Organ Sound: Authenticity and the Structure of Artistic Arguments." *Krisis. Journal for Contemporary Philosophy* 2009 (1): 5–19.

Peters, Peter Frank, and Darryl Cressman. 2016. "A Sounding Monument: How a New Organ Became Old." *Sound Studies: An Interdisciplinary Journal* 2 (1): 21–35.

Peters, Peter Frank. 2019. "Crafting Baroque Sound." In *Dialogues Between Artistic Research and Science and Technology Studies*, edited by Henk Borgdorff, Peter Frank Peters, and Trevor Pinch, 125–136. London and New York: Routledge.

Petzold, Denise. 2023. "Archives of Change: An Art Conservation Studies Approach to Innovating Classical Music." PhD diss., Faculty of Arts and Social Sciences, Maastricht University.

Piekut, Benjamin. 2014. "Actor-Networks in Music History: Clarifications and Critiques." *Twentieth-Century Music* 11 (2): 191–215.

Pinch, Trevor. 1993. ""Testing-One, Two, Three . . . Testing!": Toward a Sociology of Testing." *Science, Technology & Human Values* 18 (1): 25–41.

Pinch, Trevor. 2016. ""Bring on Sector Two!" The Sounds of Bent and Broken Circuits." *Sound Studies* 2 (1): 36–51. https://doi.org/10.1080/20551940.2016.1215595.

Pinch, Trevor. 2020. "Coda: Testing and Why It Matters." In *Testing Hearing: The Making of Modern Aurality*, edited by Viktoria Tkaczyk, Mara Mills, and Alexandra Hui, 359–374. Oxford: Oxford University Press.

Pinch, Trevor, and Frank Trocco. 2002. *Analog Days: The Invention and Impact of the Moog Synthesizer*. Harvard: Harvard University Press.

Pinch, Trevor, and Karin Bijsterveld. 2003. ""Should One Applaud?" Breaches and Boundaries in the Reception of New Technology in Music." *Technology and Culture* 44 (3): 536–559.

Pinch, Trevor, and Karin Bijsterveld. 2004. "Sound Studies: New Technologies and Music." *Social Studies of Science* 34 (5): 635–648.

Pinch, Trevor, and Karin Bijsterveld. 2012. *The Oxford Handbook of Sound Studies*. Oxford: Oxford University Press.

Pinch, Trevor, and Karin Bijsterveld. 2015. "Instruments and Innovation." In *The Routledge Reader on the Sociology of Music*, edited by John Sheperd and Kyle Devine, 301–308. London and New York: Routledge.

Reich, Steve. 2002. *Writings on Music 1965–2000*. Oxford: Oxford University Press.

Reichling, Mary J. 1997. "Music, Imagination, and Play." *Journal of Aesthetic Education* 31 (1): 41–55.

Rioux, Vincent. 2001. *Sound Quality of Flue Organ Pipe: An Interdisciplinary Study on the Art of Voicing*: Chalmers University of Technology.

Robbins, Edward. 1994. *Why Architects Draw*. Cambridge, MA: The MIT Press.

Salmen, Walter, and Markus Zepf. 2007. *Calcanten und Orgelzieherinnen: Geschichte eines "niederen" Dienstes*. Baden-Baden: Olms.

Schuiling, Floris. 2020. "(Re-)Assembling Notations in the Performance of Early Music." *Contemporary Music Review* 39 (5): 580–601. https://doi.org/10.1080/07494467.2020.1852802.

Schwab, Michael. 2014. "Artistic Research and Experimental Systems." In *Artistic Experimentation in Music: An Anthology*, edited by Bob Gilmore Darla Crispin, 111–121. Leuven: Leuven University Press.

Schweitzer, Albert. 1908. *J.S. Bach*. Leipzig: Breitkopf & Härtel.

Segurado, João. 2015. "Never Heard Before: A Musical Exploration of Organ Voicing." PhD diss., Academy of Music and Drama, Faculty of Fine, Applied, and Performing Arts, University of Gothenburg.

Semmerling, Linnea. 2020. "Listening on Display: Exhibiting Sounding Artworks 1960s-now." PhD diss., Faculty of Arts and Social Sciences, Maastricht University.

Sennett, Richard. 2008. *The Craftsman*. New Haven & London: Yale University Press.

Shannon, John R. 2009. *Understanding the Pipe Organ: A Guide for Students, Teachers and Lovers of the Instrument*. Jefferson, NC: McFarland.

Shapin, Steven, and Simon Schaffer. 1985. *Leviathan and the Air-Pump: Hobbes, Boyle and the Experimental Life*. Princeton: Princeton University Press.

Small, Christopher. 2011. *Musicking: The Meanings of Performing and Listening*. Middletown, CT: Wesleyan University Press.

Smith, Neil Thomas. 2024. "Changing Rooms: A Diary of Spatial Innovation." In *Classical Music Futures: Practices of Innovation*, edited by Neil Thomas Smith, Peter Frank Peters, and Karoly Molina, 277–298. Oxford: Open Book Publishers.

Snyder, Kerala J., ed. 2002. *The Organ as a Mirror of Its Time: North European Reflections, 1610–2000*. Oxford: Oxford University Press.

Speerstra, Joel. 2003a. "An Introduction to the North German Organ research Project." In *The North German Organ Research Project at Göteborg University*, edited by Joel Speerstra, 15–20. Göteborg: The Göteborg Organ Art Center/Göteborg University.

Speerstra, Joel, ed. 2003b. *The North German Organ Research Project at Göteborg University*. Göteborg: The Göteborg Organ Art Center.

Spronck, Veerle. 2022. "Listen Closely: Innovating Audience Participation in Symphonic Music." PhD diss., Faculty of Arts and Social Sciences, Maastricht University.

Suchman, Lucy, Randall Trigg, and Jeanette Blomberg. 2002. "Working Artefacts: Ethnomethods of the Prototype." *British Journal of Sociology* 53 (2): 163–179. https://doi.org/10.1080/00071310220133287.

Taruskin, Richard. 1995. *Text and Act: Essays on Music and Performance*. Oxford: Oxford University Press.

Théberge, Paul. 2016. "Musical Instruments as Assemblage." In *Musical Instruments in the 21st Century: Identities, Configurations, Practices*, edited by Till Bovermann, Alberto de Campo, Hauke Egermann, Sarah-Indriyati Hardjowirogo, and Stefan Weinzierl, 59–66. Singapore: Springer.

Thomas, Mark J. 2018. "Gadamer and the Hermeneutics of Early Music Performance." *Research in Phenomenology* 48 (3): 365–384.

Tkaczyk, Viktoria, Mara Mills, and Alexandra Hui, eds. 2020. *Testing Hearing: The Making of Modern Aurality*. Oxford: Oxford University Press.

Torre, Giuseppe, and Kristina Andersen. 2016. "Instrumentality, Time and Perseverance." In *Musical Instruments in the 21st Century: Identities, Configurations, Practices*, edited by Till Bovermann, Alberto de Campo, Hauke Egermann, Sarah-Indriyati Hardjowirogo and Stefan Weinzierl, 127–136. Singapore: Springer.

Valeriani, Simona. 2011. "Facts and Building Artefacts: What Travels in Material Objects?" In *How Well do Facts Travel?: The Dissemination of Reliable Knowledge*, edited by Peter Howlett and Mary S. Morgan, 43–71. Cambridge: Cambridge University Press.

Vogel, Harald. 2003. "Organ Restoration in the Twentieth Century." In *The North German Organ Research Project at Göteborg University*, edited by Joel Speerstra, 341–345. Göteborg: Göteborg Organ Art Center.

Vries, Gerard de. 2016. *Bruno Latour, Key Contemporary Thinkers*. Cambridge: Polity Press.

Warnke, Georgia. 2013. *Gadamer: Hermeneutics, Tradition and Reason*. Cambridge: Polity Press.

Watkins, Glenn E. 1965. "Schoenberg and the Organ." *Perspectives of New Music* 4 (1): 119–135.

Watson, John R. 2010. *Artifacts in Use: The Paradox of Restoration and the Conservation of Organs*. Richmond, VA: OHS Press.

Werff, Ties van de, Imogen Eve, and Veerle Spronck. 2023. "Experiments in Participation: Performing Audiences in the Symphony Orchestra Practice." In *The Routledge Companion to Applied Musicology*, edited by Chris Dromey, 275–284. London: Routledge.

Wilson, Nick. 2014. *The Art of Re-enchantment: Making Early Music in the Modern Age*. Oxford: Oxford University Press.

Winner, Langdon. 1999. "Do Artifacts have Politics?" In *The Social Shaping of Technology*, edited by Donald MacKenzie and Judy Wajcman, 28–40. Buckingham [etc.]: Open University Press.

Wolff, Christoph. 2001. *Johann Sebastian Bach: The Learned Musician*. New York: WW Norton & Company.

Wolff, Christoph. 2020. *Bach's Musical Universe: The Composer and His Work*. New York: W.W. Norton & Company, Inc.

Wolff, Christoph, and Markus Zepf. 2012. *The Organs of J.S. Bach. A Handbook*. Urbana, Chicago and Springfield: University of Illinois Press.

Yearsley, David. 2009. "Treading the Organ Bellows." *Early Music* 37 (1): 119–121.

Yearsley, David. 2010. "The Organ-Building of Munetaka Yokota." *Counterpunch*, December 3, 2010.

Yokota, Munetaka and Pamela Ruiter-Feenstra. 2003. "Historical Pipe Metal Casting Techniques in Seventeenth Century North Germany." In *The North German Organ Research Project at Göteborg University*, edited by Joel Speerstra, 165–186. Gothenburg: Göteborg Organ Art Center / Göteborg University.

Young, James. 2002. "The Concept of Authentic Performance." In *Arguing about Art: Contemporary Philosophical Debates*, edited by Neill A. and A. Ridley, 69–81. London: Routledge.

INDEX

Page numbers followed by an *f* denote a figure; *t* denotes a table.

Inside Technology Series

Edited by Wiebe E. Bijker and Rebecca Slayton

Peter F. Peters, *Electronic Baroque: Building a Historical Organ for the Present*

John Downer, *Rational Accidents: Reckoning with Catastrophic Technologies*

Jesse Olszynko-Gryn, *A Woman's Right to Know: Pregnancy Testing in Twentieth-Century Britain*

Dagmar Schäfer, Annapurna Mamidipudi, and Marius Buning, editors, *Ownership of Knowledge: Beyond Intellectual Property*

Tamar Novick, *Milk and Honey: Technologies of Plenty in the Making of a Holy Land*

Kristin Asdal and Tone Huse, *Nature-made Economy: Cod, Capital, and the Great Transformation of the Ocean*

Cyrus C. M. Mody, *The Squares: US Physical and Engineering Scientists in the Long 1970s*

Brice Laurent, *European Objects: The Troubled Dreams of Harmonization*

Florian Jaton, *The Constitution of Algorithms: Ground-Truthing, Programming, Formulating*

Kean Birch and Fabian Muniesa, *Turning Things into Assets*

David Demortain, *The Science of Bureaucracy: Risk Decision-Making and the US Environmental Protection Agency*

Nancy Campbell, *OD: Naloxone and the Politics of Overdose*

Lukas Engelmann and Christos Lynteris, *Sulphuric Utopias: The History of Maritime Fumigation*

Zara Mirmalek, *Making Time on Mars*

Joeri Bruyninckx, *Listening in the Field: Recording and the Science of Birdsong*

Edward Jones-Imhotep, *The Unreliable Nation: Hostile Nature and Technological Failure in the Cold War*

Jennifer L. Lieberman, *Power Lines: Electricity in American Life and Letters, 1882–1952*

Jess Bier, *Mapping Israel, Mapping Palestine: Occupied Landscapes of International Technoscience*

Benoît Godin, *Models of Innovation: The History of an Idea*

Stephen Hilgartner, *Reordering Life: Knowledge and Control in the Genomics Revolution*

Brice Laurent, *Democratic Experiments: Problematizing Nanotechnology and Democracy in Europe and the United States*

Cyrus C. M. Mody, *The Long Arm of Moore's Law: Microelectronics and American Science*

Tiago Saraiva, *Fascist Pigs: Technoscientific Organisms and the History of Fascism*

Teun Zuiderent-Jerak, *Situated Interventions: Sociological Experiments in Healthcare*

Basile Zimmermann, *Technology and Cultural Difference: Electronic Music Devices, Social Networking Sites, and Computer Encodings in Contemporary China*

Andrew J. Nelson, *The Sound of Innovation: Stanford and the Computer Music Revolution*

Sonja D. Schmid, *Producing Power: The Pre-Chernobyl History of the Soviet Nuclear Industry*

Casey O'Donnell, *Developer's Dilemma: The Secret World of Videogame Creators*

Christina Dunbar-Hester, *Low Power to the People: Pirates, Protest, and Politics in FM Radio Activism*

Eden Medina, Ivan da Costa Marques, and Christina Holmes, editors, *Beyond Imported Magic: Essays on Science, Technology, and Society in Latin America*

Anique Hommels, Jessica Mesman, and Wiebe E. Bijker, editors, *Vulnerability in Technological Cultures: New Directions in Research and Governance*

Amit Prasad, *Imperial Technoscience: Transnational Histories of MRI in the United States, Britain, and India*

Charis Thompson, *Good Science: The Ethical Choreography of Stem Cell Research*

Tarleton Gillespie, Pablo J. Boczkowski, and Kirsten A. Foot, editors, *Media Technologies: Essays on Communication, Materiality, and Society*

Catelijne Coopmans, Janet Vertesi, Michael Lynch, and Steve Woolgar, editors, *Representation in Scientific Practice Revisited*

Rebecca Slayton, *Arguments that Count: Physics, Computing, and Missile Defense, 1949–2012*

Stathis Arapostathis and Graeme Gooday, *Patently Contestable: Electrical Technologies and Inventor Identities on Trial in Britain*

Jens Lachmund, *Greening Berlin: The Co-Production of Science, Politics, and Urban Nature*

Chikako Takeshita, *The Global Biopolitics of the IUD: How Science Constructs Contraceptive Users and Women's Bodies*

Cyrus C. M. Mody, *Instrumental Community: Probe Microscopy and the Path to Nanotechnology*

Morana Alač, *Handling Digital Brains: A Laboratory Study of Multimodal Semiotic Interaction in the Age of Computers*

Gabrielle Hecht, editor, *Entangled Geographies: Empire and Technopolitics in the Global Cold War*

Michael E. Gorman, editor, *Trading Zones and Interactional Expertise: Creating New Kinds of Collaboration*

Matthias Gross, *Ignorance and Surprise: Science, Society, and Ecological Design*

Andrew Feenberg, *Between Reason and Experience: Essays in Technology and Modernity*

Wiebe E. Bijker, Roland Bal, and Ruud Hendricks, *The Paradox of Scientific Authority: The Role of Scientific Advice in Democracies*

Park Doing, *Velvet Revolution at the Synchrotron: Biology, Physics, and Change in Science*

Gabrielle Hecht, *The Radiance of France: Nuclear Power and National Identity after World War II*

Richard Rottenburg, *Far-Fetched Facts: A Parable of Development Aid*

Michel Callon, Pierre Lascoumes, and Yannick Barthe, *Acting in an Uncertain World: An Essay on Technical Democracy*

Ruth Oldenziel and Karin Zachmann, editors, *Cold War Kitchen: Americanization, Technology, and European Users*

Deborah G. Johnson and Jameson W. Wetmore, editors, *Technology and Society: Building Our Sociotechnical Future*

Trevor Pinch and Richard Swedberg, editors, *Living in a Material World: Economic Sociology Meets Science and Technology Studies*

Christopher R. Henke, *Cultivating Science, Harvesting Power: Science and Industrial Agriculture in California*

Helga Nowotny, *Insatiable Curiosity: Innovation in a Fragile Future*

Karin Bijsterveld, *Mechanical Sound: Technology, Culture, and Public Problems of Noise in the Twentieth Century*

Peter D. Norton, *Fighting Traffic: The Dawn of the Motor Age in the American City*

Joshua M. Greenberg, *From Betamax to Blockbuster: Video Stores tand the Invention of Movies on Video*

Mikael Hård and Thomas J. Misa, editors, *Urban Machinery: Inside Modern European Cities*

Christine Hine, *Systematics as Cyberscience: Computers, Change, and Continuity in Science*

Wesley Shrum, Joel Genuth, and Ivan Chompalov, *Structures of Scientific Collaboration*

Shobita Parthasarathy, *Building Genetic Medicine: Breast Cancer, Technology, and the Comparative Politics of Health Care*

Kristen Haring, *Ham Radio's Technical Culture*

Atsushi Akera, *Calculating a Natural World: Scientists, Engineers and Computers during the Rise of U.S. Cold War Research*

Donald MacKenzie, *An Engine, Not a Camera: How Financial Models Shape Markets*

Geoffrey C. Bowker, *Memory Practices in the Sciences*

Christophe Lécuyer, *Making Silicon Valley: Innovation and the Growth of High Tech, 1930–1970*

Anique Hommels, *Unbuilding Cities: Obduracy in Urban Sociotechnical Change*

David Kaiser, editor, *Pedagogy and the Practice of Science: Historical and Contemporary Perspectives*

Charis Thompson, *Making Parents: The Ontological Choreography of Reproductive Technology*

Pablo J. Boczkowski, *Digitizing the News: Innovation in Online Newspapers*

Dominique Vinck, editor, *Everyday Engineering: An Ethnography of Design and Innovation*

Nelly Oudshoorn and Trevor Pinch, editors, *How Users Matter: The Co-Construction of Users and Technology*

Peter Keating and Alberto Cambrosio, *Biomedical Platforms: Realigning the Normal and the Pathological in Late-Twentieth-Century Medicine*

Paul Rosen, *Framing Production: Technology, Culture, and Change in the British Bicycle Industry*

Maggie Mort, *Building the Trident Network: A Study of the Enrollment of People, Knowledge, and Machines*

Donald MacKenzie, *Mechanizing Proof: Computing, Risk, and Trust*

Geoffrey C. Bowker and Susan Leigh Star, *Sorting Things Out: Classification and Its Consequences*

Charles Bazerman, *The Languages of Edison's Light*

Janet Abbate, *Inventing the Internet*

Herbert Gottweis, *Governing Molecules: The Discursive Politics of Genetic Engineering in Europe and the United States*

Kathryn Henderson, *On Line and On Paper: Visual Representation, Visual Culture, and Computer Graphics in Design Engineering*

Susanne K. Schmidt and Raymund Werle, *Coordinating Technology: Studies in the International Standardization of Telecommunications*

Marc Berg, *Rationalizing Medical Work: Decision Support Techniques and Medical Practices*

Eda Kranakis, *Constructing a Bridge: An Exploration of Engineering Culture, Design, and Research in Nineteenth-Century France and America*

Paul N. Edwards, *The Closed World: Computers and the Politics of Discourse in Cold War America*

Donald MacKenzie, *Knowing Machines: Essays on Technical Change*

Wiebe E. Bijker, *Of Bicycles, Bakelites, and Bulbs: Toward a Theory of Sociotechnical Change*

Louis L. Bucciarelli, *Designing Engineers*

Geoffrey C. Bowker, *Science on the Run: Information Management and Industrial Geophysics at Schlumberger, 1920–1940*

Wiebe E. Bijker and John Law, editors, *Shaping Technology / Building Society: Studies in Sociotechnical Change*

Stuart Blume, *Insight and Industry: On the Dynamics of Technological Change in Medicine*

Donald MacKenzie, *Inventing Accuracy: A Historical Sociology of Nuclear Missile Guidance*

Pamela E. Mack, *Viewing the Earth: The Social Construction of the Landsat Satellite System*

H. M. Collins, *Artificial Experts: Social Knowledge and Intelligent Machines*

http://mitpress.mit.edu/books/series/inside-technology

Publisher contact:
The MIT Press
Massachusetts Institute of Technology
77 Massachusetts Avenue, Cambridge, MA 02139
mitpress.mit.edu

EU Authorised Representative:
Easy Access System Europe, Mustamäe tee 50,
10621 Tallinn, Estonia
gpsr.requests@easproject.com

Printed by Integrated Books International,
United States of America